Magnetic Resonance in Food Science

Magnetic Resonance in Food Science

Edited by

P. S. Belton
AFRC Institute of Food Research, Norwich, UK

I. Delgadillo
University of Aveiro, Portugal

A. M. Gil
University of Aveiro, Portugal

G. A. Webb
University of Surrey, Guildford, UK

THE ROYAL SOCIETY OF CHEMISTRY

Proceedings of the Second International Conference on Applications of Magnetic Resonance in Food Science, University of Aveiro, Portugal, 19–21 September, 1994

Special Publication No. 157

ISBN 0-85404-725-5

A catalogue record for this book is available from the British Library

Published by The Royal Society of Chemistry,
Thomas Graham House, Science Park, Milton Road,
Cambridge CB4 4WF, UK

Printed and bound by Bookcraft (Bath) Ltd

Preface

This volume covers the oral presentations made at the Second International Conference on Applications of Magnetic Resonance in Food Science. The material given in seventy five posters is additional. The Conference was held at the University of Aveiro, Portugal between the 19 and 21 September 1994 and was attended by 130 scientists from 20 countries with representatives from each continent.

The conference consisted of Invited and Submitted lectures grouped into four symposia. The order in which the lectures were given is the same as that of the chapters in this volume.

Symposium A covered Applications of Magnetic Resonance: The Developing Scene. The first eight chapters are based upon this symposium. The following six chapters belong to Symposium B which was devoted to Analysis and Authentication. Chapters fifteen to eighteen relate to Magnetic Resonance and Nutrition which was the theme of Symposium C. Symposium D dealt with Magnetic Resonance in the Study of Biopolymers and Complex Systems; the final five chapters relate to this symposium.

Contents

Applications of Magnetic Resonance: The Developing Scene

Basic Principles and Applications of Magnetic Resonance

L. H. Sutcliffe

CHEMISTRY DEPARTMENT, UNIVERSITY OF SURREY, GUILDFORD, SURREY GU2 5XH, UK

1 INTRODUCTION

Since the discovery of both electron and nuclear magnetic resonance 50 years ago, tremendous benefits from the phenomena have accrued in all branches of science and medicine. Because of the enormous amount of information that has accumulated, this brief review can only touch lightly on principles and applications. In this article, the following aspects will be dealt with in turn:

1. The determination of molecular structure
2. Molecular dynamics
3. Chemical analysis
4. Imaging

Throughout the article, attention will be drawn to the ways in which electron magnetic resonance (EMR - usually referred to as electron spin resonance (ESR) or electron paramagnetic resonance (EPR)) and nuclear magnetic resonance (NMR) make their own specific contributions. On the one hand, NMR requires the presence of a magnetic nucleus and there are plenty of these to be found in the Periodic Table[1]: on the other hand, EMR requires the presence of an unpaired electron; while this may appear to be a severe restriction on the usefulness of EMR, it is not because there are many organic and inorganic free radicals and many paramagnetic transition metal compounds[2-4] that fulfil this condition. Both types of spectroscopy depend upon the lifting of degeneracy of the nuclear or electron spins by an external magnetic field (the Zeeman effect). The fundamentals of the magnetic resonance phenomenon for nuclei and for electrons can be described both in quantum mechanical and in classical terms - the latter is particularly useful when considering spin dynamics and pulse methods. Thus, for a particle having a spin value of 1/2, two energy levels are produced separated by an energy, ΔE, given by:

For the electron, $$\Delta E = g_e \mu_B B_o \quad (1)$$

where the g_e has a value of 2.0023: μ_B is the Bohr magneton (= 9.274×10^{-24} JT^{-1}) and B_o is the applied magnetic field.

For the nucleus of spin 1/2, $$\Delta E = g_n \mu_n B_o \quad (2)$$

where the g factor is a property of the nucleus (it has a value of 5.5856 for the proton): μ_n is the Bohr nucleon (= 5.051×10^{-27} JT^{-1} for the proton). Transitions can occur between the two levels with a frequency given by $\Delta E = h\nu$.

For EMR, it is customary to use a spectrometer having a frequency of about 9.5 GHz (X-band radar): thus a magnetic field of about 340 mT is required: for proton NMR a mid-field spectrometer operates at 400 MHz and hence a field of 9.3 T is used. These requirements tell us that the energy levels are much farther apart for the electron than they are for a magnetic nucleus. Hence, because of the Boltzmann condition, EMR is intrinsically more sensitive by a factor of about 600 than is proton NMR. Another set of properties, depending upon the rate at which spins in the upper levels return to the lower levels (*relaxation*), are also very different for NMR and EMR; generally being much shorter for the latter. The monitoring and manipulation of relaxation processes provide magnetic resonance spectroscopy with much of its power.

2 THE DETERMINATION OF MOLECULAR STRUCTURE

When bond angles and bond lengths are required, and when good single crystals are on offer, the technique of choice is X-ray crystallography but, without doubt, NMR spectroscopy is the most powerful tool available for the elucidation of structures - even for large protein molecules.[5] In addition, magnetic resonance spectroscopy can give types of information from crystals difficult or impossible to get by X-ray crystallography, namely NMR can measure hydrogen distances and EMR can provide information on the distribution of unpaired electrons in a molecule. When single crystals are not available, then NMR rules supreme - indeed, it is hard to imagine how any chemical research laboratory could function without a NMR spectrometer.

2.1 NMR Spectroscopy

The basic pulse NMR experiment is carried out by switching on and off the resonant radio frequency ν; the pulse length τ determines the frequency range covered, that is, $\nu \pm 1/\tau$. Thus, for $\tau = 25$ μs the range covered is $\nu \pm 40{,}000$ Hz. The intensity of the decaying signal (the free induction decay - FID) following the pulse is digitised and recorded in a computer as a function of time. The FID time-domain signal is converted into a recognisable spectrum by Fourier transformation (FT). Normally, the FIDs from many pulses are recorded before the Fourier transformation is carried out.[6] From spectra obtained in this way, two useful parameters can be measured, these are chemical shifts and spin-spin coupling constants. The chemical shift determines the position of a line in the spectrum in the absence of spin-spin coupling and it is dependent upon the electronic environment of the resonant nuclei. For a rapidly-tumbling molecule, the effective magnetic field at a nucleus is given by:

$$B = B_0 (1 - \sigma) \qquad (3)$$

where σ is the isotropic shielding constant. Thus the signal position is dependent on the applied magnetic field. Because of this and because we cannot use the "bare" nucleus as a reference, the chemical shift is defined:

$$\delta = [\nu\,(\text{sample}) - \nu\,(\text{reference})]/\nu_0 \qquad (4)$$

where ν_0 is the spectrometer frequency in MHz and the other frequencies are in Hz; thus δ is dimensionless in units of ppm and is independent of spectrometer frequency. As well as providing important structural information, chemical shifts can be used to study minor perturbations such as hydrogen bonding, solvent effects and conformational changes. Spin-spin coupling arises when two magnetic nuclei interact with one another. The coupling constant is labelled J and it has the units of Hertz. Simple splitting patterns can arise from nuclei of spin ½, for example, when nucleus A is coupled to n equivalent nuclei of type B then at the chemical shift position for A there will be $n + 1$ lines having intensities given by the coefficients of the binomial expansion $(1 + x)^n$. The patterns observed can be very complicated if a large number of such nuclei are present and an additional complexity is introduced when (i) magnetic non-equivalence is present and (ii) spectra are second order due to the magnitude of the coupling constant being comparable with that of the chemical shift. Indeed, computer simulation of the spectra may be necessary in order to assign complex spectra. In one-dimensional NMR spectroscopy, the complexity from spin-spin coupling can be reduced by introduction of a second irradiating frequency - spin decoupling. There are many ways of decoupling[6,7] and, in addition to simplifying spectra, it is vital to the elucidation of chemical structure.[8] For structural studies of organic compounds, the most useful magnetic nuclei are those of hydrogen-1, carbon-13, fluorine-19, phosphorus-31 and nitrogen-15. Although carbon-13 and nitrogen-15 are only present in low abundance (1.1 and 0.37 atom% respectively) they can give good spectra in small molecules without recourse to enrichment. Figures 1 and 2 show examples of proton and carbon-13 NMR spectra. In the latter example, no spin-spin coupling between carbon-13 and protons is to be seen because it has been removed by "noise" or broadband decoupling of the protons. The nuclear Overhauser effect (NOE) produces an increase in signal intensity so that the carbon-13 spectrum benefits from this as well as from the removal of line splittings.

In the absence of relaxation processes, spins would not be able to return rapidly to the ground state and then no signals could be observed - if this happens when relaxation is not sufficiently efficient then "saturation" is said to occur. Usually relaxation does take place and there are two main types of process[9] (i) spin-lattice (longitudinal) relaxation, characterised by a time T_1, which occurs when the spins lose their energy to the surrounding molecules, (ii) spin-spin (transverse) relaxation, characterised by a time T_2, caused by the transfer of energy from one spin to another. The spectral line width is determined by the magnitude of T_2, according to $\nu_{1/2} = \pi T_2$, where $\nu_{1/2}$ is the half-height line width; for rapidly tumbling molecules, $T_1 = T_2$. Although relaxation times can give structural information they play a more important part in the timing of pulse sequences in many experiments and they are very useful in the study of molecular dynamics.

2.2 Multidimensional NMR Spectroscopy

The analysis of the NMR spectra of small molecules hinges on the assignments of chemical shifts and coupling constants. The process can be assisted considerably by resorting to 2D methods.[8,10,11] A 2D spectrum is defined as one in which both the abscissa and the ordinate are frequency axes. There are two distinct types of 2D experiment (i) J-resolved (ii) correlated spectroscopy; in the former, one axis contains coupling constant information while the other has the chemical shift data: in the latter, both axes contain chemical shift information correlated by spin-spin coupling, NOE or by exchange processes. Since the spin-spin coupling may be dipolar, information on the geometric structure of molecules may be obtained. The most commonly used of the 2D techniques will now be discussed.

The proton-proton chemical shift correlation spectrum (COSY - correlated spectroscopy) is probably the most commonly applied 2D technique. The spectrum is presented as a contour plot, having proton chemical shifts along each axis, showing intensities greater than a preselected level. Contours along the diagonal correspond to the 1D spectrum. The off-diagonal peaks correspond to spin-spin coupling constants thus giving in one spectrum the information that would require a series of 1D decoupling experiments.

Another common 2D experiment is homonuclear J-resolved spectroscopy which is employed when the 1D spectrum contains overlapping signals. Here, chemical shifts are presented along one axis and spin-spin coupling constants are presented along the other. A variation of this experiment is the heteronuclear version - usually done with carbon-13 shifts along one axis and with proton coupling constants along the other, the main purpose being to determine the number of protons attached to a particular carbon atom.

A very useful 2D technique is that of proton-carbon-13 chemical shift correlation (HETCOR) (see Figure 3). The contour peaks show which protons are attached to a

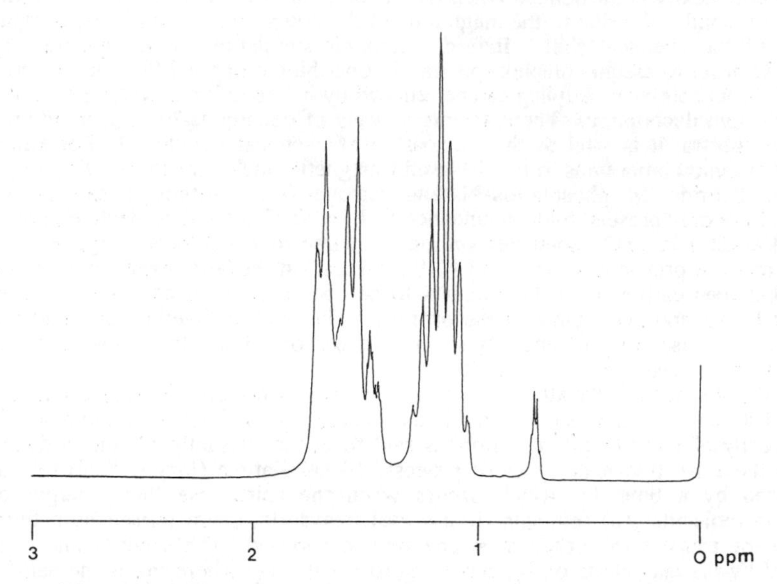

Figure 1 *The 300.1 MHz proton NMR spectrum of triicyclohexylmethane in $CDCl_3$.*

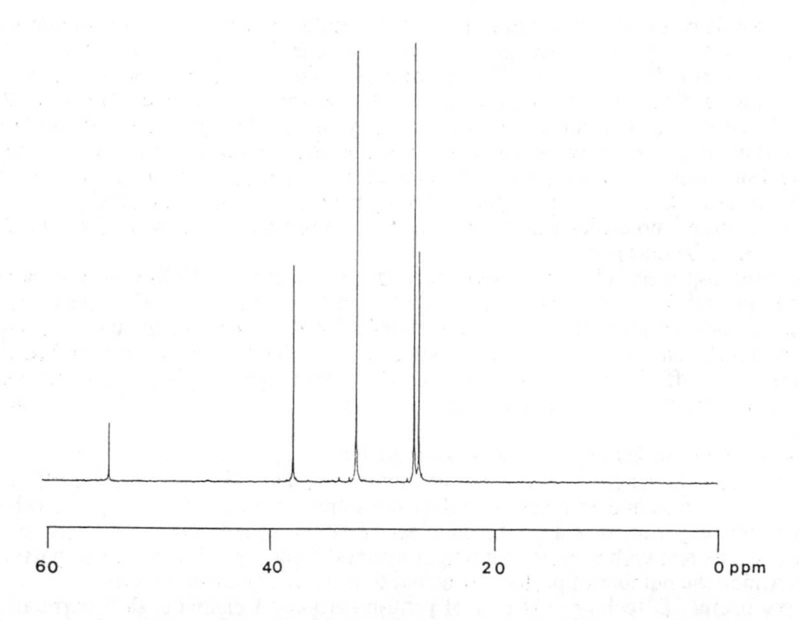

Figure 2 *The proton-decoupled 75.47 MHz carbon-13 NMR spectrum of tricyclohexylmethane in $CDCl_3$.*

given carbon atom but, of course, correlation cannot be observed for quaternary carbon atoms.

Structural studies of very large molecules are assisted considerably by 2D NOESY (nuclear Overhauser and exchange spectroscopy). Here, the contours on the diagonal correspond to the 1D spectrum while the off-diagonal contours arise from through-space proton-proton interactions between non-bonded protons. If a system under study has a dominant conformation and there are sufficient NOE-based distance constraints, the quality of the structures generated can be as good as those produced by X-ray crystallography.[12] If a complex molecule such as a biopolymer gives complex proton or other resonances, simplification can be achieved by a type of "spin labelling" namely by introducing tritium[13] or fluorine.[14] While 2D techniques require considerable spectrometer time, the benefits more than outweigh this drawback and they are used routinely. Indeed, the even more time-consuming 3D and 4D techniques have been developed. Other means of obtaining structural information include the use of lanthanide shift reagents[15] and chiral solvents.[16]

2.3 Solid-state NMR Spectroscopy

The spectra of randomly-oriented crystalline solids are broad and usually featureless due to (i) dipolar coupling between nuclear spins (ii) chemical shift anisotropy (iii) electric fields (when quadrupolar nuclei are present) (iv) scalar coupling anisotropy (v) lifetime broadening: large values of the relaxation time T_1 compound the difficulties requiring long times between pulses thus lowering sensitivity. In the liquid state, rotational averaging causes the effects (i) and (iii) to vanish. When the nucleus under examination is in low abundance, for example carbon-13, most of the broadening from dipolar coupling comes from the surrounding protons. This broadening can be eliminated by proton decoupling but, because the interactions are of the order of 50 kHz, very high radiofrequency powers of about 100 watts are required: the residual line width is about 10 Hz. Cross-polarization is used to increase the sensitivity by using the abundant spins to provide spin polarization which can be transferred to the rare spins as required. Using the decoupling and cross-polarization techniques an amorphous solid would give a "powder pattern" characterized by the anisotropy of the shielding tensors, α_{xx}, α_{yy} and α_{zz}.

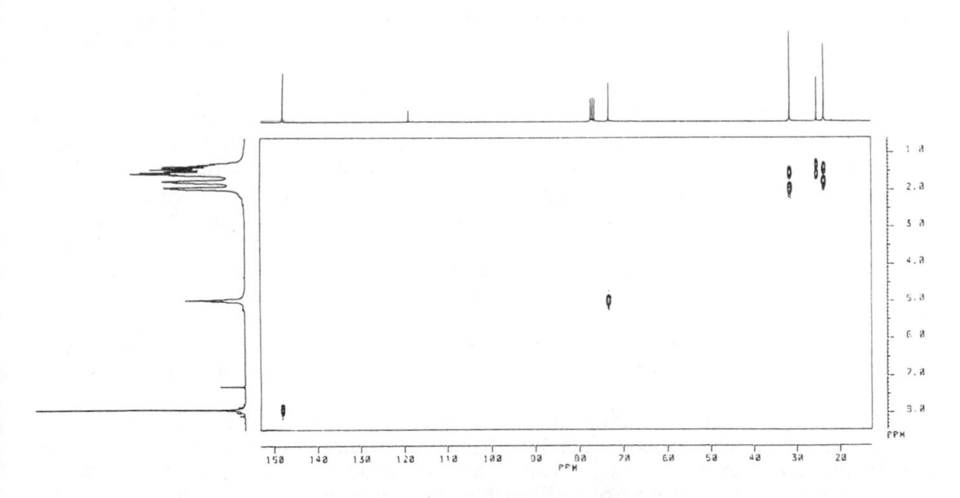

Figure 3 *The ^{1}H (300 MHz) - ^{13}C (proton-decoupled) 2D chemical shift correlation spectrum of dicyclohexyl-3, 4-furandicarboxylate in $CDCl_3$*

Sometimes it is possible to grow a sufficiently large crystal to allow these tensors to be evaluated by orienting the crystal along the x, y and z axes with respect to B_0. Where there are only two different shielding tensors (axial symmetry), the orientation dependence of the shielding is given by :

$$\sigma(\theta) = \sigma_i + (3\cos^2\theta - 1)(\sigma_{\parallel} - \sigma_{\perp})/3 \quad (5)$$

where θ is the angle between the symmetry axis and B_0 and σ_i is the isotropic shielding constant (= $(2\sigma_{\perp} + \sigma_{\parallel})$). If $3\cos^2\theta - 1$ is made equal to zero then the anisotropic terms disappear and the powder pattern will reduce to a single line. In practice, this is done by rotating the sample rapidly at an angle of 54°44' - *the magic angle*. The rotation speed clearly depends upon the extent of the chemical shift anisotropy and the operating field of the spectrometer. For 75 MHz carbon-13 spectroscopy, a typical rotation speed is 5 kHz. With all these conditions fulfilled, a solid-state carbon-13 spectrum of a simple carbohydrate is almost as well resolved as the corresponding solution spectrum. Multidimensional experiments can be carried out in similar manner to those for solutions .

It is obviously important to be able to study solids because (i) they may be insoluble (ii) the solid state itself is interesting, for example celluloses[17,18] and food materials.[19]

2.4 EMR Spectroscopy

Since EMR spectra are only observed when unpaired electrons are present, "structure" here relates to the orbital distribution of unpaired electrons in a molecule. Ideally, a single crystal should be available for study but amorphous solids and solutions can provide a wealth of information. A simple spectrum is seen when a free radical of the nitroxyl type is observed in a low viscosity medium. Here, the isotropic g factor (which corresponds to chemical shift in NMR spectroscopy) has a value of 2.006 which is close to the free electron value: most free radicals have g factors of about 2. In the absence of a magnetic nucleus, a single line would be observed having a line width of about 0.04 mT; however, in nitroxyl radicals there is a hyperfine interaction with the nitrogen-14 nucleus: thus the observed spectrum comprises a 1:1:1 triplet as may be seen in Figure 4.

Since the unpaired electron distribution is unlikely to be spherically symmetric, the radical actually has anisotropic g and hyperfine tensors that have been averaged out by rapid molecular motion. A single crystal of the radical would allow the three values of each (g_{xx}, g_{yy}, g_{zz}, A_{xx}, A_{yy} and A_{zz}) to be evaluated: in order to achieve this the radical would have to be placed in a diamagnetic isomorphous host lattice to avoid inter-radical interactions which would cause the spectrum to appear as a single line. The crystal is oriented in turn with each of its three axes aligned with the main magnetic field direction - for each direction there would be a simple spectrum of just three lines. Similar information can be obtained from a frozen solution (preferably using a deuterated solvent to reduce dipolar interactions) of a radical but now the spectrum observed is that of a "powder", namely an average of all the orientations of the radical is seen. It is customary to display an EMR spectrum as a first derivative, and occasionally a second derivative, in order to enhance the resolution. Figure 5 shows such a powder spectrum for the simple 1,3,2-dithiazoyl radical. From the anisotropic g and A values the symmetry of the orbitals occupied by the unpaired electron can be deduced together with the s and p orbital contributions to the unpaired spin density. Transition metal and rare earth compounds generally give much more complex spectra [20,3] but structural aims are similar to those for free radicals.

For a full structural examination, all the nuclear hyperfine interactions have to be evaluated, in addition to those originating from the magnetic nucleus upon which the unpaired electron is largely centred. When these are similar to or less than the line width, other techniques have to be invoked. Probably the most important of these is Electron-Nuclear Double Resonance (ENDOR) spectroscopy.[21] This can be regarded as EMR

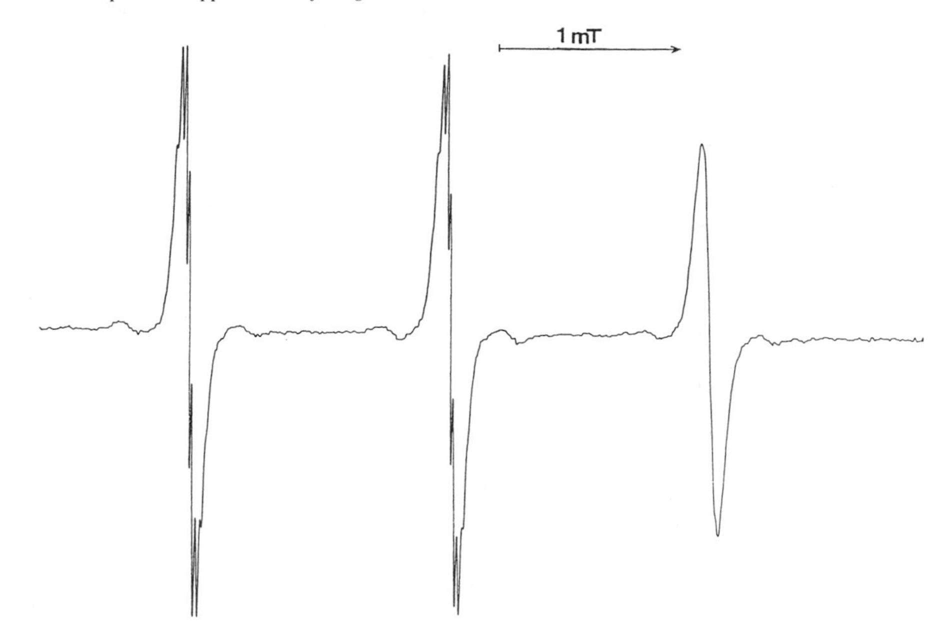

Figure 4 *The X-band EMR spectrum if the 1,1,3-tetramethylisoindolin-2-yloxyl radical in toluene. The fine structure seen in the nitrogen-14 lines arises from proton hyperfine interactions of the methyl groups.*

detection of NMR - in the continuous wave (CW) experiment an EMR transition is saturated and the NMR frequencies are swept through. Figure 6 shows a typical ENDOR spectrum. Pulse techniques provide a complementary means of obtaining ENDOR spectra.[22]

Another means of measuring very small isotropic hyperfine interactions is to resort to NMR spectroscopy. Typically, a proton NMR spectrum of a free radical is compared with that of its precursor and from the chemical shift differences the isotropic *a* value of nucleus *i* can be calculated from the equation:

$$a^i = -\Delta\delta/[\gamma_e/\gamma_i)(g\beta B_o/4kT)] \tag{6}$$

where $\Delta\delta$ is the chemical shift difference (ppm) between the radical and its diamagnetic precursor (δ_P - δ_D), γ_e and γ_i are the magnetogyric ratios of the electron and nucleus *i*. This method is particularly useful for measuring carbon-13 hyperfine interactions without the need for isotopic enrichment. Assignments can be a problem but 2D NMR spectroscopy can be of considerable assistance for this purpose.[23]

3 MOLECULAR DYNAMICS

3.1 Dynamic NMR Spectroscopy

This is one of the more important areas of NMR spectroscopy[24-28] due to the fact that it is possible to extract rate constants and activation parameters from reactions that are too slow for optical spectroscopic methods but too fast for classical chemical methods, that is reactions with rates in the region 10^{-1} to 10^3 s^{-1}. The method is based upon the coalescence of peaks over a temperature range resulting from averaging of either chemical

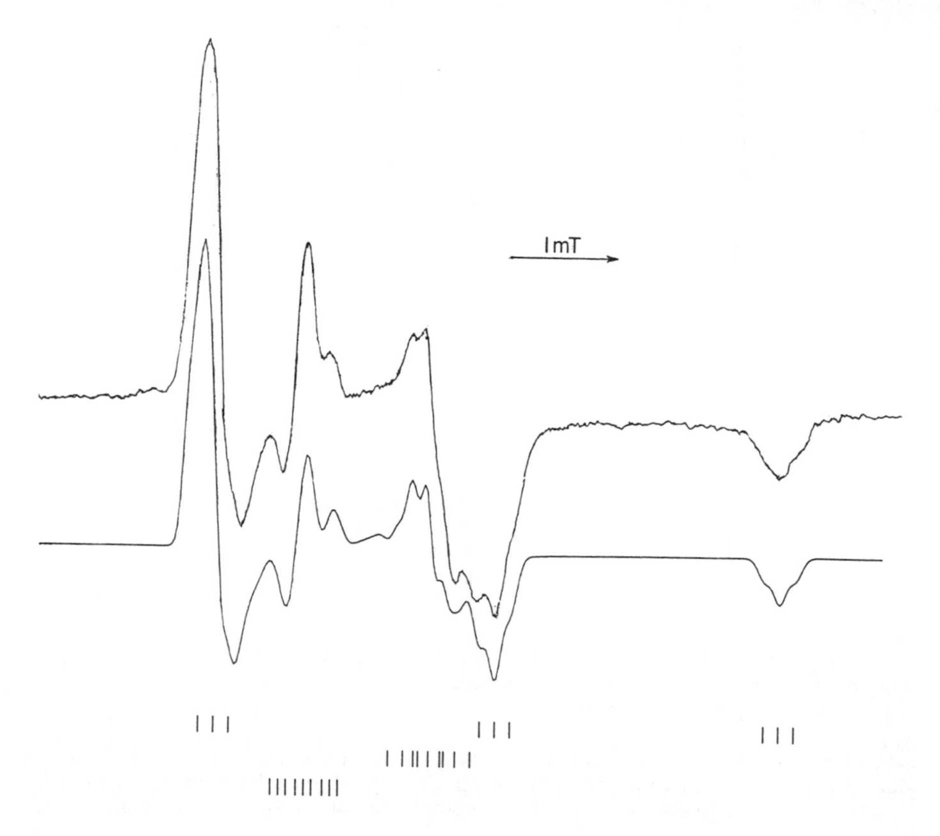

Figure 5 *The X-band EMR spectrum of the 1,3,2-dithiazol-2-yl radical in deuteriotoluene at 95 K. The lower spectrum is a computer simulation.*

shifts or of spin-spin coupling: the processes causing the averaging may be either intra- or inter-molecular. The kinetic data are obtained from the spectra either by complete band-shape analysis[29] or by measuring the coalescence temperature. An example of the latter is the evaluation of the rate constant, k_c, at the coalescence temperature from the equation:

$$k_c = 2.22(\Delta\nu^2 + 6J^2_{AB})^{1/2} \qquad (7)$$

This equation applies to exchange processes taking place between two nuclei A and B coupled by the spin-spin interaction J_{AB}, and having a chemical shift difference of $\Delta\nu$. Typical applications are (i) rotation about single CC bonds (ii) rotation about partial double bonds (iii) inversion at nitrogen or phosphorus atoms (iv) ring inversion (v) valence tautomerism (vi) keto-enol tautomerism (vii) proton exchange.

3.2 Dynamic EMR Spectroscopy

In order to obtain high resolution EMR spectra of free radicals it is necessary to work with concentrations less than 10^{-3} M otherwise electron exchange between the radicals will cause line broadening - known as Heisenberg broadening. The effect can be put to good use in measuring translational diffusion coefficients of a spin probe in, for

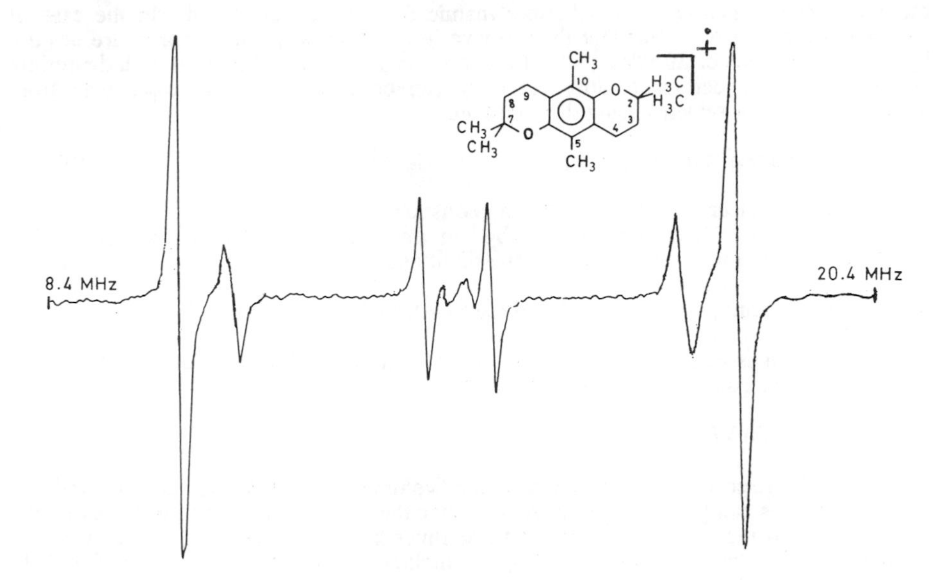

Figure 6 *The ENDOR spectrum of a cation radical in CH_2Cl_2 at 180 K. The spectrum is characterized by four proton hyperfine interactions.*

example, the lipids of a membrane.[30] When dynamic processes are present the hyperfine structure of an EMR spectrum collapses in the same way as it does for an NMR spectrum, as described above, and a number of mechanisms can be responsible in addition to electron spin exchange, namely, (i) electron transfer between a paramagnetic and a diamagnetic species (ii) cation exchange (iii) conformational equilibria.[31] ENDOR can also be used to investigate such dynamic processes.[21]

3.3 NMR Determination of Rotational Correlation Times

Carbon-13 is a particularly suitable nucleus for relaxation studies of molecular dynamics because (i) at natural abundance there are no complications from spin-spin interactions (ii) directly-attached protons can be used to provide NOE enhancements (iii) spin-lattice relaxation times are easy to measure (iv) the analysis of NOE and T_1 data is relatively simple and gives the spectral densities[32] (v) the range of chemical shifts is large. Lipari and Szabo[33] introduced a "model-free" approach to allow internal (τ_e) and overall (τ_c) motions to be separated giving a spectral density;

$$J(\omega) = S^2\left\{\frac{2\tau_c}{(1+\omega_c^2\tau_c^2}\right\} + (1-S^2)\left\{\frac{2\tau}{(1+\omega^2\tau^2)}\right\} \tag{8}$$

where $\tau^{-1} = \tau^{-1}{}_c + \tau^{-1}{}_e$ and S^2 is a generalised order parameter which is a measure of the spatial restriction of the internal motion. In a long-chain hydrocarbon the internal motions arise from *trans-gauche* isomerizations and from fluctuations of the local director. From the measurements of the motion of each individual carbon, the overall motion of the molecule can be obtained along with a measure of the spatial restriction from S^2 (the higher the value the more restricted the motion). From the temperature dependence of these parameters, the associated thermodynamic data can be evaluated. In the case of hydrocarbon chains it is clear that they behave in a similar way whether they are in pure hydrocarbons, lipids or micelles.[34-36] If a hydrogen atom can be replaced with deuterium in the molecule under examination, then τ_c can be evaluated in a simple way from deuterium T_1 measurements using the equation:

$$T_1^{-1} = 3\pi^2\kappa^2\tau_c/[10I^2(I + 1)] \quad (9)$$

where κ is the nuclear quadrupolar coupling constant.

All the NMR experiments described in this section can be carried out at high pressures[37,38] in order to study the compressibility of molecules.

3.4 EMR Determination of Rotational Correlation Times

The Stokes-Einstein-Debye equation gives the rotational correlation time, τ_{SED}, for spherical molecules obeying classical mechanics:

$$\tau_{SED} = V\eta/kT \quad (10)$$

where η is the viscosity and V is the molecular volume. Thus measurement of rotational correlation times can provide a measure of either the microviscosity or the radius of the molecule. NMR methods of determining these times have been discussed above: it can be done using EMR spectroscopy for isotropic tumbling with rates in the range 10^9 to 10^{11} s^{-1} by measuring the line widths W:

$$W(m_I) = A + Bm_I + C(m_I)^2 \quad (11)$$

where m_I is the magnetic quantum number of a line; A, B, and C depend upon the signs and magnitudes of the g and hyperfine tensors. The B and C parameters are given by:

$$B = 4b(\Delta\gamma)B_o\tau_c/15 \quad (12)$$

$$C = b^2\tau_c/8 \quad (13)$$

where:

$$b = 4\pi[A_{zz} - (A_{xx} + A_{yy})/2]/3 \quad (14)$$

$$\Delta\gamma = -\beta_e h^{-1}[g_{zz} - (g_{xx} + g_{yy})/2] \quad (15)$$

For tumbling rates in the range 10^9 to 10^6, band shape analysis can be used to measure τ_c.[39] Slower rates of rotation require the use of saturation transfer methods[30]. Pulse EMR techniques can also be used to measure τ_c. The practical consequences of the dependence of asymmetric line broadening on τ_c are enormous: the whole area of spin labelling depends on the phenomenon. This is a powerful means of studying biochemical and biophysical characteristics of lipids, proteins and amino-acids. One of the principal advantages of the approach is the sensitivity of detection; it is possible to work with radical concentrations as low as 10^{-7} M. The key to the method is to synthesize a free radical (usually a nitroxyl) having a functional group which will couple selectively to the biological system whilst causing minimum perturbation of biological activity. The importance of imaginative synthesis cannot be overstressed; an example is the labelling of carbohydrates.[40] To date thousands of spin labels have been synthesized.[41] A less demanding application is to use a small stable free radical (again it is usually a nitroxyl) as a spin probe to measure microviscosities; this can be done for the fluid inside a biological cell.

3.5 NMR Measurement of Translational Diffusion Coefficients

The experimental method normally used is the Fourier-transform-pulsed-field-gradient-spin-echo (FT PGSE) technique[42-44] which enables self-diffusion coefficients, D, to be measured for each species exhibiting an absorption line.[45] The method is based on the attenuation of a spin echo arising from diffusion dephasing under the influence of a steady applied field gradient. The attenuation is given by:

$$S(g)/S(O) = \exp[-\gamma^2 g^2 \delta^2 D(\Delta - \delta/3)] \quad (16)$$

where g is the magnitude of the gradient, δ is the pulse duration and Δ is the pulse separation.

Other NMR methods are available.[46] Typical values for D are 10^{-9} m^2s^{-1} for small molecules in a mobile liquid and $10^{-12} m^2s^{-1}$ for high polymers in solution.

3.6 EMR Measurement of Translational Diffusion Coefficients

The Heisenberg line broadening method has been described above: this measures D over molecular dimensions. Another method based on EMR imaging (EMRI), known as *dynamic imaging of diffusion*, measures diffusion over diffusive lengths of about 100 μm. In this method, large field gradients are applied to samples having an inhomogeneous distribution of spins and a special Fourier analysis is carried out.[47,48] A simple and more direct technique, when the volume element is not restricted, is to use a capillary approach which is commonly in use with radioactive tracers.[49,50] A spin probe can be prepared that is suitable for the medium under investigation, for example, a water-soluble nitroxyl radical in a polyelectrolyte gel. Progress of the radical through the gel can be monitored easily in an EMR spectrometer over a period of several days.[51] The diffusion equation is:

$$c = [c_o \exp(-x^2/4Dt)]/A(\pi Dt)^{1/2} \quad (17)$$

where c_o is the initial concentration, c is the concentration at distance x from the top of the capillary, t is the time taken to reach distance x and A is the cross-sectional area of the tube.

4 CHEMICAL ANALYSIS

4.1 NMR Spectroscopy

It is obvious that NMR spectroscopy provides a very powerful means of identifying a very wide variety of substances and it has the advantage of being non-destructive. There are limitations of sensitivity but these are becoming less important as spectrometers operating at proton frequencies of 750 Mhz and above become available.

Quantitative NMR analysis is simple because there are no extinction coefficients to consider - thus the integrated signal intensity is directly proportional to the number of nuclei giving rise to the signal. However, careful attention has to be paid to the spectrometer settings in order to get accurate intensity relationships, for example there must be complete relaxation between successive pulses for all types of nuclei at resonance.

A spectacular application of quantitative analysis is isotopic content determination. One of these is site-specific natural isotope fractionation (SNIF) NMR of deuterium.[52] The natural abundance of the latter at different sites in a natural product, such as ethanol in wine, can be measured giving information on its biological origin.

All the NMR techniques referred to so far in this review are based on pulse methods. However, for some applications there are advantages in returning to the original methods of NMR, namely continuous-wave (CW) spectroscopy. The latter has advantage of cost and of allowing a very small region of the spectrum to be scanned: thus strong solvent peaks can be avoided. Furthermore, CW spectra can be scanned rapidly so there is no time penalty.[53,54] The method has been applied to the analysis of ethanol in the range 0.01 to 15 % in wines.

Simple proton resonance pulse instruments are available for measuring the oil and water contents of materials such as foods.

4.2 EMR Spectroscopy

Whilst most analyses using EMR spectroscopy concern free radicals, there are a few oxidation states of transition metals that can be used analytically, namely, Mn(II), V(IV), Fe(III) and Cu(II). An example is the study of non-porphyrin vanadyl complexes in petroleum crudes.[55]

Short-lived free radicals can be detected by a technique known as "spin trapping".[56] In this method a stable diamagnetic compound ("spin trap"), T, is used which will react with a reactive free radical, $R^\bullet$, to give a stable free radical adduct, $A^\bullet$:

$$R^\bullet + T \rightarrow A^\bullet \tag{18}$$

In favourable cases the original free radical can be identified from the EMR spectrum of $A^\bullet$. Probably the most important application of spin trapping is the study of radicals in biological systems.

Most EMR analyses are quantitative and these cover a very wide range of substances; some examples are (i) free radicals in irradiated foods such as seafood, fruit, spices and meat (ii) the effect of additives on food deterioration[57] (iii) naturally-occurring radicals such as melanins (iv) chars in caramels (v) defects in rocks allowing dating to be carried out from the concentration of irradiation defects. An interesting analytical technique is that devised for measuring the concentration of oxygen, known as *oximetry*. This can be done by measuring the broadening caused by oxygen of an EMR line of either a soluble stable free radical (normally a nitroxyl)[58] or of a small crystal of a paramagnetic solid such as lithium phthalocyanine[59] which has a line width of less than 0.002 mT in the absence of oxygen. These materials enable oxygen concentrations as low as 10^{-7} M to be measured in cells, plants, animals, foods, etc. The response time of the method is of the order of milliseconds, hence rapid changes can be monitored easily.

5 IMAGING

5.1 NMR Imaging

There have been enormous developments in the medical applications of NMR imaging - it is generally known as MRI.[60] Proton resonance is normally used in order to get as large a signal as possible , differences in chemical shifts for polar (water) and non-polar substances (fat) or for their T_1 or T_2 values are used to produce an image. Spatial encoding is carried out by applying field gradients in the x, y and z directions. Field strengths up to 3 T are used for large objects and up to 12 T for small objects; resolution for microscopy can be about 10 μm.[44,61] Apart from their application to medical body scanners, the former magnets can be used for "on-line" searching for foreign bodies in large objects, such as cheeses. MRI microscopy can be applied to a large variety of problems in material science, biology and medicine. The application of MRI to food research has been reviewed recently.[62]

5.2 EMR Imaging

By analogy with MRI it is useful to use the abbreviation EMRI for this methodology. Unlike MRI, for EMRI it is usually necessary to introduce something to image, that is to place a paramagnetic substance in the system since both naturally-occurring free radicals and transition metal compounds are not likely to be present in sufficiently large concentrations: however, peroxy radicals have been imaged in coffee beans and chestnuts.[63] Thus spin labels/probes, such as those based on nitroxyl radicals need to be introduced. While it is usual to carry out EMR experiments at microwave frequencies this cannot be done for samples larger than a few mm in diameter if they are "lossy", for example if they contain a large fraction of water. Thus it is found that X-band frequencies are suitable for microscopy but it is necessary to use radiofrequencies of about 300 MHz for samples about 10 cm in diameter.

References

1. R K Harris and B E Mann, "NMR and the Periodic Table", Academic Press,
2. F E Mabbs and D Collison, "Electron Paramagnetic Resonance of Transition Metal Compounds", Elsevier, Amsterdam, 1992.
3. J R Pilbrow, "Transition Ion Electron Paramagnetic Resonance", Clarendon Press, Oxford, 1990.
4. A Abragam and B Bleaney, "Electron Paramagnetic Resonance of Transition Ions", Clarendon Press, Oxford, 1970.
5. D Neuhaus and P A Evans in *Methods in Molecular Biology*, Volume 17, *Spectroscopic Methods and Analyses.*, eds. C Jones, B Mulloy and A D Thomas, Humana Press, Totowa, New Jersey, 1993.
6. A E Derome, "Modern NMR Techniques for Chemical Research", Pergamon Press, Oxford, 1987.
7. J M Saunders and B K Hunter, "Modern NMR Spectroscopy", Oxford University Press, Oxford, 1987.
8. E Breitmaier, "Structure Elucidation By NMR in Organic Chemistry", Wiley, Chichester, 1993.
9. R K Harris, "Nuclear Magnetic Resonance Spectroscopy", Longman Scientific and Technical, Harlow, 1986.
10. R R Ernst, G Bodenhausen and A Wokaun, "Principles of Nuclear Magnetic Resonance in One and Two Dimensions", Clarendon Press, Oxford, 1992.
11. H Friebolin, "Basic One- and Two-dimensional NMR Spectroscopy", VCH, Weinheim, 1991: W R Croasmun and R M K Carlson, eds. "Two-dimensional NMR Spectroscopy", VCH, New York , 1987.
12. M Billeter, *Quart. Rev. Biophys.*, 1992, **25**, 325.
13. T M O'Connell, P G Williams and J T Gerig, *J. Amer. Chem. Soc.*, 1993, **115**, 3048.
14. J T Gerig, *Progress in Nuclear Magnetic Resonance Spectroscopy.*, 1994, **26**, 293. Eds, J W Emsley, J Feeney and L H Sutcliffe.
15. T C Morill, ed., "Lanthanide Shift Reagents in Stereochemical Analysis", VCH, New York, 1986.
16. W H Pirkle and D J Hoover, *NMR Chiral Solvating Agents*, in *Topics in Stereochemistry*, E L Eliel and N L Allinger, eds., 1982, 13, 263.
17. C A Fyffe, "Solid State NMR for Chemists", CFC Press, Guelph, 1983.
18. M Mehring in "NMR: Basic Principles and Progress", Volume 11, Eds. P Diehl, E Fluck and R Kosfeld, Springer-Verlag, Berlin, 1976.
19. M J Gidley, *Trends in Food Science & Technology.*, 1992, **3**, 231.
20. J A Weil, M K Bowman, J R Morton and K F Preston, eds. "Electronic Resonance of the Solid State", Canadian Society for Chemistry, Ottawa, 1987.
21. H Kurreck, B Kirste and W Lubitz, "Electron Nuclear Double Resonance Spectroscopy of Radicals in Solution", VCH, Weinheim, 1988.
22. A Schweiger, *Agnew. Chem. Int. Ed. English*, 1982, **30**, 265.
23. D G Gillies, L H Sutcliffe and X Wu, *J. Chem. Soc. Perkin Trans.2.*, 1993, 2049.
24. M Oki, ed., "Applications of Dynamic NMR Spectroscopy to Organic Chemistry", VCH, Deerfield Beach, 1985.
25. J Sandström, "Dynamic NMR Spectroscopy", Academic Press, New York, 1982.

26. J I Kaplan and G Fraenkel, "NMR of Chemically-exchanging Systems", Academic Press, New York, 1980.
27. A Steigel and H W Spiess, "Dynamic NMR Spectroscopy", in "NMR: Principles and Progress", Vol. 15, Springer-Verlag, Berlin, 1978.
28. L M Jackman and F A Cotton eds., "Dynamic Nuclear Magnetic Resonance Spectroscopy", Academic Press, New York, 1975.
29. D S Stephenson and G Binsch, *J. Magn. Reson.*, 1978, **30**, 145; 1978, **32**, 145.
30. D Marsh and L I Horváth, "Spin Label Studies of the Structure and Dynamics of Lipids and Proteins in Membranes" in "Advanced EPR: Applications in Biology and Biochemistry", ed. A J Hoff, Elsevier, Amsterdam, 1989.
31. J A Weil, J R Bolton and J E Wertz, "Electron Paramagnetic Resonance: Elementary Theory and Practical Applications", Wiley-Interscience, New York, 1994.
32. D Doddrell, V Glushko and A Allerhand, *J. Chem. Phys.*, 1972, **56**,3683.
33. G Lipari and A Szabo, *J. Amer. Chem. Soc.*, 1982, **75**, 4546.
34. D G Gillies, S J Matthews, L H Sutcliffe and A J Williams, *J. Magn. Reson.*, 1990, **86,** 371.
35. P J Bratt, D G Gillies, L H Sutcliffe and A J Williams, *J. Phys. Chem.*, 1991, **94**, 2727.
36. P J Bratt, D G Gillies, A M L Krebber and L H Sutcliffe, *Magn. Reson. Chem.*, 1992, **30,** 1000.
37. J Jonas, *High Pressure NMR Studies of the Dynamics in Liquids and Complex Systems* in *NMR: Basic Principles and Progress: High Pressure NMR*, eds. P Diehl, E Fluck, H Günther, R Kosfeld and J Seelig, Volume 24, Springer-Verlag, Berlin, 1991.
38. D G Gillies, S J Matthews and L H Sutcliffe, *Magn. Reson. Chem.*, 1991, **29**, 823.
39. D J Schneider and J H Freed, *Calculating Slow Motional Magnetic Resonance Spectra*, in *Biological Magnetic Resonance*, Volume 8, ed. L J Berliner, Plenum, New York, 1989.
40. T Gnewuch and G Sosnovsky, *Chem. Rev.*, 1986, **86**, 203.
41. D Marsh, *Tech. Life Science*. B4/11, 1982, **B426**, 1.
42. P T Callaghan, "Principles of Nuclear Magnetic Resonance Microscopy", Clarendon, Oxford, 1993.
43. T L James and G C McDonald, *J. Magn. Reson.*, 1973, **11**, 58.
44. P T Callaghan, C M Trotter and K W Jolley, *J. Magn. Reson.*, 1980, **37**, 247.
45. P Stilbs, *Progress in Nuclear Magnetic Resonance Spectroscopy*, 1987, **19**, 1. Eds. J W Emsley, J Feeney and L H Sutcliffe.
46. J Karger, H Pfeiffer and W Heink, *Adv. Magn. Reson.*, 1988, **12**, 1.
47. J K Moscicki, Y K Shin and J H Freed, *J. Magn. Reson.*, 1989, **84**, 554.
48. J K Moscicki, Y K Shin and J H Freed, in "EPR Imaging and In Vivo EPR", eds. G R Eaton, S S Eaton and K Ohno, CRC Press, Boa Boca, Florida, 1991.
49. S F Patil, N S Rajurka and A V Borhade, *J. Chem. Soc. Faraday Trans.*, 1991, **87**, 3405.
50. L Johansson and J-E Löfroth, *J. Colloid Interface Sci.*, 1991, **142,** 116.
51. C Beadle, D G Gillies, V Iyer, L H Sutcliffe and X Wu, to be published.
52. G J Martin, X Y Sun, C Guillou and M L Martin, *Tetrahedron*, 1985, 3285.
53. H Barjat, P S Belton and B J Goodfellow, *Analyst*, 1993, **118**, 73.
54. H Barjat, P S Belton and B J Goodfellow, *Food Chemistry*, 1993, **48**, 307.
55. N M Atherton, S A Fairhurst and G J Hewson, *Magn. Reson. Chem.*, 1987, **25**, 829.
56. E G Janzen and D L Haire in "Advances in Free Radical Chemistry", Vol. 1, ed. D D Tanner, JAI Press, London, 1990.
57. "Free Radicals and Food Additives", eds. O I Aruoma and B Halliwell, Taylor and Francis, London, 1991.
58. H Hu, G Sosnovsky and H M Swartz, *Biochim. Biophys. Acta*, 1992, **1112**, 161.

59. X S Tang, M Moussavi and G C Dismukes, *J. Amer. Chem. Soc.*, 1991, **113**, 5914.
60. F W Wehrli, D Shaw and J B Kneeland, "Biomedical Magnetic Resonance Imaging", VCH, New York, 1988.
61. "Magnetic Resonance Microscopy", eds. B Blümich and W Kuhn, VCH, Weinheim, 1992.
62. P S Belton, I J Colquhoun and B P Hills, *Ann. Rep. NMR Spectroscopy,* Vol. 26, ed. G A Webb, Academic Press, 1993.
63. A Hochi, M Furasawa and M Ikeya, *Appl. Radiat. Isot.*, 1992, **45**, 33.

NMR in Heterogeneous Systems

P. S. Belton

INSTITUTE OF FOOD RESEARCH, NORWICH RESEARCH PARK, COLNEY, NORWICH, NR4 7UA, UK

1. INTRODUCTION

Simple visual examination of almost any foodstuff will show that, on the microscopic scale, different parts of the food are different from one another. Food, therefore, in general is inhomogeneous. This inhomogeneity typically persists over a variety of length scales. Figure 1 demonstrates this in a piece of bread. In the whole loaf inhomogeneity over the length scale of centimetres is apparent. The crust has a different texture and appearance to the crumb and it is chemically distinct, being lower in water content and partially carbonised. On the scale of millimetres it is apparent that the crumb is a foam consisting of air bubbles supported in a matrix of lipid starch and gluten. Consideration of the matrix shows that on the scale of micrometres a further range of inhomogeneities are present: starch granules are surrounded by a gluten matrix which is in turn coated with a lipid layer. Finally, on a scale of nanometres, local chemical inhomogeneities start to appear. Starch consists of branched amylopectin and linear amylose, gluten is a mixture of different proteins, and so on. At smaller scale (not shown in the diagram) further inhomogeneity becomes apparent, the gluten proteins contain a variety of amino acids; each acid is not a continuous uniform whole, but consists of distinct atomic nuclei surrounded by an inhomogeneous electron density. Nuclei themselves contain different particles. Even at the level of empty space the Uncertainty Principle ensures that elementary particles are constantly created and annihilated.

The example given illustrates some of the major factors that need to be considered when examining the problem of inhomogeneity. The first of these is the principle of course graining in space. For the systems of interest to food science we may define this as:

"provided measurements are carried out on a suitable time scale homogeneity of a system can only be defined with respect to a particular length scale. For any given length scale, greater than sub atomic, which appears homogeneous, there will always be a shorter length scale on which inhomogeneities are apparent".

In essence a system appears homogeneous only when its properties are averaged over some length scale.

The second of the factors to be considered is that of the means of detecting inhomogeneity. Visual inspection gives one view. On the large scale colour and reflectivity differences can give an indication of inhomogeneity, but as the scale is contracted different methods will be required.

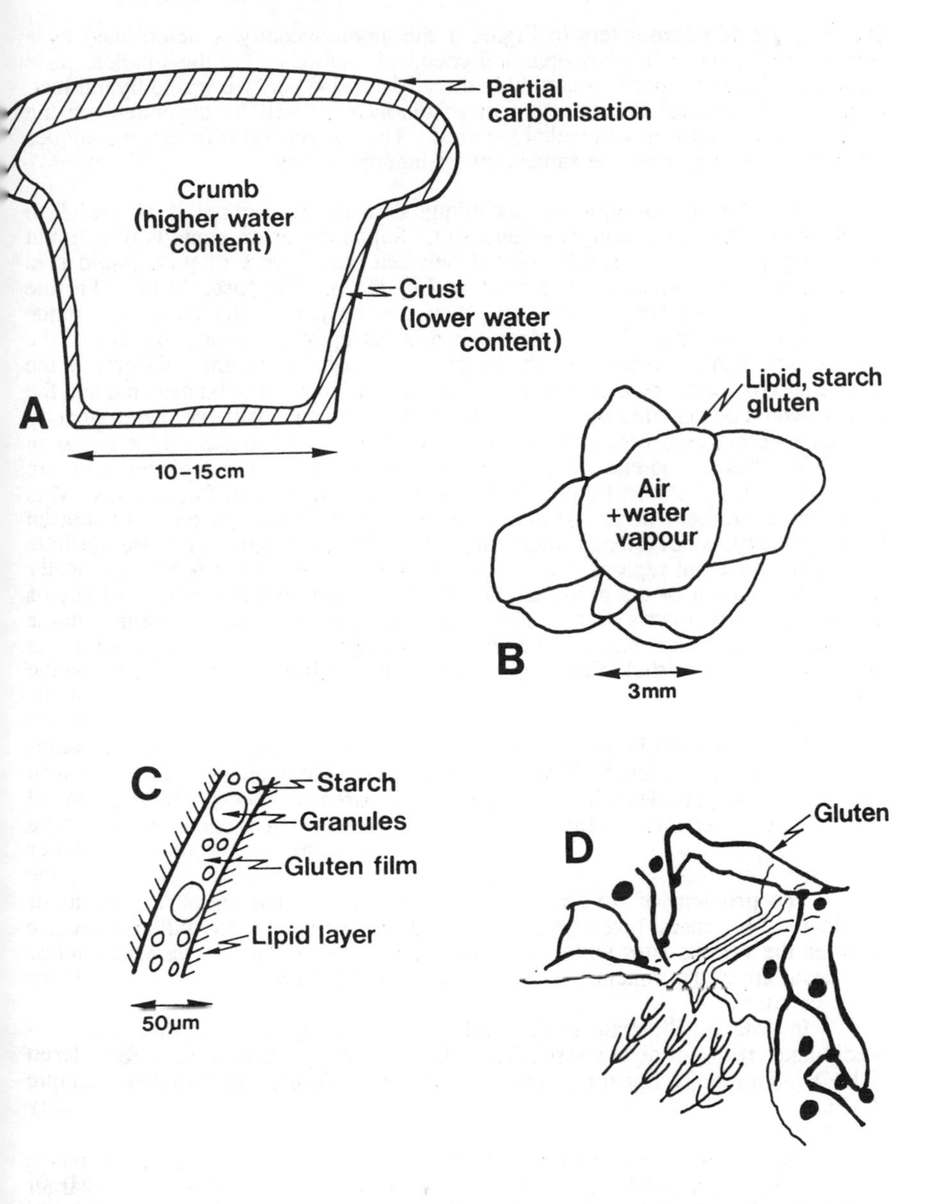

Figure 1. The heterogeneity of a slice of bread on a decreasing length scale. Part D is not strictly to scale but represents biopolymers with dimensions of nanometres.

On the scale of micrometers in Figure 1 the inhomogeneity is determined by a collection of electron microscopic and chemical methods. On the smallest scale the inhomogeneity is partly measurable directly but also partly inferred by indirect methods. In general the detection of inhomogeneity will be dependent on the experimental procedure and technique used. Thus in general different techniques will not necessarily give the same view of inhomogeneity.

In order to consider the technique dependence further it is useful to consider the following thought experiment. Suppose that an annulus of a liquid containing particles is carefully placed between two layers of pure liquid in a tube. This is the situation illustrated in Figure 2a. Suppose, further that the nature of the particles is such that they can only be detected by a unique spectroscopic method. An experiment is now set up to determine the size of the annulus and hence characterise the heterogeneity of the system. Clearly if the unique spectroscopic method is not used heterogeneity will not be detected and the system would be declared homogeneous. If the method is available, heterogeneity will only be detected provided its spatial resolution is adequate. If it can scan quickly and has a good enough spatial resolution the situation is depicted in Figure 2a. The intensity plot in Figure 2a shows evidence of random fluctuations. This is always a problem in real systems and is another limitation on resolution of inhomogeneity, since clearly there must be sufficient signal to noise ratio to distinguish different regions. In the case illustrated here it arises because of the spatial distribution of the particles. There are random and therefore will have a random intensity distribution. This illustrates the principle of spatial course graining. The figure has two levels of inhomogeneity, on the large scale the distribution of the particle filled liquid and on the smaller scale the distribution of the particles.

Even if the spatial resolution is high, there is still the problem of temporal resolution to be considered. If the measuring time is long enough that significant diffusion of the particles takes place during measurement then the boundaries of the region will become increasingly indistinct. This is illustrated in Figure 2b, c and d.

The problem of time resolution is the third of the major factors to be considered. In general heterogeneous systems are dynamic: there will be exchange between the various sites within the sample and the sites themselves will only be time invariant if the system is at thermodynamic equilibrium.

In order to illustrate these points consider a crystalline hydrate which is precipitated from an aqueous solution. For simplicity we assume a formula of $X.H_2O$. Figure 3a illustrates the situation immediately after precipitation of $X.H_2O$.

There is exchange between X in solution and X in solid as well as exchange of water. On a longer time scale (Figure 3b) Ostwald ripening ensures that larger particles will grow at the expense of smaller particles until surface free energy is minimised.

The effectiveness of any method in determining the existence and nature of inhomogeneity will thus depend on its ability to detect those characteristics which differ between different regions and its ability to deal with and characterise the length and timescale over which the inhomogeneity exists.

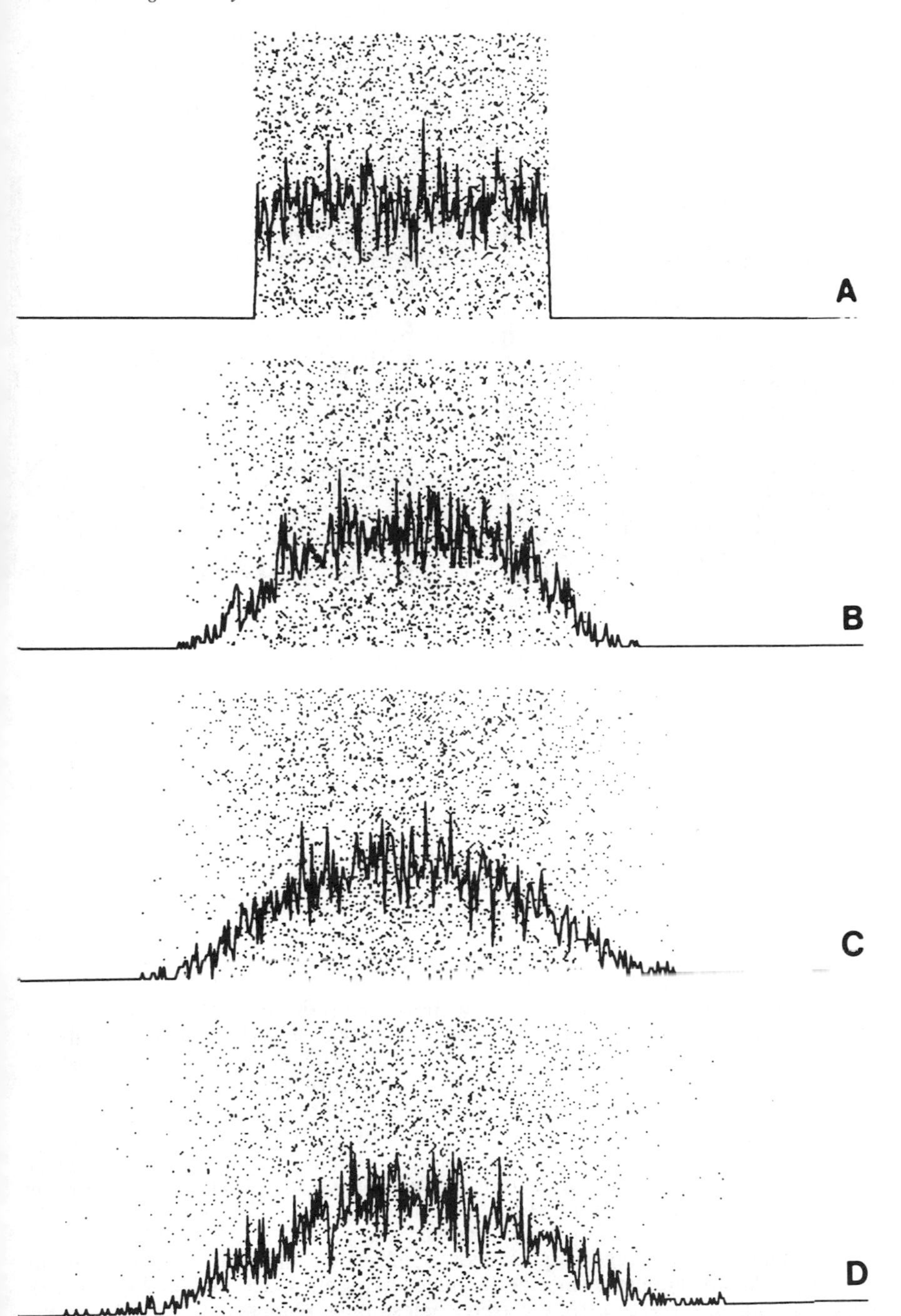

Figure 2. The diffusion of an initially homogeneous distribution of particles. A is the distribution at time zero. B, C, D are at increasing times after time zero.

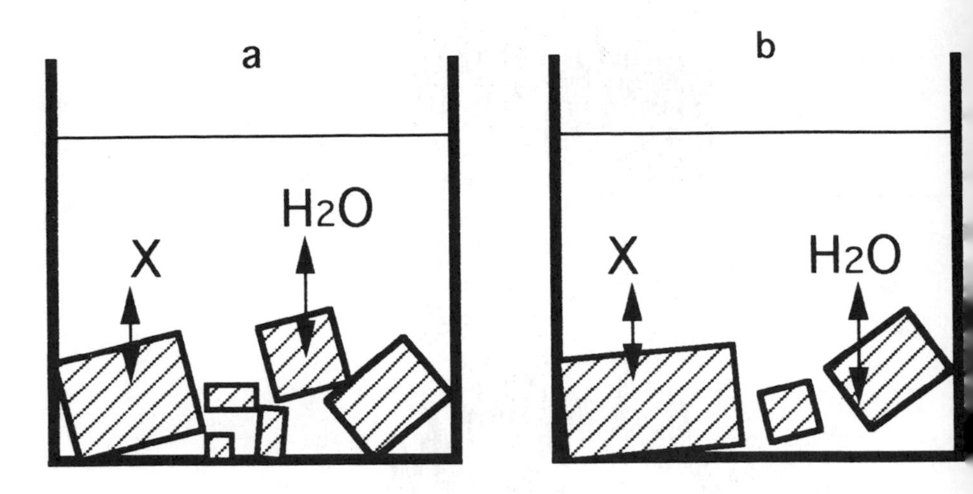

Figure 3. Recrystallisation and exchange in the solid X.H_2O.a is immediately after precipitation, b is after sometime has elapsed.

2. THE USE OF NMR TO DETECT INHOMOGENEITY

The net result of an NMR experiment is always a set of measurements of macroscopic magnetisation. These may be as a function of frequency, or of time, or of both. For many purposes the distinction is not relevant since the process of Fourier Transformation allows interconversion of the time and frequency domains.

Two types of experiments may be distinguished however: those in which the time average frequency is measured and those in which the time dependence of the frequency is measured.

In the first class fall measurements such as chemical shift, fine structure arising from static dipolar or quadrupolar interactions and NMR imaging methods using field gradients. In all of these inhomogeneity is inferred from variations in signal intensity at different frequencies. Some care and subtlety is required in order to make this inference since at the molecular scale different atoms within the molecule will resonate at different frequencies, thus inhomogeneity at this scale will always be apparent. Typically however it is inhomogeneities at a longer length scale that are of interest. Such inhomogeneities may be inferred by a variety of observations. Often the differences in magnetic environment between different regions are such that two or more distinct resonances can be observed. In biological systems different cellular compartments may have different pH's. In this case resonances from species which have pH dependent shifts may be used as indicators of heterogeneity. The phosphorus resonance of orthophosphate is a good example[1]. An example from a gel is shown in Figure 4; this is of potassium resonance in a kappa carrageenan gel[2]. At high temperatures the gel system is homogeneous and in a sol state, as it is cooled a gel is formed and potassium chloride solution is expelled by syneresis. The expelled solution has a narrow line. The potassium in the gel is bound to the carrageenan and thus has a shifted and broadened resonance. In this case that the observed effects are due to heterogeneity can be easily demonstrated by removal of the supernatant liquid and the subsequent disappearance of the sharp line. Such convenience is not usually observed and it is often very difficult to demonstrate conclusively that observed effects do arise from an inhomogeneous distribution of material. One way in which this can be done is by the use of shift reagents whose chemical or physical properties are such that they cannot enter one compartment or region of the system. Such reagents have been used to great effect in biological systems where rare earth complex of tripolyphosphate have been shown not to cross cell membranes[3]. These materials are excellent shift reagents for alkali and alkaline earth metal ions and can be used to distinguish between inter and extra cellular metal ions[3,4].

Static interactions are usually associated with short range, homo or heteronuclear interactions and thus have been used for measuring internuclear distances on the molecular distance scale[5]. However, where there is anisotropic motion the residual interactions may only be averaged out over much longer distances[6,7] by diffusive sampling of a variety of environments. Distances on the scale of a micrometres may be sampled.

A recent paper by Warren and coworkers[8] has suggested that in the presence of field gradients long range dipolar interactions over distances of 10 micrometres may be examined. This offers very interesting possibilities for examining heterogeneity in novel way.

So far all the techniques discussed allow the inference of heterogeneity and may allow some estimate of the size and density of the differing regions. They do not, however, give an indication as to the location of the different regions. They do not therefore allow the creation of maps. This is because the properties measured are related to the local properties that distinguish the different regions. The coordinate system and the metric on which the measurement is based is therefore a local one. In order to examine the distribution and location of regions within the sample a whole sample coordinate system and metric is required. This is supplied by the NMR imaging experiment. In this a gradient of magnetic field is applied across the sample as a whole. The local field and hence the resonance frequency will depend on the nucleus in the sample. In principle any NMR experiment may be carried out, so that heterogeneity may be detected by any NMR method and both the size and the location of the region may be estimated. Very often a simple measurement of intensity is sufficient to identify heterogeneity. Figure 5 shows a cross section of the intensity profile of a powdered sample in which water has filled the interstices between the powder particles. The overall shape is shown by the outline and local variations in proton density are shown by the smaller intensity variations. The large peak arises from a marker in the sample. Note the similarity between this example and the simulation in Figure 2.

The second class of experiments to detect inhomogeneity are those in which the time dependence of frequency is measured. These measurements are essentially dynamic ones in which some kind of relaxation process is examined. The time dependent fluctuations of local frequencies or fields give rise to the relaxation process. Different parts of the sample having different local fluctuation rates will be distinguishable from each other by different relaxation rates and thus heterogeneity is detected.

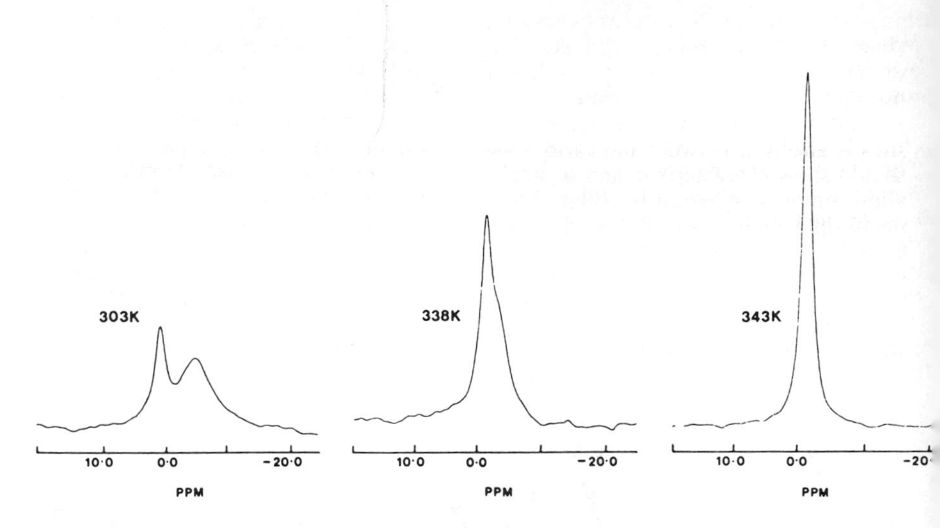

Figure 4. Syneresis in a kappa - carrageenan gel. ^{39}K *spectra are depicted at the temperatures indicated.*

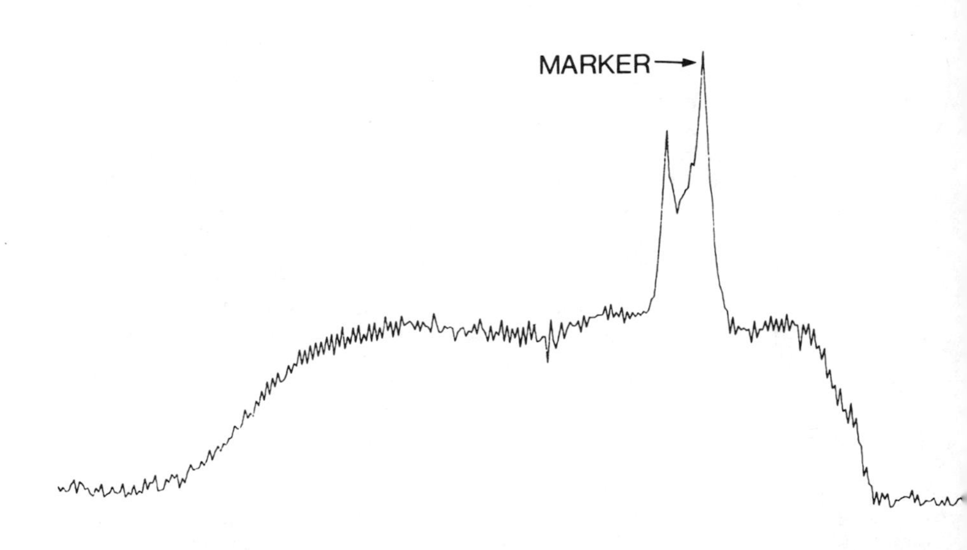

Figure 5. NMR Intensity profile of a water filled powder.

A diagram representing two sites with different properties is shown in Figure 6. Relaxation times in each site are characterised by a time T_i, diffusion coefficient by a time D_i and resonance frequencies by ω. The subscript i denotes the fact that a number of species exists in each site. The relaxation time T can represent any of the measurable relaxation times. If there is no exchange between sites A and B or between the various species within sites then measurements of T_i should show characteristic and separate relaxation processes for A and B and thus allow simple distinction between them. Similar considerations apply for measurements of ω and D. As discussed previously shift reagents may be used to enhance differences in ω. In a similar way relaxation agents may be used to enhance differences in relaxation times.

In practice however the processes of intra-and intersite exchange complicate the situation. Consideration of these is given in the next section.

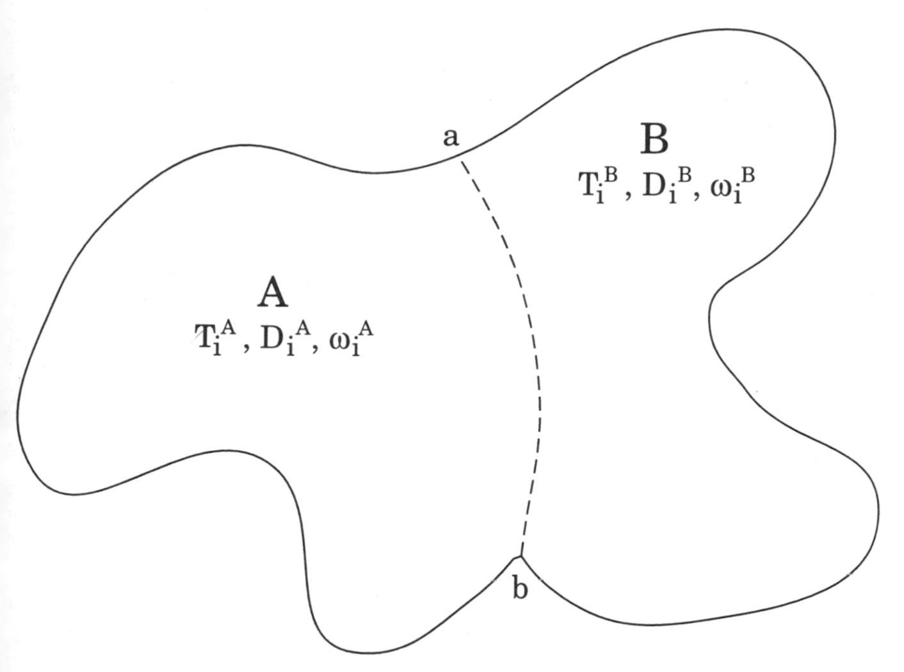

Figure 6. Reaction times, diffusion coefficients and resonance frequencies in a two compartment system.

3. THE EFFECT OF EXCHANGE

When the parameters of interest are sufficiently distinct, the NMR measurement, in the limit of a slow exchange rate, will result in a set of intensities which correspond to the populations of various sites. Such a measurement will contain no information about the exchange rates between the sites. As the rate of exchange increases a point will be reached such that, on the time scale of the experiment, spins from one site will experience the environment of another site. Information about the two sites will be lost in the sense that the distinctness of the sites has been reduced. However, information about the sites is now convoluted with information about the exchange rates between the sites, so that, in principle another parameter of the system becomes available. When exchange between sites becomes very fast all information about the separateness of the sites disappears and only one single observable parameter is available.

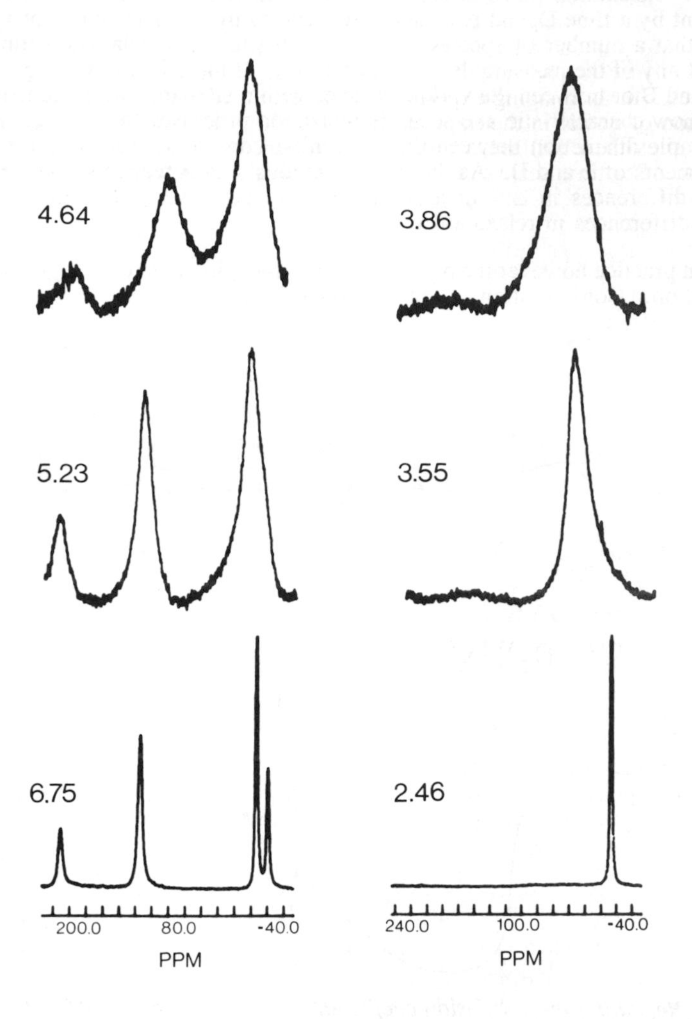

Figure 7. The variation of ^{31}P spectra of a dysprosium tripolyphosphate complex with pH. The pH values at which the spectra were obtained are given by the side of each spectrum.

Figure 7 shows an example of these effects on line-shape[9]. The sample is a dysprosium III tripolyphosphate complex in solution; the spectrum is a ^{31}P spectrum. Initially at low pH, 4 well resolved lines are observed (Fig. 7a). These correspond to complexed tripolyphosphate and free tripolyphosphate. As pH is decreased there is exchange between free and bound tripolyphosphate. This results first of all, in a broadening of lines, then a coalescence into a broad lines, then a coalescence into a broad lines and, finally, the formation, at least for one set of peaks, of single narrow line. In the example shown, in fact, there are some further complications and a more detailed analysis is required[9]. However the general principle of information loss is clear; on going from slow exchange to fast exchange there is a general loss of information and hence structure in the spectrum. In this example structure in the spectrum resulted from chemical differences. In an imaging experiment differences in frequency arise because of the position of nuclei in the imposed field gradient. If there is diffusion between different sites there will be a mixing of frequencies and hence a loss of spatial resolution. The time scale over which the mixing and loss of information occurs is determined by the experimental timescale. In the case of the Figure 7 exchange rates of the order of the frequency differences between the peaks will cause line broadening and loss of resolution. Exchange much faster than this will cause the total loss of information about separate resonances and result in a narrow single line. For relaxation measurements the relaxation time is the criterion by which slow and fast are judged.

When the exchange process is fast compared to relaxation rates only a single averaged process is observed. As exchange slows, two or more processes may be observed depending on the details of the exchange mechanism. However the rates and relative populations of these processes are not in general a simple function of the rates and populations in the various regions [1,10]. Finally in the limit of very slow exchange the relative populations and rates of the different processes represent their rates in the various regions.

To some extent relaxation times are experimentally determined parameters. T_1 is dependent upon the magnetic field strength and $T_{1\rho}$ is dependent on the radio frequency field strength and T_2 when measured by the Carr-Purcell-Meiboom-Gill (CPMG) sequence is pulse spacing dependent[10].

The use of the pulse spacing dependence of the CPMG sequence has proved invaluable in compartmental systems[10]. Even when only one relaxation process is observed the pulse spacing dependence can still be used to obtain evidence of exchange. This is illustrated in Figure 8. When the spacing between pulses is very short compared to the exchange time the mixing sites of different chemical shift is very limited. Thus the mixing of frequencies is small; as the pulse spacing is decreased mixing during the time between pulses increases. Uncertainty as to frequency increases and thus relaxation rate, which is proportional to line width, increases. Finally when the pulse spacing is very long, maximum mixing takes place and the greatest relaxation rate is observed. In this way the dispersion curve marked A in Figure 8 can be readily explained. However, the data in Figure 8 is derived from a sample of Sephadex beads in water[11]. The dispersion marked A results from exchange between water protons and hydroxyl protons on the polysaccharide units of the Sephadex. As the pulse spacing is increased another dispersion becomes apparent, this arises from water outside the beads diffusing into a bead and then undergoing exchange. The rate determining step in this case is not the chemical exchange rate but the diffusion rate. Thus, in this example, chemical heterogeneity becomes apparent by the chemical exchange process but spatial heterogeneity becomes apparent by the diffusive exchange process. The conclusion to draw from this is that rate determining steps may arise from a variety of causes and that it is important to distinguish them in order to properly characterise the system.

In a spatially heterogeneous system such as that shown in Figure 6 translational motion from regions A to B has the potential to mix information from both regions. The boundary a-b may be a membrane, or particle edge, such as the edge of an oil droplet in water, or it may be less well defined such as the edge of

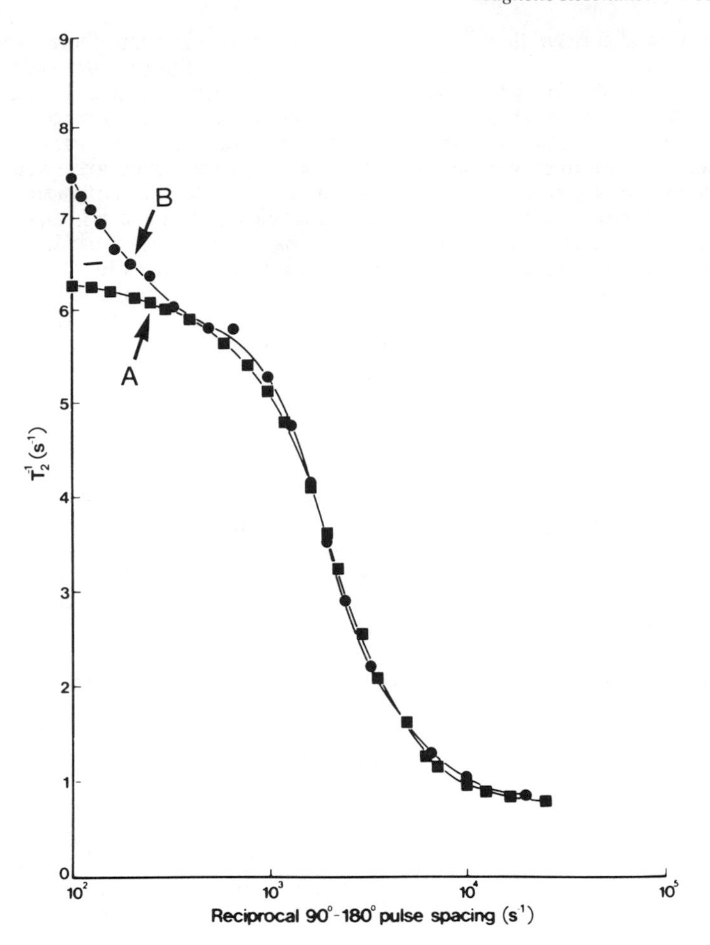

Figure 8. The variation of apparent relaxation rate (T_2^{-1}) with reciprocal pulse spacing in a CPMG pulse sequence. For a discussion of A and B see text.

a gelled particle. Diffusion across it may therefore range from being totally restricted to free. The nature of the diffusion across the boundary will therefore effect the ability to resolve different regions.

In solids translational motion is very restricted but information can still be exchanged between sites by the process of spin-diffusion[12]. Most commonly this is observed in systems which are dense in protons; solid biopolymers are typical examples. The process is a quantum mechanical one which occurs by transfer of magnetic polarisation from nucleus to nucleus. The effect of this is mix polarisation between different sites and therefore to mix relaxation processes which measure polarisation such as T_1 and $T_{1\rho}$.

Whether diffusion is of matter or is quantum mechanical in origin the mathematical expressions describing the mixing are similar and similar phenomena are observed. Quantitatively the effects may be seen by assuming that there is free diffusion across the boundary a-b in Figure 6. Exchange between spins close to a-b will be very easy since they are spatially very close to one another. In this region therefore mixing will be almost complete and averaged relaxation parameters will be observed. Further away from a-b mixing will be less complete and spins very far from a-b will not experience any exchange and will thus retain their own intrinsic relaxation parameters. In practice this implies that for a simple two site system many relaxation processes could be observed. The formal mathematical expression [13] is

$$M_t = \sum_{j=1}^{\infty} A_j \exp(-b_j t)$$

where M_t is the magnetisation at time t and A_j and b_j are coefficients.

The effect of diffusion is create a new level of inhomogeneity. The zone where diffusive mixing takes place rapidly is distinguished from the distant zones because it has some average relaxation time which is distinguishable from the extremes. As the exchange rate increases, or the relative dimensions of the compartments decreases the relative size of the mixed zone increases and at some point will include the whole sample. When exchange rates decrease or the compartmental size increases, the relative size of the mixed zone decreases. In the limit of very large compartments or very slow exchange the size of this zone will become negligible and only relaxations corresponding to the two compartments will be observed.

In general there is no simple way of directly inferring the nature of the heterogeneity of a system from NMR measurements other than imagining. It is usually necessary to set up a model and then test the model against the observed data to evaluate it. Great care must be taken to ensure that the model is physically reasonable and does not contain so many parameters that almost any data set could be fitted. There is a strong analogy between the problems of fitting data where resolution is lost through exchange to problems of the analysis of data where resolution is lost because of instrumental or other effects. One way of approaching this resolution loss is to assume that there is some underlying well resolved data which is broadened by a point spread function. The problem then is to deconvolute the signal from this and the noise. Information theoretic methods have proved particularly valuable in this context [14] and may prove valuable in the analysis of exchange broadened data.

4. THE USE OF TRANSLATIONAL MOTION TO EXPLORE INHOMOGENEITY

From the foregoing discussion it is clear that translational motion has a profound effect on the NMR signal. A very useful way of measuring translational motion is by the application of a magnetic field gradient [15]. The general principle is dealt with in some detail in reference 15. Here only the salient points are considered. At some time a pulse of magnetic field of gradient strength g is applied for a time. This results in the spin experiencing a position dependent resonance frequency given by

$$\omega = \gamma g r + \gamma B o$$

where ω is the resonance frequency, r is the position of the spin in the gradient and Bo is the main magnetic field.

For convenience we redefine ω as the difference in frequency from that when there is no gradient and write

$$\omega = \gamma \mathbf{gr}$$

As the gradient is applied for a time δ the phase angle accumulated is given by

$$\theta = \gamma g r \delta = \omega \delta$$

If, at some later time the sign of the field gradient were reversed and the pulse reapplied the signs of θ and ω would be reversed and provided remained unchanged no net phase shift would result. In practice the field gradient is kept constant and θ is reversed by the application of a 180° pulse[15]. The result of this procedure is the formation of a spin echo; if the value of r for the spin remains unchanged throughout this process the spin echo would have a maximum equivalent to the intensity of magnetisation at the start of the experiment, provided there are no effects of relaxation. If translational motion has taken place then the spin will arrive at a new position r^1 when the second field gradient pulse is applied. The phase shift resulting from this pulse will be given by

$$\theta^1 = \gamma g r^1 \delta$$

and the net change of phase will be

$$\theta^1 - \theta = \gamma g \, (r^1 - r) \, \delta$$

Experimentally this will be observed as a change in echo intensity resulting from translational motion.

The experiment is thus one in which information about position is translated into information about phase shifts. This is analogous to the situation in X-ray scattering where the phase of the electromagnetic wave depends upon the crystal plane from which it was reflected. In the same way as it is possible to interpret the diffraction pattern in terms of a structure so it is possible to interpret NMR signals in a field gradient.

An intuitive way in which to see how motion can explore structure is to consider diffusion in open and constrained systems. Figure 9 shows the path of a randomly diffusing particle in a medium where the diffusion coefficient is spatially uniform. Also shown is the effects of barriers on constraining diffusion. The displacement of the particle is clearly affected by both the presence of the barriers and their shape. The random motion of the spins explores the geometry of the regions in which they are constrained and their response in a field gradient reflects these constraints. In a particularly elegant application [16] of this response it has proved possible to explore the porosity of an array of polystyrene spheres. Hills and Snaar [17] have pointed out that for many systems of interest much structural information can be obtained by light or electron microscopy and that one of the value of the diffusion approach is to exploit it for examining the permeability of the barriers within the system and exchange between different regions.

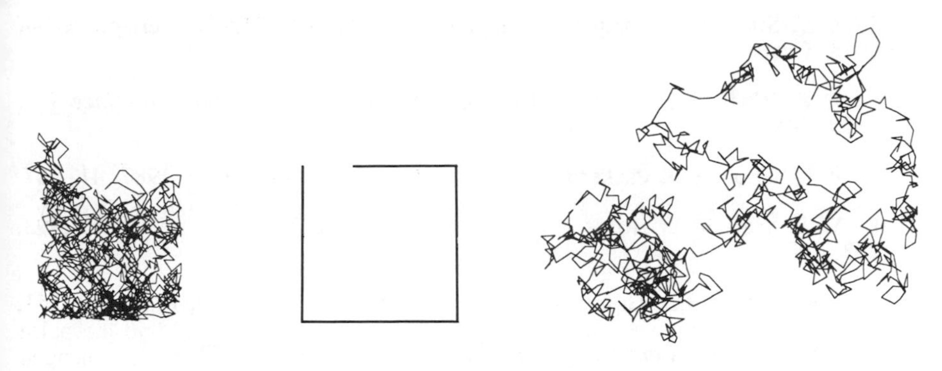

Figure 9. The right hand figure shows a simulation of an unconstrained random walk. The left hand figure shows a similar walk constrained by the barriers shown in the centre.

Random diffusion is an incoherent process. Random motion within a field gradient is a process which causes spins to experience different frequencies. It is therefore analogous to an exchange process between sites with different chemical shifts. In effect the field gradient experiment is one in which the net information loss is measured, information about the time course of the phase of an individual spin is not available. This situation has its analogy in X-ray scattering. In this experiment the intensity of the X-ray signal is measured and information about the phase of the scattered radiation is lost. This is the well known "phase problem" of X-ray diffraction.

In contrast to X-ray methods alternative approaches are available to NMR spectroscopy. One interesting possibility is that of using flow rather than diffusion as the translational probe for structure. Flow is a coherent process and in which locally molecules move in the same direction. This coherence is reflected in the behaviour of magnetisation of a flowing system in a field gradient and may have the potential for dealing with the phase problem.

5. CONCLUSION

NMR is a powerful and versatile method for exploring heterogeniety. Sample preparation is not required and the method is non-invasive. However the effects of heterogeneity are subtle and require subtle interpretation. Recent developments in the interpretation of the effects of translational motion in a field gradient offer exciting opportunities for development.

References

1. P.S. Belton and R.G. Ratcliffe, *Prog. NMR. Spectrosc.*, 1985, **17**, 241.

2. P.S. Belton, V.S. Morris and S.F. Tanner, *Macromolecules*, 1986, **19**, 1618.

3. M.M. Pike, S.R. Simon, J.A. Balschi and P.S. Springer, *Proc. Nat. Acad. Sci.*, 1982, **79**, 810.

4. J.C. Veniero and R.K. Gupta, *Ann. Reps. NMR Spectrosc*, 1992, **24**, 219.

5. C.P. Slichter, "Principles of Magnetic Resonance" Springer Verlag, Berlin, 1978, Chaper 3, p.71.

6. D.E. Woessner, B.S. Snowden and G.M. Meyers, *J. Colloid Interface Sci.*, 1970, **3443.**

7. P.S. Belton, K.J. Packer and T.E Southon, *J. Sci. Fd. Agric.* 1987, **41**, 267.

8. W.S. Warren, W. Richter, A.M. Andreatti and B.T. Farmer, *Science*, 1993, **262**, 2005.

9. S.M. Anson, R.B. Homer and P.S. Belton, *Inorg. Chim. Acta*, 1987, **138**, 241.

10. P.S. Belton. *Comments Agric. Food Chem.*, 1990, **2**, 179.

11. B.P. Hills, K.M. Wright and P.S. Belton, *Molec. Phys.* 1987, **67**, 1309.

12. V.J. McBrierty and K.J. Packer, " Nuclear Magnetic Resonance in Solid Polymers", Cambridge University Press, Cambridge, 1993, Chapter 6, p195.

13. P.S. Belton and B.P. Hills, *Molec. Phys.*, 1987, **61**, 999.

14. B. Buck and V.A. Macaulay "Maximum Entropy in Action", Clarendon Press, Oxford, 1991.

15. P.T. Callaghan, "Principles of Nuclear Magnetic Resonance Microscopy" OUP, Oxford 1991 Ch6.

16. P.T. Callaghan, A. Coy, D. MacGowan, K.J. Packer and F.O. Zelaya, *Nature*, 1991, **351**, 467.

17. B.P. Hills and J.E.M. Snaar, *Molec. Phys.* 1992, **76**, 979.

The Role of Water Mobility in Promoting *Staphylococcus aureus* and *Aspergillus niger* Activities

J. Lavoie and P. Chinachoti

DEPARTMENT OF FOOD SCIENCE, UNIVERSITY OF MASSACHUSETTS, AMHERST, MA 01003, USA

1 INTRODUCTION

Food microbiologists agree that the minimum water activity (a_w) at which microorganisms grow can vary greatly with the type of food, the types and concentration of the solute used to adjust a_w (1), (2), (3), (4). There have been a number of factors proposed to regulate microbial osmotic stress, including accumulation of intracellular solutes (e.g., L-proline) and other metabolic response (5), (6), (7). Osmotic stress, whether brought about by increasing the solute concentration, decreasing the water content, or both, is normally accompanied by changes in other influencing parameters, such as significant decreases in pH, dissolved oxygen and perhaps nutrient availability. Thus, microbial response to limited moisture environment is a complex one and much is likely to do with specific microorganism metabolic pathways controlled by specific genetic make-ups. Nevertheless, food safety and quality is empirically controlled, with respect to water, by using the a_w term.

The issue about whether a_w is the primary factor controlling the water availability, has been raised over the years as scientists has observed that the minimum a_w for microbial growth is always in a range rather than any exact value. Low molecular weight solutes can have large differences in minimum a_w for microbial growth (1). Franks (8), (9) pointed out that it is a simplistic belief that the a_w (or osmotic pressure) is the key determinant for all the various processes. This is due to the fact that osmotic pressure being dependent on the quantity rather than quality of the particles in solution. Additionally, van den Berg and Bruin (10) raised some concerns that most foods are complex and their deviation from ideal solution could result in the inability to reach a true equilibrium. Thus, using a_w to describe kinetically controlled reactions being inappropriate.

1.1 Mold Germination

Questions have been raised about the role of a_w on the ability of a mold spore to germinate (11), (12), (13), (14). An Aspergillus parasiticus germination experiment by Lang (11) suggested that the type of the solid substrate (and not a_w) determines its spore germination. In this experiment, a starch substrate at 0.86 a showed no germination at all, while a sucrose substrate at same a_w did. Similar results suggested to be a result of different water binding energies (12), (13), (14). Slade and Levine (15) proposed that mechanical relaxation of food components determined water translational mobility and this can be better understood, explained, and predicted by the glass transition concept. They also suggested potential use of terms, such as Tg', Wg' and Tm/Tg, in predicting mold germination. Bolton et al. (16) found that bacterial vegetative cells and spores could not grow in a glassy gelatin medium but they and their toxin remained viable for a significant period of time.

1.2 Water Mobility

Many physical methods, such as proton NMR and dielectric relaxation, have been employed to determine water mobility. More recent literature shows that water is perturbed in mobility by solid surfaces but the majority is far from being tightly bound, i.e., its mobility is closer to bulk water rather than to ice (17), (18). Thus, even though a_w of a given system is reduced, water molecules in the food system can still be dynamically active and available. Thus, from the reasons above, it might be useful to investigate the role of water mobility in governing microbial activity (by affecting transport of moisture, O_2, and nutrients, etc.). If so, studying water mobility in the media that the microorganisms grow on might lead to a better understanding regarding to the effect of water.

NMR water mobility is used in this work. The preliminary results from Aspergillus niger spore germination have been used to compared to ^{17}O NMR data. This nuclei stands out compared to ^{1}H and ^{2}H mainly because ^{17}O NMR data is not complicated by the cross-relaxation and proton/deuterium exchange involving non-water nuclei, that contribute to ^{1}H and ^{2}H data (19), (20), (21), (22), (23). It is probably the most powerful nucleus for studying water (24), (25), (26), (27), (28).

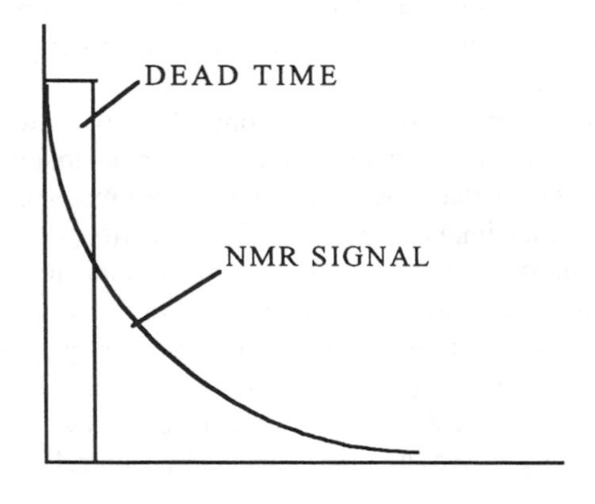

Figure 1 *Free induction decay of NMR signal.*

Chinachoti and Stengle (27) have shown that, in sucrose-starch-water mixtures, varying sucrose, starch or water content resulted in a drastic change in ^{17}O NMR line width and the signal intensity. Not all ^{17}O nuclei were detected by the spectrometer as determined from the total intensity of a given sample and pure water. The weight fraction of water that was detected increased with moisture and sucrose content.

The undetected water was the water that relaxed rapidly (short relaxation time) and slowly exchanged with the rest of the water within the time frame of the experiment (200 msec deadtime). Thus, the water already relaxed before the signal recording started (T_2 less than 200 msec) became detected water (Figure 1). LW of the NMR spectra thus only indicated the average mobility of the detected, mobile water. Such drastic changes in both the detected water and the line width with sucrose at a given a_w, showed that the water mobility and the exchange rate were modified with changing composition even though a_w was kept constant.

The germination time of A. niger was studied by Kou et al. (29) and was found to vary significantly with sucrose and water content. When plotted against a_w, it was clear that a_w failed to predict mold germination in media of varying composition. When plotted against the NMR linewidth and relative intensity (calculated detected water), the relative intensity showed a far better correlation eventhough the relationship was found curvilinear (Figure 2)

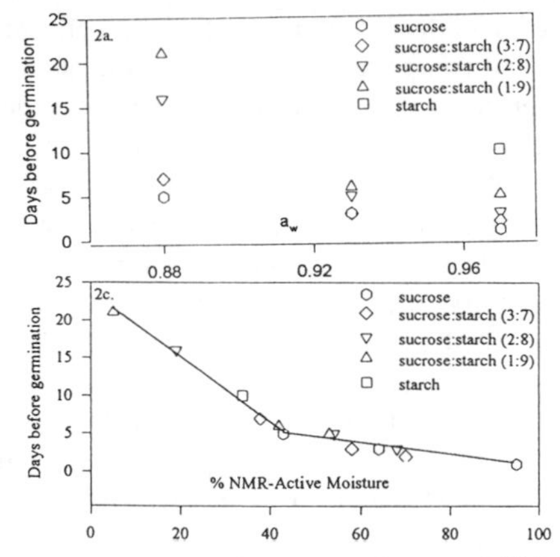

Figure 2. *Mold germination time as function of a_w and ^{17}O NMR relative intensity expressed as NMR active (detected) water.*

It seems from the demonstration above that a_w is a not very good predictor of mold germination if the food composition is to change drastically and it seems that NMR intensity detected may indicate the presence of bulk or free water and thus a better indicator. However, one must be careful before such conclusion can be made since this is only a preliminary data on starch-sucrose media systems which are not designed to take all microbiological factors into consideration. For instance, there is a vast difference in nutrients available to A. niger among the media containing varying amount of sucrose:starch ratio. Thus, the difference observed here could reflect the difference due to both moisture and carbon sources. Nevertheless, the movement of sucrose in the media depends on its ability to be dissolved in water and mobility of the water may be a good indicator.

There is a serious question whether the results shown above is rather an effect of sucrose as a more accessible carbon source than starch, thus, in samples of higher sucrose always resulted in a faster germination and growth. Additionally, how much other changes in physical and chemical properties of the media play a role here; these include the viscosity effect, oxygen, and pH. It is most likely that changing one parameter would affect many others.

There are many situations in food when, in practice, food composition is not allowed by producers to change dramatically or when selected solutes used are similar in functionality that a_w may be found to be very well correlated with the microbial activity (30) and thus a_w in this case is quite useful.

1.3 Bacterial Growth

The minimum a_w for growth of Staphylococcus aureus varies vastly with the types of the solutes used to lower a_w (2), (4), (31), (32), (33) and the mode of sorption (3). a_w alone is not the only factor controlling growth of the toxin producing bacteria (4), (33), (34), (35), (36). Thus, if it continues to be used as a quality control parameter in food products, a simple modification in food ingredient may lead to a possibility of serious food poisoning and intoxication.

For decades scientists have been using the a_w concept to explain how water molecules are bound to solid components in food. This is quite a simplistic model and a_w is being extended to mean more than what it was originally defined (i.e., the ability to escape to vapor). Nevertheless, a_w has been important in regulating moisture in food, processing criteria and shelf-life. However, the very mechanism that a cell respond to moisture still need to be clarified. There has been many hypotheses proposed but practically cellular response to dryness can be divided into two categories, 1) a cell must maintain its membrane integrity, and 2) a cell must prevent moisture loss effect, either by losing less water from the cell or accumulating some osmoregulatory or compatible solutes to maintain and protect organelle integrity (37). The fact that most solutes in food do not readily penetrate into and plasmolyse the cells, makes it reasonable to picture their effect could be much rather external and osmolarity of the media is expected to give some indication of the osmotic stress of the cells. The question is whether a cell actually senses a_w of the surrounding per se (8).

The objective of this work is to show preliminary results exploring the possible role of water mobility in bacterial activity in model food systems using ^{17}O NMR relaxation to provide useful information about the linewidth (relaxation time) and the signal intensity of detected water.

2 MATERIALS AND METHODS

2.1 Staphylococcus aureus

Brain Heart Infusion (BHI) was used as the growing mediums containing varying amount of raffinose, glycerol or NaCl keeping a constant solute/BHI ratio and varying only the water content. Staphylococcus aureus was maintained on nutrient agar slants in a 4 °C incubation chamber and transferred weekly. 100 ml of sterilized media was inoculated with the culture (acclimatized to 30°C for several hours prior to inoculation). After an incubation period (overnight at 30°C), the log phase cells were inoculated into the

a_w adjusted media (0.1 ml inoculum/30 ml media) before incubated at 30 °C in shaking flasks. Cell counts were obtained by absorbance at 600 nm with a Beckman spectrophotometer (with predetermined standard curve of absorbance versus cell number) and by actual cell count by pouring plate technique. All absorbance data was confirmed by the values obtained from the plate count technique.

2.2 NMR Analysis

^{17}O spectra of the samples were obtained at 25°C under a Varian XL-300 spectrometer (Varian Instruments, Palo Alto, CA) operating at 40.662 MHz. A 90° pulse width (15 µsec), proton decoupling, and spectral width of 20 KHz were used. Maximum sensitivity was obtained by fast acquisition (acquisition time, AT=0.02sec) and a large number of repetitive accumulations (number of scans, NT = 4000- 60,000). A 200 µsec delay (dead time) between the pulse and the acquisition was also introduced to suppress the rolling base line effect (27). T_2 relaxation time was calculated as $1/(\Delta\nu)$, where $\Delta\nu$ was the linewidth at the spectral half-height. The total intensity of the observed ^{17}O NMR signal was determined from intensity integration. Signal intensities of known amounts of liquid water were used as the standard curve. A comparison of intensity data with known moisture levels of the samples allowed calculation of weight of NMR-active moisture, and the fraction of total moisture which contributed to the NMR signal (27). All NMR and water sorption data gave experimental error within 3 percent.

3 RESULTS AND DISCUSSION

3.1 Staphylococcus aureus

Growth curves for Staphylococcus aureus were found to indicate a significant decrease in the growth rate at a decreasing moisture content as expected. Raffinose containing media showed a double exponential curve (Figure 3) in 65-83% moisture range, indicating a complex metabolic response to the media. Since raffinose is not

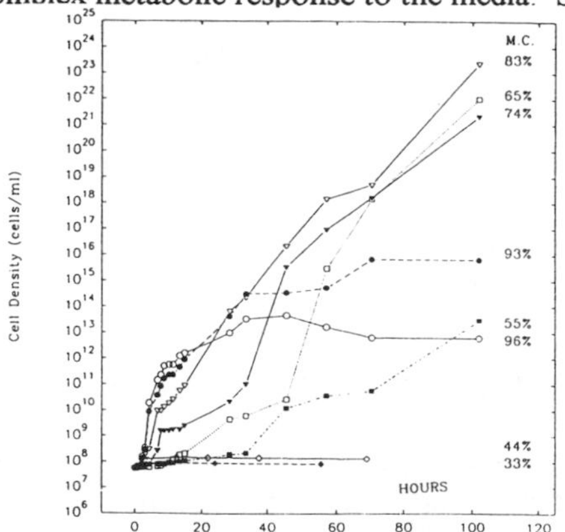

Figure 3. Growth *curves for Staphylococcus aureus in raffinose containing media.*

supposed to be utilized by Staphylococcus aureus, this could not possibly be the response due to an increase in utilization of raffinose. pH measurement showed no sign of dramatic reduction in pH indicating no fermentation of raffinose. This has made the determination of the growth rate and generation time complicated and thus, only the cell counts are used here. Other media gave a single exponential growth.

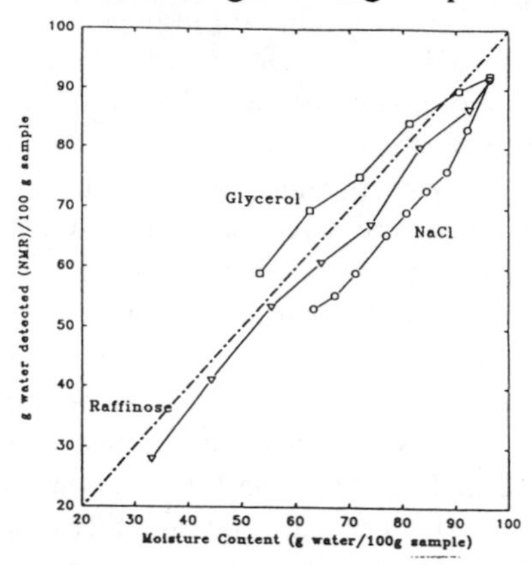

Figure 4. *Amount of detected water observed by NMR in Staphylococcus aureus media at various moisture content.*

The media used for Staphylococcus aureus resulted in strong NMR signal intensity which was found to increase with moisture content. Compared to control bulk water signal, the signal obtained from all samples represented almost all of the water present (Figure 4). This means that the majority of the water was highly mobile.

Linewidth plotted against moisture content showed linebroadening at lower moisture content for all media, depending on the media composition (Figure 5).

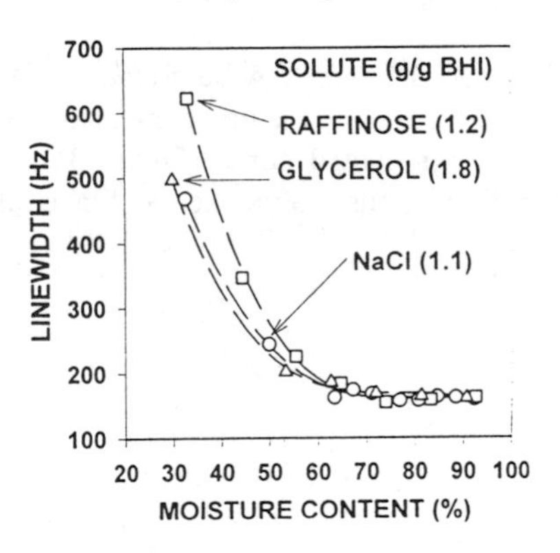

Figure 5 *^{17}O NMR linewidth versus moisture content in the media used.*

Comparing line width with the growth data, we observed some discrepancies. For instance, the raffinose samples showed the greatest line broadening effect at 30-50% moisture content when there was little or no growth observed. On the other hand, The NaCl containing media showed no line broadening effect but a significant inhibitory effect at<70% moisture content. Glycerol showed similar result over the moisture content studied.

The time during the lag phase is also an important parameter. It was found that the lag time was correlated curvilinearly with a_w and the intensity data (Figure 6). The interesting phenomena is that the raffinose curve and glycerol and NaCl curves switched their position (Figure 6a vs 6b). NaCl and glycerol resulted in a longer lag period than raffinose of the same NMR detected water. Or, in order to have the same lag phase effect, glycerol and NaCl are more inhibitory and thus the media did not have to be lowered in the amount of water as much as raffinose. Figure 6a shows that glycerol and NaCl have a stronger colligative property and thus were able to lower a_w further than raffinose.

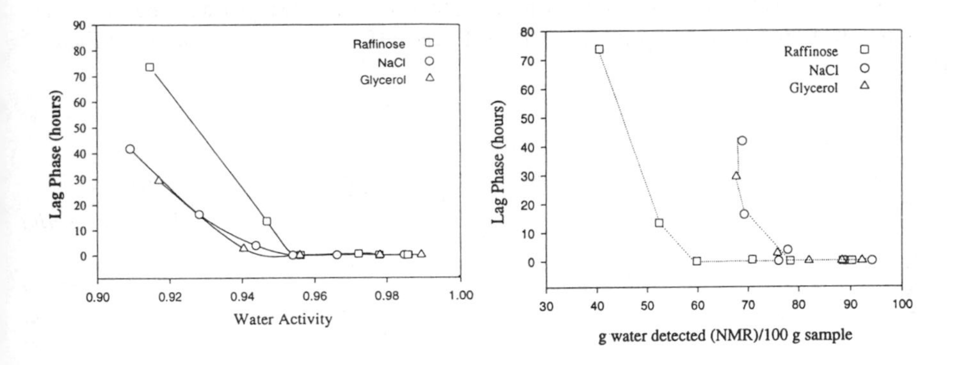

Figure 6 *Lag time for Staphylococcus aureus plotted against a_w and NMR intensity (detected water).*

The viable cell density of three media containing NaCl (1.1:1 NaCl:BHI ratio), raffinose (1.2:1 raffinose:BHI ratio), and glycerol (1.8:1 glycerol:BHI ratio) are plotted against incubation time in Figure 7. All positive growth curves showed different initial growth rate and a short stationary phase before a drop in cell density. At some lower moisture content, the there was a rapid death in the microorganisms; this occurred at different moisture content. The cell density values after 24 hours of incubation at 37°C was plotted against moisture content, a_w, linewidth, and NMR intensity in Figure 7. The data shows that while there was no correlation found with the linewidth data, there was some correlation found with the other parameters. Within the experimental variation, one can conclude that there was a solute dependence of these curves for all cases (Figure 7a, 7b, 7d). However, for the case of detected intensity, the curves for NaCl, raffinose and glycerol were not very far apart and may be found not very significantly different. For the case of raffinose (Figure 7d), at <60% water detected value when the microorganism stopped from growing, the presence of raffinose some how allowed some but small growth. This small growth extended far to as low as 25 % detected water.

It should be considered that these three different solutes are markedly different in their physiological nature, i.e., NaCl and glycerol can penetrate into the cell but raffinose cannot, glycerol can be utilized by Staph. aureus but raffinose cannot, and raffinose resulted in a much more viscous media than the other solutes. Additionally, it was found that only glycerol resulted in a very significant drop in pH (due to fermentation of glycerol by the microorganisms) and thus the effect of solutes in this case is also influenced by the change in pH. With these factors in mind, the ^{17}O NMR intensity data in Figure 7d gave a reasonably good correlation in particularly at higher detected water content.

It is interesting to note that, if we base the safety criteria on the critical point where there is no growth in the media, a_w is a fairly acceptable empirical parameter since the x-axis intercept where there is no growth fell within 0.84-0.89 range.

Raffinose behaved differently than the other two solutes probably because the microorganism responded to raffinose very differently (from the growth curves in Figure 3).

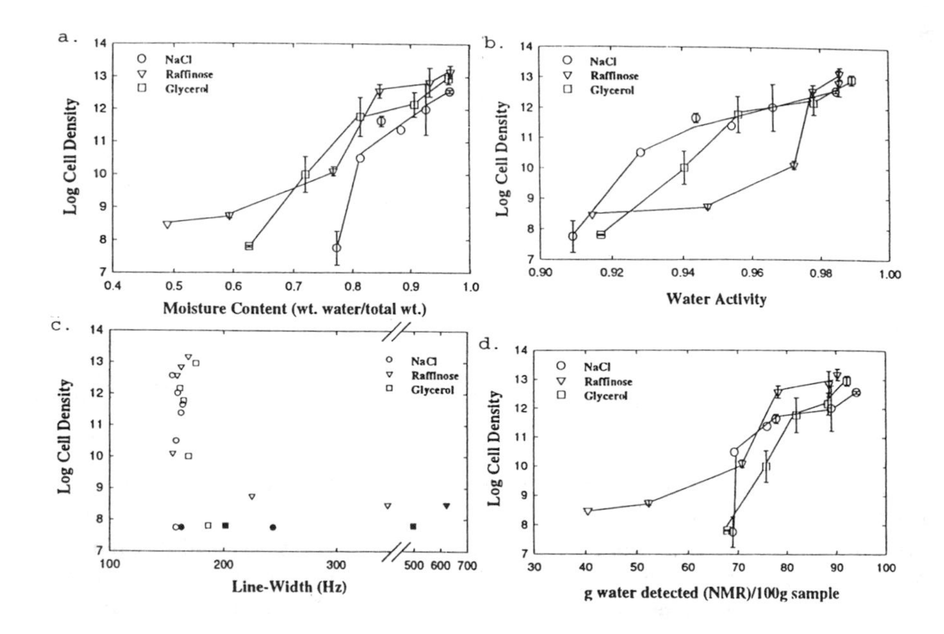

Figure 7 *Cell density after 24 hours of incubation at 37 °C plotted against the four parameters studied.*

4 SUMMARY

This investigation demonstrates the powerful application of ^{17}O NMR technique to help understand the role of water mobility in microbial activity. It can be seen from this preliminary work that the types of solutes play a critical role and the use of water parameters alone may not be adequate.

Germination of molds was found to be better correlated with the NMR detected intensity although more results are needed in real food systems and in varying species. Staphylococcus aureus ability to grow a low moisture systems seemed to be solute specific and complicated by the change in water mobility due to different solutes. However, due to the complication of different metabolic response to solutes, this work could not lead to specific conclusion on how critical the water NMR linewidth or intensity influences the bacterial growth.

The use of the NMR can help monitor the water availability factor as related to the microbial activity in food. This could have an impact on moisture control to assure food quality and safety. If found useful, this technique will serve as a more accurate measure of water availability rather than a_w alone. Already, our preliminary data shows that it can be applied to A. niger germination in carbohydrate systems. Thus mold spoilage is a possible application which could help minimize many spoilage problems in food and food crops, such as corn, wheat, and processed food. This would help lower waste caused by molds. In addition, a better control of toxin producing mold would enhance the safety control in the food supply. If the technique will be proven applicable to predict bacterial growth, it could have a major impact in the food safety area preventing food-borne poisoning related to low and intermediate moisture foods.

References

1. L.B. Rockland and G.F. Steward, 'Water Activity-Influences on Food Quality', Academic Press, NY, 1981, p. 825.
2. M. Plitman, Y. Park, R. Gomez, and A. J. Sinskey, J. Food Sci., 1973, 38, 1004.
3. K. M. Acott and T. P. Labuza, J. Food Sci., 1975, 40, 137.
4. G. Vaamonde, G. Scarmato, J. Chirife and J. L. Parada, IWT (Lebensmittel-Wissenchat und-Technologic), 1986, 19, 403.
5. J. Bae and K. J. Miller, Appl. and Envir. Microbiol., 1992, 58, 471.
6. J.Chirife, C.F. Fontan, O.Z. Scorza, J. Food Technol., 1980,15,383.
7. J. C. Measures, Nature, 1975, 25, 398.
8. F. Franks, Cereal Foods World, 1982, 27, 403.
9. F.Franks, Trends in Food Science and Technology,1991, March, 68.
10. L.B. Rockland and G. F. Stewart, ' Water Activity - Influences on Food Quality', Academic Press, NY, 1981, p1.
11. K. W. Lang, 'Physical, chemical and microbiological characteristics of polymer and solute bound water', Ph.D. thesis, Dept. of Food Science, Univ. of Illinois at Urbana-Champaign,1980.
12. F.-F. Chu, Relationship of water content and composition of food to mold spore germination. M.Sc. Thesis, Rutgers University, New Brunswick, NJ, 1984.
13. S. W. Paik, 'The state of water in food components related to germination of mold spores', Ph.D. Thesis, Rutgers University, New Brunswick, NJ, 1984.
14. G. Charalambous, 'Shelf-life of foods and Beverages', Elsvier Sci. Publ. Amsterdam, 1986, 791.
15. L. Slade and H. Levine, Pure & Appl. Chem, 1988, 60, 1841.
16. J. M. V. Blanshard and P. J. Lillford, 'The Glassy State in Foods', Nottingham University Press, UK, 1992

17. L. B. Rockland and L.R. Beuchat, Water Activity: Theory and Applications to Foods', 1987, chapter 11, 235.
18. P. Chinachoti, Food Technology, 1993, January,134.
19. H. T. Edzes and E. T. Samulski, J. Magn. Reson., 1978., 31, 307.
20. W. B. Wise and P. E. Pfeffer, Macromolecules, 1987, 20, 550.
21. J.Bremer,G. L. Mendz and W. J. Moore, J. Am. Chem. Soc., 1984, 106, 4691.
22.W.M. Shirley and R.G. Bryant, J. Am. Chem. Soc.,1982,104, 2910.
23. S. H. Koenig, R. G. Bryant, K. Hallenga and G. S. Jacob, Biochem., 1978, 17, 4348.
24.F.Franks, Biophysics of water, John Wiley&Sons Ltd., NY, 1982.
25. L. Picullel, J. Chem. Soc. Faraday Trans., 1986, 82, 387.
26. A. Mora-Gutierrez, I. C. Baianu, J. Agric. Food Chem.,1989, 37, 1459.
27. P. Chinachoti and T. R. Stengle, J. Food Sci., 1990, 55, 1732.
28. P. Chinachoti, V. A. White, L. Lo and T. R. Stengle, Cereal Chem., 1991, 68, 238.
29. Y. Kou, S. J. Schmidt and P. Chinachoti, Unpublished Results.
30. J. Chirife, J. of Food Engr., 1994, 22, 409.
31. J. Chirife, G. Vaamonde, G. and G. Scarmato, J. Food Sci., 1982, 47, 2054.
32. G.Vaamonde, J.Chirife and O.C. Scorza, J. Food Sci.,1982,47, 1259..
33. G.Vaamonde, J.Chirife and G.Scarmato, J. Food Sci.,1984,49,296.
34. S. Notermans and C. J. Heuvelman, J. Food Sci., 1983, 48, 1832.
35. D. Simatos and J. L. Multon, 'Properties of Water in Foods in Relation to Quality and Stability,' Martinus Nijhoff Publishers, Boston., 1985, 247.
36. S. Ewald and S. Notermans, Inter. J. of Food Micro., 1988, 6, 25.
37, C.C. Seow, 'Food Preservation by Moisture Control', Elsvier Applied Science, New York, 1988, 43.

Acknowledgement Support from Massachusetts Agricultural Experiment Station number MAS 00709 and USDA Grant number 93-37500-9252. Special thanks to Yang Kou for his preliminary work and Profs. Ronald Labbe and Robert Levin for their advise on microbiological aspect of this work. Use of the NMR facility and close collaboration with Prof. T. R. Stengle of Chemistry is appreciated.

Analysis of Molecular Motions in Food Products

Marcus A. Hemminga

DEPARTMENT OF MOLECULAR PHYSICS, WAGENINGEN AGRICULTURAL UNIVERSITY, PO BOX 8128, 6700 ET WAGENINGEN, THE NETHERLANDS

1 INTRODUCTION

Many food systems are multi-component, multiphase systems with complicated structures that are generally not in equilibrium. In such systems, a variety of changes often occurs, which at best are only partly understood. These macroscopic changes are induced on a molecular level by structural changes in the molecules and by molecular motions. An example is the glass-rubber transition of solid food materials, which is of importance for the processing, quality, and storage stability. Of special concern in this respect is the effect of water and small organic molecules that act as plasticizers of the food material.

Electron Spin Resonance (ESR) spectroscopy and saturation transfer (ST) ESR are powerful magnetic resonance tools to obtain information about the rotational mobility of spin-labeled molecules in food materials over a very large motional range of eight decades from 10^{-3} to 10^{-11} s. For glassy systems, this information is related to the presence of molecular cavities, solvent interactions, and arrangement of the hydrogen-bonded network in the matrix.

Another example of the application of ESR spectroscopy is found in the study of wheat proteins. The bread making quality of flours is related to the quantity and nature of the glutenins and gliadins, which are water-insoluble proteins. These form a protein network, called gluten, that is established upon hydration and subsequent mixing. ESR and ST-ESR spectroscopy are used here to provide information about the protein network by specifically labelling the cysteine residues of the proteins with maleimide spin probes.

The present paper gives an overview of work carried out in collaboration with food scientists in France (Jeremy Hargreaves and Martine Le Meste from the "Ecole Nationale Supérieure de Biologie Appliquée à la Nutrition et à l'Alimentation" (ENSBANA) in Dijon, and Yves Popineau from the "Institut National de la Recherche Agronomique" (INRA) in Nantes) and The Netherlands (Marcel Roozen, Cornelius van den Berg, and Pieter Walstra from the Department of Food Science, Wageningen Agricultural University).

2 ESR SPECTRAL ANALYSIS

A survey of the various motional regions and characteristic ESR and ST-ESR spectra for isotropic motion of nitroxide spin labels is shown in Figure 1. In ESR spectroscopy the effect of molecular motion is to average the anisotropic magnetic interactions that arise from the g-factor tensor and hyperfine tensor of the nitroxide. The molecular motion is generally expressed by a rotational correlation time τ_R. If the motion is very fast, i.e. $\tau_R < 10^{-11}$ s, the ESR spectrum of nitroxide spin probes consists of three sharp lines of equal height and a spacing of about 1.5 mT. As the motion becomes progressively slower, there is a differential broadening of the lines in the spectrum, while the line positions remain constant. This is the fast motional region. For values of $\tau_R > 3\text{x}10^{-9}$ s a distortion of the line positions and line shape is observed. This region is called the slow motional region. In the very slow motional region ($\tau_R > 10^{-6}$ s) the rigid powder spectrum is reached for conventional ESR, in which the full effects of the anisotropic magnetic interactions are observed. In this region, however, saturation transfer ESR results in spectra that are still sensitive to molecular motions up to a final limit of $\tau_R \approx 10^{-3}$ s.

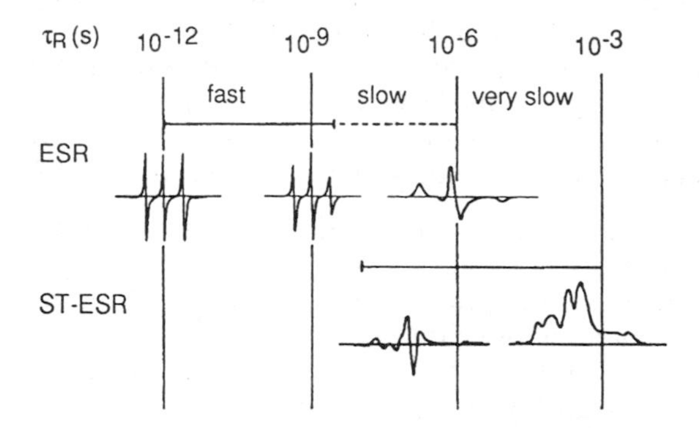

Figure 1 Survey of the various motional regions and the corresponding characteristic ESR and saturation ESR spectra for isotropic motion of a nitroxide spin label with rotational correlation time τ_R (Courtesy Ref. 1)

In the fast motional region, the rotational correlation time (in s) may be obtained directly from the spectra from the following equation:[1]

$$\tau_R = 6.5\text{x}10^{-6}\, B_0 \left(\sqrt{h_H/h_C} - 1\right), \tag{1}$$

where h_H and h_C are the heights of the high field and central lines in the ESR spectra, respectively. B_0 is the linewidth of the central line in tesla (T). The rotational motion of the spin probes is assumed to be isotropic. It should be noted that Equation (1) applies for ESR spectra that have been obtained in the X-band microwave region at a microwave frequency of 9.5 GHz.

In the slow motional region of the spin probes it is possible to characterize the changes in the spectral line shape by a parameter $S = A'_z/A_z$, where A'_z is the separation of the outer hyperfine extremes in the ESR spectrum, and A_z is the rigid limiting value for the same quantity. From computer simulations[2] the following empirical relation has been obtained:

$$\tau_R = a(1 - S)^b. \quad (2)$$

The constants a and b depend on the nature of the motion of the spin probe and of the intrinsic linewidth of the spectra. For a Brownian diffusion model it follows that $a = 1.09 \times 10^{-9}$ s and $b = -1.05$.[3] This model has been employed in the analysis of ESR spectra of probes in glassy materials.[4, 5]

For rotational correlation times $\tau_R > 10^{-6}$ s, conventional ESR spectroscopy yields a so-called rigid-limit powder spectrum that is insensitive to the rate of motion. However, by carrying out ESR spectroscopy under saturation conditions of the microwave radiation, a non-linear effect arises that is usually referred to as saturation transfer (ST) ESR spectroscopy. ST-ESR spectra are sensitive in the very slow motional region (10^{-6} s $< \tau_R < 10^{-3}$ s). In ST-ESR spectroscopy the rotational correlation time τ_R is usually obtained in an empirical way from calibration curves of the line shape against the value of τ_R obtained from reference spectra of spin-labeled hemoglobin that are recorded at various temperatures in anhydrous glycerol (for reviews, see Refs. 1 and 6).

3 SPIN PROBES

In general food systems do not posses intrinsic paramagnetism and hence in the unlabelled state do not give rise to an ESR spectrum. The introduction of a stable free radical ("spin label") thus enables one to use ESR spectroscopy to study specific molecular environments within the food system. The spin label almost exclusively used is the nitroxide radical, which has a very high stability. The radical has a three-line nitrogen hyperfine structure whose splitting varies with the orientation of the nitroxide moiety with respect to the externally applied magnetic field, which is necessary for obtaining ESR spectra. This makes the ESR spectra very sensitive to rotational motions, as illustrated in Figure 1. The nitroxide group can be attached to various molecules with different functionality. The motion of the nitroxide will then directly reflect the motion of the labeled part. The spin label group must of course introduce some steric perturbation, but this effect is generally relatively small.

Two useful examples of spin labelled molecules for application in food materials are shown in Figure 2. Nitroxide spin probe I (4-hydroxy, 2,2,6,6-tetra-methyl-piperidino-oxyl (TEMPOL)) is an alcohol derivative that has been applied to monitor changes in hydrogen bond formations in carbohydrate-water systems. For spin labelling proteins, several nitroxide analogues of standard protein modifying agents can be used, e.g. the maleimide, iodacetamide, and isothiocyanate derivatives. These agents are capable of labelling nucleophilic groups by aqueous reaction, particularly sulphydrils or amino groups, with the specificity to some extent depending on the reaction conditions. It is generally observed that amino groups are not labelled very extensively at pHs below their pK_a and thus sulphydryl group labelling is normally favoured at neutral pH. In the present work results obtained with maleimido spin label II covalently attached to sulphydryl groups of gluten will be shown. Because of the high sensitivity of ESR spectroscopy, only a small amount of labelling (about 1%) is required. Spin label ESR spectroscopy can be seen as complementary to NMR spectroscopy in determining molecular motion. However, the advantage of spin label ESR is that this information is obtained from specifically labelled sites in the system.

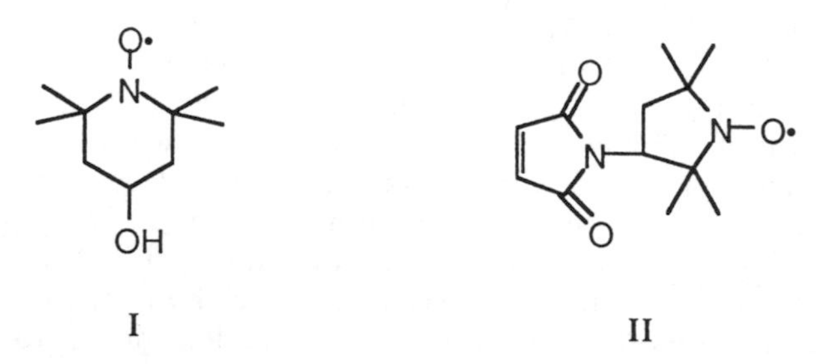

Figure 2 *Chemical structure of the spin labels used in motional analysis of food materials. Spin label I is the TEMPOL spin label, and spin label II a maleimide spin label that is used to covalently label cysteine groups of protein molecules*

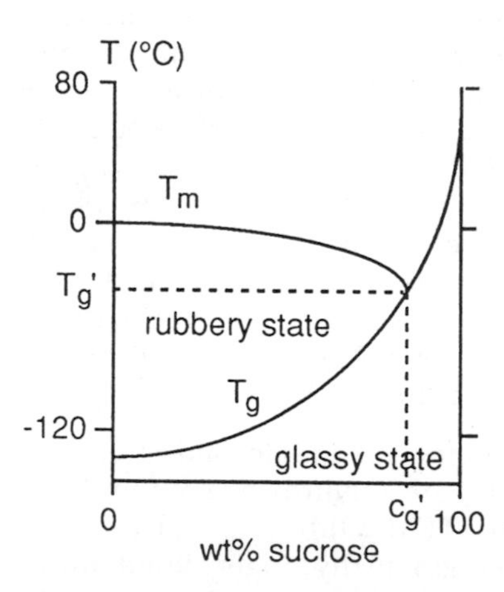

Figure 3 *Simplified state diagram of temperature against wt % sucrose for an aqueous solution of sucrose.* T_m *is the freezing curve, and* T_g *the glass transition temperature curve*

4 THE GLASS TRANSITION TEMPERATURE

Many biopolymer materials show a glass transition at which the solid phase displays a discontinuous change in the specific heat on changing the temperature. At the glass transition temperature (T_g), there is not only a sudden change in thermal and mechanical properties of the system, but also an extreme decrease in the rates of molecular motion. The glass transition temperatures vs. composition of an aqueous solution of glass-forming sucrose are shown in Figure 3. The T_g value particularly important for aqueous systems is the one attained at maximal freeze concentration, T_g' in Figure 3. T_g' is of crucial importance for food processing at low temperatures.

4.1 Malto-oligose-Water Mixtures

The rotational correlation time for spin probe I after rewarming of a rapidly cooled 20% maltotriose-water mixture is shown in Figure 4. Between -70 and -20°C log τ_R decreases almost linearly with temperature. A sudden change in the temperature dependence of the rotational mobility can clearly be observed at a temperature of -23°C. This temperature is very close to the special glass transition temperature T_g' obtained from differential scanning calorimetry[7] (T_g'= -23.5°C). Table 1 presents the results for other 20% malto-oligosaccharide-water mixtures and a comparison with the results obtained by Levine and Slade,[7] showing an excellent agreement.

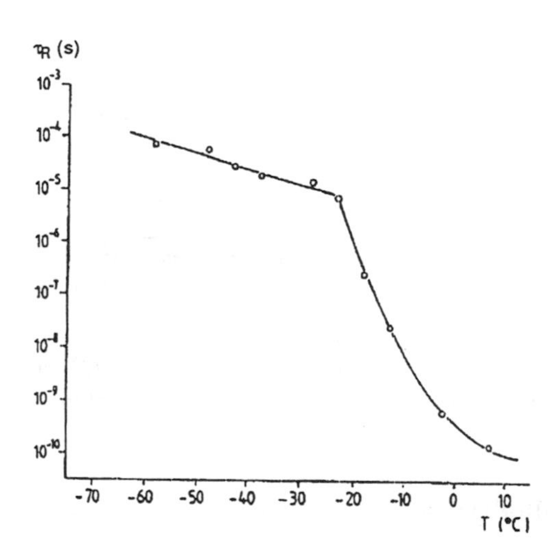

Figure 4 *Rotational correlation time τ_R of TEMPOL spin probe I as a function of temperature in a 20% maltotriose-water mixture. (Courtesy Ref. 8)*

Table 1 *The special glass-transition temperature (T_g') for rapidly cooled 20% malto-oligosaccharide-water mixtures*

Oligosaccharide	$T_g'^{a)}$	$T_g'^{b)}$
Maltotriose	-23	-23.5
Maltopentaose	18	-16.5
Maltoheptaose	-14	-13.5

[a]*Determined by spin-probe ESR spectroscopy (±1 °C)*[8]
[b]*Determined by differential scanning calorimetry by Levine and Slade*[7]

The relative decrease of τ_R, as shown in Figure 4 is much smaller than the relative change in macroscopic viscosity, which varies from about 10^{12} Pa.s at the glass transition temperature to about 10^3 Pa.s at temperatures 20°C above the glass transition. This observation indicates that the spin probe is present in cavities in the lattice of the amorphous glass.

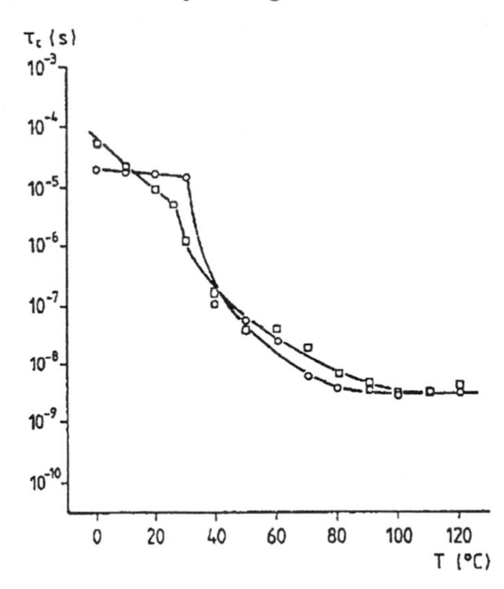

Figure 5 *Rotational correlation time τ_R of the TEMPOL spin probe I as a function of temperature in a maltodextrin-water system. (o) 12.7 wt% water; (□) 12.7 wt% water and 8 wt% glucose*

4.2 Commercial Maltodextrin

The ESR technique can also be applied to materials used in the food industry, such as enzymatically converted starch, which is a mixture of dextrin molecules of different molecular weight. In Figure 5 the rotational mobility of spin probe I in a commercially available sample of maltodextrin (average molecular weight of 8000 Da) is shown as a function of temperature.[8] The sudden change in the temperature dependence of the rotational correlation time τ_R at 30 °C represents the glass-rubber transition. It has recently been shown by Van den Berg et al.[9] that similar observations can be made of spin probe I incorporated in various water-carbohydrate (fructose, maltose, sucrose, trehalose) systems.

In Figure 5 it is observed that by adding 8% (by weight) glucose to the polymer results in a decrease of the glass transition temperature by 5°C, indicating that glucose acts as a plasticizer in maltodextrin.

4.3 Activation Energy For Molecular Reorientation

It has been shown by Roozen et al.[8] that the temperature dependency of the correlation time of spin probes in the glassy state is described by an Arrhenius type equation:

$$\tau_R = \tau_a \exp(E_a/RT) \tag{5}$$

The parameter τ_a is a temperature independent factor, R and T have their usual meaning; E_a is interpreted as an activation energy for molecular reorientation.[8] The values of E_a for the spin probe I in glassy malto-oligosaccharide-water mixtures are given in Table 2.

As discussed before, spin probe I is present in cavities in the glass, however, the values for τ_R and E_a indicate that the spin probe does not rotate freely in the cavities. The activation energy E_a corresponds not only to the number and strength of hydrogen bonds formed between the spin probe and its surrounding. but is also determined by the free volume of the system. This latter effect is observed in Figure 5 on the addition of glucose to maltodextrin: the strong increase of E_a from 8 to 60 kJ/mol is related to the decrease in the free volume.

Table 2 *Apparent activation energy E_a for rotation of TEMPOL spin probe I in glassy 20% malto-oligosaccharide-water-ice mixtures*

Oligosaccharide	*E_a (kJ/mol)*
Maltotriose	17
Maltopentaose	14
Maltoheptaose	26

5 GLUTEN SYSTEMS

Gluten consists of a mixture of glutenins and gliadins, as is illustrated in Figure 6. The visco-elastic properties of gluten depend largely on the relative proportions of the two components. Too few gliadins in proportion to glutenins results in a rigid mass, and too many gliadins make it too slack. In native gluten, the proportion of gliadins is about 40%.

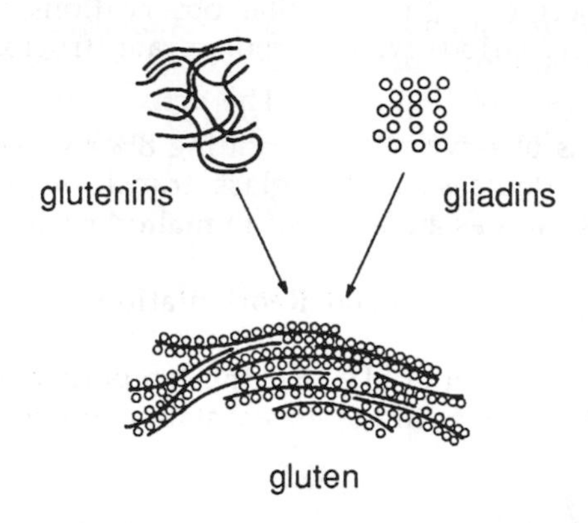

Figure 6 *Schematic illustration of a gluten system consisting of glutenins and gliadins*

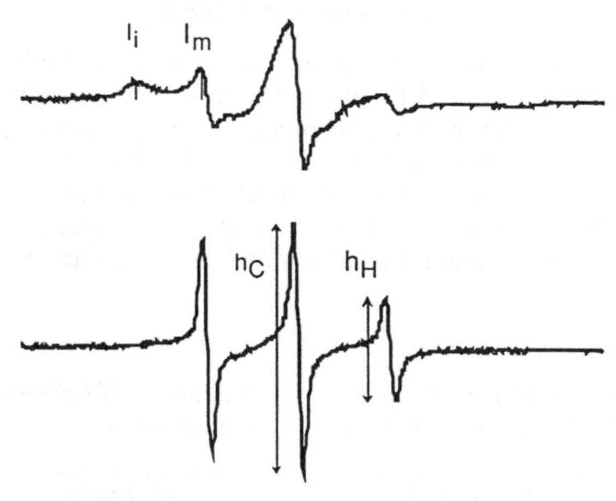

Figure 7 *ESR spectra of hydrated, functional, spin-labelled gluten., using maleimide spin label II. The R factor is given by the ratio* I_m/I_i. *The line height ratio* h_H/h_C *determines* τ_R *as given by Equation (1). The temperature is 20°C*

The ESR spectrum of a gluten system, labelled with spin label II is shown in Figure 7. The spectrum consists of a "mixture" of labels at different positions on the proteins, giving rise to composite ESR spectra in which at least two populations of spin labels largely differing in molecular mobility can be observed. Two parameters are used to describe the spectra. The factor R (see Figure 7) reflects the number of immobile with respect to mobile spin labels.

The rotational correlation time τ_R is used to describe the molecular mobility of the mobile spin label population.

The values of R and τ_R decrease with increasing temperature, indicating an increase in molecular dynamics both by a transfer of spin labels from the immobile to the mobile population and by an increased motion of the mobile spin labels. Both parameters show a linear dependence when drawn in an Arrhenius plot (Figure 8). The activation energy for the motion derived from the τ_R values is about 17 kJ/mol for all gluten systems. The ESR spectra show a monotonal variation of molecular motion (Figure 8) up to 90 °C. Also the temperature dependence of the R parameter shows no break, indicating that the local molecular motion as sensed by the maleimide spin label on the cysteine residues changes gradually. This effect is proposed to be related to a reduction of hydrogen bonds possibly corresponding to a disruption of α-helices in the gluten system.[10]

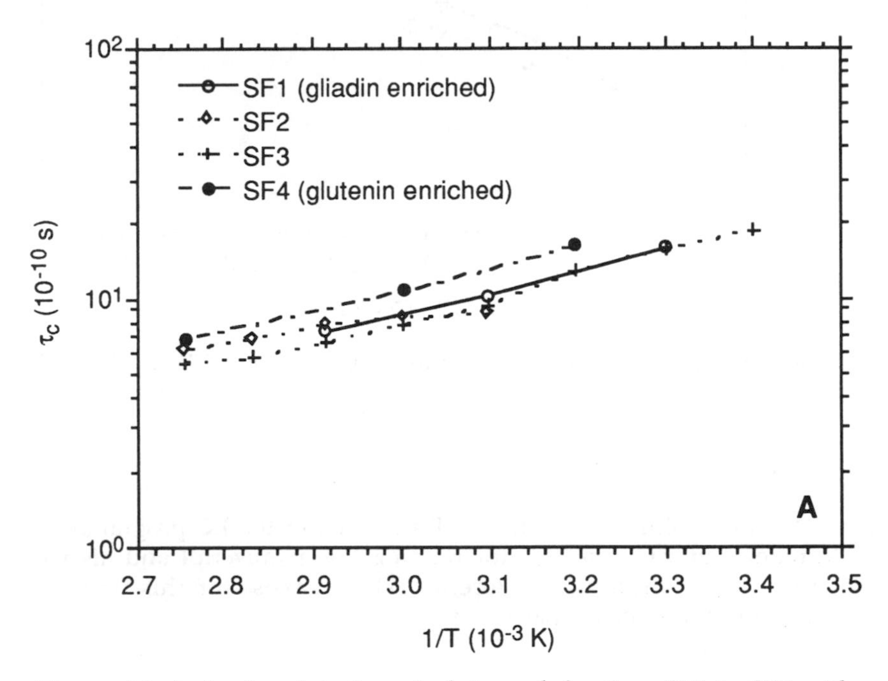

***Figure 8A** Arrhenius plot of τ_R of gluten sub-fractions SF1 to SF4 with different gluten/gliadin ratio*

6 CONCLUSIONS

The work presented here shows that ESR spectroscopy is a powerful technique to obtain information about molecular mobility in various food systems. In carbohydrate-water systems the glass transition temperatures can be obtained that agree very well with those obtained from differential scanning calorimetry measurements. In addition the technique provides information about hydrogen bonds, and presence of cavities in these systems. The ESR technique is also suitable for the characterisation of gluten proteins on a molecular basis. This information is complementary to techniques such as microscopy and rheology, which reflect the macroscopic network organisation.

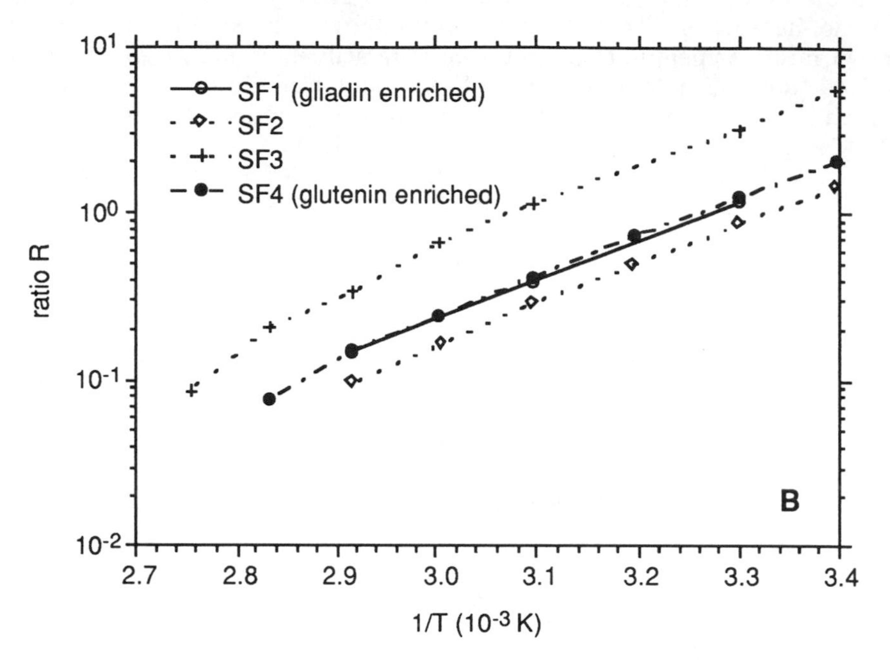

Figure 8B *Arrhenius plot of R of gluten sub-fractions SF1 to SF4 with different gluten/gliadin ratio*

Acknowledgements

This work was subsidized by Unilever Research and the EC programme AIR. I wish to thank Cor van den Berg for reading the manuscript and his valuable comments. I am grateful to Jeremy Hargreaves for his enthusiastic collaboration in the gluten project.

References

1. Hemminga, M.A. *Chem. Phys. Lipids* **32**, 323 (1983).
2. Freed, J.H. in *Spin Labeling* (eds. Berliner, L.J.) p. 53 (Academic Press, New York, 1976).
3. Hyde, J.S. *Methods Enzymol.* **49**, 480 (1978).
4. Le Meste, M. & Simatos, D. *Cryo-letters* **1**, 402 (1980).

5. Roozen, M.J.G.W. & Hemminga, M.A. *J. Phys. Chem.* **94**, 7326 (1990).

6. Hemminga, M.A. & De Jager, P.A. in *Biological Magnetic Resonance. Spin Labeling - Theory and Applications* (eds. Berliner, L.J. & Reuben, J.) p. 131 (Plenum Press, New York, 1989).

7. Levine, H. & Slade, L. *Cryo-Lett* **9**, 21 (1988).

8. Roozen, M.J.G.W., Hemminga, M.A. & Walstra, P. *Carbohydr. Res.* **215**, 229 (1991).

9. Van den Berg, C., Van den Dries, I.J. & Hemminga, M.A. *Molecular mobility around the glass transition in sugar-water systems* (Second International Conference on Applications of Magnetic Resonance on Food Science, Aveiro, Portugal, 19-21 September 1994).

10. Hargreaves, J., Popineau, Y., Le Meste, M. & Hemminga, M.A. *The effect of heat on gluten as seen by ESR spectroscopy* (Second International Conference on Applications of Magnetic Resonance on Food Science, Aveiro, Portugal, 19-21 September 1994).

Enzyme Activity in Low Water Content Biopolymer Systems Studied by ^{19}F NMR

J. J. Cappon[1], J. S. Hebblethwaite[1,2], J. M. V. Blanshard[2], and P. G. Morris[2]

[1] MAGNETIC RESONANCE CENTRE, DEPARTMENT OF PHYSICS, UNIVERSITY OF NOTTINGHAM, UNIVERSITY PARK, NOTTINGHAM NG7 2RD, UK

[2] DEPARTMENT OF APPLIED BIOCHEMISTRY AND FOOD SCIENCE, UNIVERSITY OF NOTTINGHAM, SUTTON BONINGTON CAMPUS, LOUGHBOROUGH LE12 5RD, UK

1 INTRODUCTION

1.1 Enzyme Activity in Food Systems

The knowledge and understanding of parameters that determine enzyme activity in foodstuffs are of great importance to the food industry. The action of enzymes can change the properties and quality of food during processing or storage. This effect may be desirable, for example in the manufacture of cheese and bakery products, or undesirable, leading for example to loss of flavour and freshness. In both cases, an understanding of the factors affecting enzyme activity is crucial, if better control is to be effected.

The activity of enzymes in foods cannot simply be extrapolated from their activity in aqueous solution. Many food systems are characterized by a relatively low moisture content and can be described in terms of state diagrams, ranging from the solution state, through the rubbery state and into the glassy state[1]. Even at very low water contents, enzymes exhibit activity provided the protein is sufficiently hydrated to maintain its active conformation[2,3]. The main effects on the extent of enzyme activity in food systems are probably due to decreased water content and restricted mobility. One of the major parameters that may determine enzyme activity under these conditions is the flexibility or internal motion of the enzyme, especially the motion of residues and substrates at the active site. Diffusion processes may be another limiting factor, including diffusion of substrates and products to and from the active site, and possibly diffusion of the protein itself.

To study the phenomena that are significant for enzyme activity in food systems we address in this investigation the changes in the molecular motion of the enzyme, the substrate and the enzyme-substrate complex, as the water content is gradually decreased from the solution state to the rubbery state. Nuclear magnetic resonance is ideally suited for the study of systems in which changes in molecular motions are involved, as mobility at the molecular level can be probed by relaxation time measurements. However, the restricted mobility and the complexity of food systems will reduce the resolution of the NMR spectra compared to solution systems, making their interpretation more difficult. The main limitations will be excessive line broadening and overlapping signals. One way of overcoming these limitations is to use isotopic labelling to selectively enhance specific resonances of the substrate and the protein. Specific ^{19}F-labelling in combination with ^{19}F NMR is a powerful strategy for the study of complex systems, because of the high receptivity of ^{19}F to NMR detection, its extremely wide range of frequencies and the absence of background signals[4]. Labelling of the substrate and the enzyme with ^{19}F will enable us to selectively study resonances of the substrate and the enzyme in complex food-like systems. The relaxation behaviour of these resonances as a function of matrix properties (particularly water content) will then give information about the dynamics of the enzyme-substrate system under these conditions. The primary interest is in the molecular motions at the active site because of the likely relationship with activity. Insight into the active site dynamics may be gained by studying the relaxation behaviour of the substrate bound in the active site relative to the relaxation behaviour of the free substrate. Information about the relaxation behaviour of the enzyme molecule will then allow us to correlate the active site dynamics to the overall dynamics of the enzyme.

As food systems are very diverse and complicated in their nature we decided to develop a model system that allows the study of the general features of enzyme activity in foods. Such a model system must consist of an enzyme, a substrate and a food-like matrix. Hen egg white (HEW) lysozyme has been chosen as the enzyme to examine, since it is an extensively studied and well characterized enzyme[5,6]. Its three-dimensional structure has been determined by X-ray crystallography[7,8] and by ^{1}H NMR[6]. Its catalytic mechanism is well understood[9] and it has been the subject of numerous molecular dynamics studies[10,11]. It is a relatively small, globular enzyme (15 kDa) and therefore relatively easily accessible by NMR. Lysozyme is found in many foodstuffs, particularly in milk products and eggs, and catalyzes the hydrolysis of β(1-4)-glycosidic bonds in bacterial cell wall polysaccharides and chitin[5], both consisting of *N*-acetylmuramic acid and *N*-acetylglucosamine. Since these complex polysaccharides are difficult to obtain in a well-defined form, studies of lysozyme activity have mostly concentrated on the corresponding mono- and oligosaccharide substrates. In our studies we have used monomeric derivatives of *N*-acetyl-D-glucosamine (GlcNAc, **1**, Figure 1) as substrates. These compounds act as competitive inhibitors and, consequently, their chemical structure will not be affected during the experiments. Both kinetic and structural data for these enzyme-inhibitor systems are available from the literature[5].

Specific ^{19}F-labelling of *N*-acetyl-D-glucosamine derivatives has been reported in the literature starting from the commercially available *N*-acetyl-D-glucosamine and D-glucosamine[12,13,14]. We have chosen to label the acetyl group by introduction of a monofluoroacetyl group because of the relative ease of the synthetic method, involving a one-step monofluoroacetylation of the 2-amino group of D-glucosamine[12]. This resulted in *N*-monofluoroacetyl-D-glucosamine (GlcNFAc, **2**), in both α- and β-anomeric forms. We have used this compound as a model substrate for lysozyme and probed the dynamics of the active site of the enzyme by studying the relaxation behaviour of both the free and the bound substrate. To obtain insight into the overall dynamics of this enzyme system we have also labelled the enzyme molecule and measured its relaxation parameters. The ^{19}F-labelling of lysozyme has been performed by means of trifluoroacetylation of the ω-amino groups of the lysine residues according to a method described by Adriaensens *et al.*[15]. Lysozyme contains six lysine residues which are relatively easy accessible for chemical modification because the

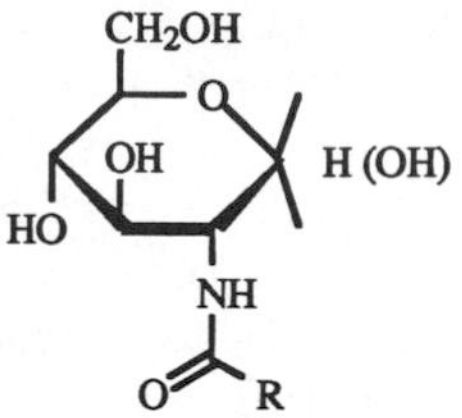

(1) GlcNAc, R = CH_3

(2) GlcNFAc, R = CH_2F

Figure 1. *Lysozyme inhibitors N-acetyl-D-glucosamine (GlcNAc, 1) and N-monofluoroacetyl-D-glucosamine (GlcNFAc, 2).*

side chains are exposed on the surface of the protein and the amino groups can be modified under mild conditions. Since this chemical modification has effect on the activity of the enzyme we have employed the labelled lysozyme only for the determination of the overall mobility of the enzyme. Investigations in which the enzyme-substrate complex was involved have been carried out using native lysozyme only.

Ultimately, we intend to study enzyme activity in food systems ranging from the solution state through the rubbery state to the glass-rubber transition. For this purpose we need a food-matrix material that displays such behaviour and allows sample preparation in the corresponding range of water contents (100-20 w/w%). Furthermore, these samples must be as homogeneous as possible to minimise inhomogeneity effects in the NMR measurements. We have chosen gelatin as a biopolymer matrix since it is a widely applied model food system that fulfils these requirements. Gelatin is based on collagen and thus has a protein structure. Polysaccharide materials, such as agarose, amylose, amylopectin and polyglucose have been considered, but, as lysozyme acts on polysaccharides, they all have the disadvantage of possible interaction with lysozyme. Thus, the model system we designed for our studies consists of lysozyme, *N*-monofluoroacetyl-D-glucosamine and a gelatin matrix.

To study the relaxation behaviour of the enzyme-substrate system as a function of its matrix, we need to consider the theory of chemically exchanging systems that applies to NMR of enzyme-substrate systems.

1.2 Theory of Exchanging Enzyme-Substrate Systems

As a result of the binding of the substrate to the active site of the enzyme, the substrate experiences two environments in which its NMR parameters, including chemical shift (δ) and spin-spin relaxation rate ($1/T_2$), are different. The effect of the exchange process depends on its rate relative to the difference in NMR parameters. If the substrate exchanges at the active site at a rate greater than the resonance frequency difference between the two environments, then we are in the fast exchange limit. In this regime the NMR experiment shows a single averaged resonance.

To define the timescale of the exchange process and the nature of the NMR parameters we must consider the enzyme kinetics. For second order enzyme kinetics, the following equations apply[16,17]:

$$K_s = \frac{k_d}{k_a} = \frac{[E]\,[S]}{[ES]} \tag{1}$$

$$[ES] = \frac{1}{2}\{[E]_0 + [S]_0 + K_s - \sqrt{([E]_0 + [S]_0 + K_s)^2 - 4[E]_0[S]_0}\} \tag{2}$$

in which [E], [S] and [ES] are the enzyme, substrate and enzyme-substrate complex concentrations at equilibrium, $[E]_0$ and $[S]_0$ the total concentrations of enzyme and substrate, K_s the equilibrium constant, and k_a and k_d the association and dissociation rate constants.

Looking at the chemical shift values under fast exchange conditions we can say:

$$\Delta_{obs} = P_{bound}\,\Delta_{bound} \tag{3}$$

where Δ_{bound} is the difference between the chemical shift of the bound substrate and of the free substrate ($\delta_{bound} - \delta_{free}$), Δ_{obs} the difference between the observed chemical shift and the chemical shift of the free substrate ($\delta_{obs} - \delta_{free}$), and P_{bound} is the fraction of substrate bound to the enzyme. Substitution of

$$P_{bound} = \frac{[ES]}{[S]_0} \tag{4}$$

and equation 2 into equation 3 gives

$$\Delta_{obs} = \frac{\Delta_{bound}}{2[S]_0}\{[E]_0 + [S]_0 + K_s - \sqrt{([E]_0 + [S]_0 + K_s)^2 - 4[E]_0[S]_0}\}. \tag{5}$$

Thus, measurement of the chemical shifts (Δ_{obs}) for a range of different substrate concentrations ($[S]_0$) at a given enzyme concentration ($[E]_0$) and subsequent fitting of the above equation to the data using non-linear regression, will afford values for the principal parameters K_s and Δ_{bound}.

The fast exchange regime is defined as follows[18]:

$$\tau \ll \frac{1}{|\delta_{bound} - \delta_{free}|} \quad ; \quad \tau \ll \frac{1}{|1/T_{2bound} - 1/T_{2free}|} \tag{6a,b}$$

where

$$\frac{1}{\tau} = \frac{1}{\tau_{bound}} + \frac{1}{\tau_{free}} \tag{7}$$

in which τ is the lifetime of a state and therefore the reciprocal of the rate constant:

$$\tau_{bound} = \frac{1}{k_d} \quad ; \quad \tau_{free} = \frac{1}{k_a\,[E]}\,. \tag{8a,b}$$

In aqueous solution, association is diffusion limited, and the association rate constant k_a is of the order of 10^7-10^8 $M^{-1}s^{-1}$ [17]. From this figure we can calculate k_d, as K_s can be determined from the experimental data. This allows us to estimate the lifetime τ of the exchange process and show that the substrate is in fast exchange with the enzyme in solution with respect to chemical shift and relaxation parameters.

The observed spin-spin relaxation rates ($1/T_{2obs}$) under fast exchange conditions are, like the chemical shifts, averaged values of the relaxation parameters of the bound and free substrate ($1/T_{2bound}$, $1/T_{2free}$). In 'moderately' fast exchange, as opposed to 'very' fast exchange, additional broadening may be observed due to an exchange contribution[18]:

$$\frac{1}{T_{2obs}} = \frac{P_{bound}}{T_{2bound}} + \frac{1-P_{bound}}{T_{2free}} + \frac{(1-P_{bound})^2 P_{bound} \Delta_{bound}^2}{k_d} \quad (9)$$

In very fast exchange, when $k_d \gg \Delta_{bound}^2$, this last term will be negligible and the equation simplifies to

$$\frac{1}{T_{2obs}} = \frac{1}{T_{2free}} + \left(\frac{1}{T_{2bound}} - \frac{1}{T_{2free}}\right) P_{bound} \quad (10)$$

T_{2obs} can be measured by spin-echo experiments and P_{bound} can be calculated from $[S]_0$, $[E]_0$ and the previously calculated K_s-value using equation 2 and 4. Consequently, fitting one of these equations to experimental values of $1/T_{2obs}$ as a function of P_{bound} will allow the calculation of values for T_{2free} and T_{2bound}.

2 MATERIALS AND METHODS

Hen egg white lysozyme, salt-free and albumin-free, was obtained from ICN. D-Glucosamine hydrochloride, sodium monofluoroacetate and dicyclohexylcarbodiimide were obtained from Aldrich, *S*-ethyl trifluoroacetate, trifluoroacetone, and silica gel 60 from Fluka, and gelatin (bovine skin, G-9382) from Sigma. GC-MS was performed on a Hewlett Packard 5890 gas chromatograph equipped with an HP-1 capillary column (12 m x 0.2 mm x 0.33 μm) and connected to a Hewlett Packard 5970 mass spectrometer.

Lysozyme was fluorinated following the method described by Adriaensens *et al.*, involving treatment with *S*-ethyl trifluoroacetate, dialysis, purification by ion-exchange chromatography, further dialysis and freeze-drying[15]. This resulted in a trifluoroacetyl derivative of lysozyme in which the lysine residues (Lys-1, 13, 33, 96, 97 and 116) were trifluoroacetylated at the ω-amino position. ^{19}F NMR (470.5 MHz, D_2O, pH 7.3): δ -76.77, -76.79, -76.99, -77.00, 77.06, -77.11 ppm.

For the synthesis of *N*-monofluoroacetyl-D-glucosamine (GlcNFAc, **2**), the original method described by Dwek *et al.* (19% yield)[12] was slightly modified to increase the yield. D-Glucosamine hydrochloride (21.6 g, 0.1 mol) and sodium monofluoroacetate (11.0 g, 0.11 mol) were dissolved in 100 ml of water. A solution of dicyclohexylcarbodiimide (35 g, 0.17 mol) in 300 ml of pyridine was added and the mixture was stirred for 24 hours at room temperature. After adding 500 ml of water the mixture was stirred for another hour, filtered and extracted with diethyl ether (4 x 150 ml). The aqueous layer was evaporated to dryness and the residue was redissolved in 200 ml of methanol. After removing the remaining solids by filtration, the filtrate was concentrated to about half the volume and loaded onto a silica gel column (30 x 3 cm, eluent: dichloromethane/methanol (4/1)). After the product was eluted, the solvent was evaporated and the product was crystallized from dry methanol. This resulted in 9.84 g (41% yield) of **2**. ^1H NMR (500 MHz D_2O): δ 3.50, 3.64, 3.74-4.02

(multiplets, 6 H), 4.95 (dd, 2H, $^2J_{HF}$ 45.8 Hz, α- and β-CH_2F), 5.24 (d, 1H, $^2J_{HH}$ 3.5 Hz, H1) ppm. ^{19}F NMR (470.5 MHz, D_2O): δ -228.21 (t, $^2J_{HF}$ 45.8 Hz, β-CH_2F), -228.65 (t, $^2J_{HF}$ 45.8 Hz, α-CH_2F) ppm (Figure 2a). Assignment according to Butchard *et al.*[19]. MS (tetra-trimethylsilylether, GC-MS) *m/z* 512 (M-15, <5%), 422 (<5%), 332 (5%) 277 (10%), 191 (100%).

GlcNFAc, lysozyme/GlcNFAc and [TFA-Lys]-lysozyme NMR samples were prepared in D_2O solution and in a series of gelatin matrices with different D_2O contents, 85%, 75% and 65%. The lysozyme concentration was constant throughout one series of experiments and in the range of 3.5 - 7 m*M*. The concentration of the less soluble [TFA-Lys]-lysozyme was 0.25 m*M*. For each different matrix including the D_2O solution the following range of GlcNFAc concentrations prepared were typical, 2, 7, 25, 50, 100 and 200 m*M* ([α-GlcNFAc] = [β-GlcNFAc] = 0.5 [GlcNFAc]) Concentrations in the gelatin matrices were determined as concentrations in the volume of the D_2O fraction, the weight ratio of D_2O to gelatin being constant for each series. Gelatin was weighed into a 5 mm NMR tube and the appropriate solution of lysozyme and GlcNFAc was added and mixed thoroughly at a maximum temperature of 70°C (lysozyme denatures at 75°C) until the gelatin was in solution and the mixture appeared homogeneous. The samples were sonicated at this elevated temperature to remove air interfaces, trifluoroacetone was added as reference and the samples were allowed to cool forming a homogeneous gel.

All NMR experiments were performed using a Bruker AMX 500 spectrometer equipped with an Oxford instruments magnet and a Bruker 5 mm ^1H probe, which could be retuned to ^{19}F. ^{19}F NMR spectroscopy was performed at 470.5 MHz. Trifluoroacetone was used as internal reference and set at -87.83 ppm (scaling relative to $CFCl_3$: 0.00 ppm). Chemical shifts of both the α- and β-anomers of GlcNFAc were measured. Spin-echo experiments with a set of 10 to 15 variable delays from 500 μs to 600 ms were executed in a single experiment. Phase parameters set using the spectrum with the shortest delay time were applied to the following spectra in the series. An automatic baseline correction was performed and the T_2 was calculated from the signal decay employing the Bruker T_2-calculation routine. The accuracy of the results is largely dependent on the reproducibility of the sample preparations. The given error margins are estimates based on duplicate experiments.

3 RESULTS

We have studied the molecular motions of lysozyme, the lysozyme inhibitor GlcNFAc and the lysozyme-GlcNFAc complex by analysis of their spin-spin relaxation behaviour in aqueous solution and in low-water-content gelatin gels. To improve the poor resolution of these complex and immobilized systems we have specifically labelled the inhibitor and lysozyme with ^{19}F-nuclei and studied these compounds with ^{19}F NMR. We have first determined the kinetic characteristics of this enzyme system and the timescale of the exchange process by measuring the chemical shifts of the two anomeric forms of the substrate in the presence of the enzyme as a function of substrate concentration. Under the same conditions we have measured the corresponding T_2-values for the substrate by spin-

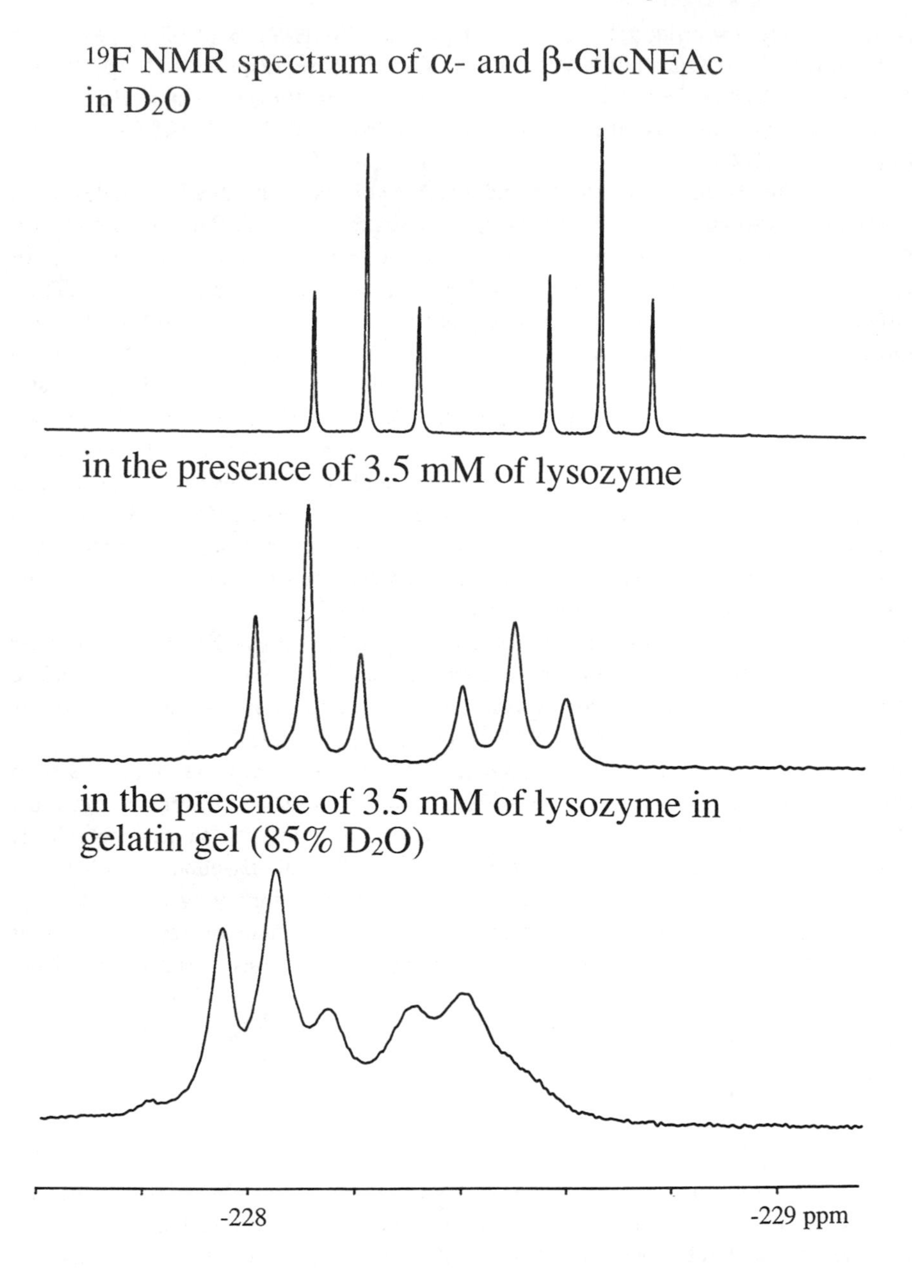

Figure 2. *^{19}F NMR spectra of α- (right signal) and β-GlcNFAc (left signal); a. 25 mM in D_2O (above), b. 25 mM in the presence of 3.5 mM lysozyme in D_2O (middle), c. 25 mM in the presence of 3.5 mM lysozyme in gelatin/D_2O (15/85) (below).*

echo experiments and calculated the relaxation parameters of the free and the bound substrate in the solution state. These experiments have then been repeated in gelatin matrices at different water contents and the relaxation parameters have been compared with those in solution. Finally, we have measured the relaxation parameters of the ^{19}F-labelled lysozyme in solution and in the same gelatin matrices.

3.1 Kinetics and Exchange Rate of the Enzyme System

The ^{19}F NMR spectrum of GlcNFAc (**2**) shows two resolved triplets at -228.21 and -228.65 ppm with equal integration values representing the β- and α-anomer, respectively (Figure 2a). Upon addition of lysozyme, both signals shift downfield due to the interaction of GlcNFAc with the active site of lysozyme (Figure 2b). The signals appear as the average of the signals of the substrate free in solution (δ_{free}) and in the enzyme-substrate complex (δ_{bound}), which indicates that the exchange rate is fast on the NMR timescale. In this regime the change in chemical shift will vary with the substrate concentration as outlined in equation 5. We have measured the chemical shifts of both anomers as a function of substrate concentration at two enzyme concentrations (3.5 and 7.0 m*M*). The results (as $\Delta_{obs} = \delta_{obs} - \delta_{free}$) for one series of experiments are depicted in Figure 3. The experimental data were subjected to non-linear regression analysis applying equation 5, which results in the curves shown in Figure 2. From this analysis we calculated values for Δ_{bound} (= $\delta_{bound} - \delta_{free}$) and K_s. For the α-anomer we found K_s = 28 (±2) m*M* and Δ_{bound} = 792 (±12) Hz (1.68 ppm) and for the β-anomer K_s = 47 (±5) m*M* and Δ_{bound} = 723 (±36) Hz (1.54 ppm). These results mean that both anomers interact in a different way with the enzyme and that the α-anomer has a slightly higher affinity for the active site. This corresponds to results reported in the literature for α- and β-GlcNAc (**1**)[5]. X-Ray data of lysozyme crystallized as a complex with these inhibitors show that both anomers of GlcNAc occupy different sections of the active site. The equilibrium constants for α-GlcNAc reported in the literature vary

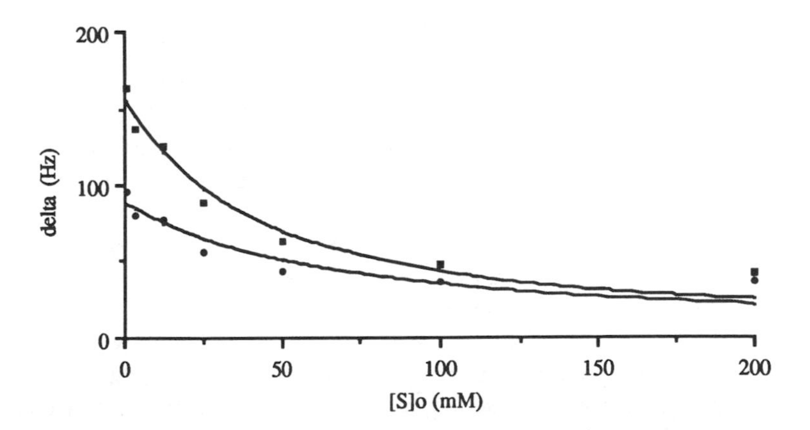

Figure 3. *Chemical shift change (Δ_{obs}) for α- (squares) and β-GlcNFAc (circles) in the presence of lysozyme (7 mM) relative to the free substrate at different substrate concentrations ($[S]_0$). The curves are obtained by fitting equation 5 to the experimental data.*

from 15 to 30 m*M*, and for β-GlcNAc from 23 to 30 m*M*, which indicates that introduction of a fluorine atom on the acetyl group slightly decreases the affinity of the substrate for the active site.

From the obtained equilibrium constants we can now estimate the timescale of the exchange process. The association rate constant k_a in aqueous solution is of the order of 10^7-10^8 $M^{-1}s^{-1}$ [17]. Using equation 1 and the obtained values for K_s (10^{-2} M) we can calculate that k_d is of the order of 10^5-10^6 s^{-1}. The lifetimes for the enzyme-substrate complex (τ_{bound}) and free substrate (τ_{free}) can be determined as 10^{-5}-10^{-6} s and 10^{-4}-10^{-5} s, respectively, according to equations 8a and 8b. These figures show that the timescale of the exchange process is determined by the lifetime of the enzyme-substrate complex (equation 7). Comparing the time constant of the exchange process τ (10^{-5}-10^{-6} s) with the reciprocal value of the frequency difference Δ_{bound} (10^{-3} s) proves that, in solution, the substrate is in fast exchange with the enzyme with regard to the chemical shift changes (equation 6a).

3.2 Relaxation Behaviour of the Substrate in the Presence of the Enzyme in Aqueous Solution and in Gelatin Matrices

Besides changes in chemical shift the ^{19}F NMR signals of GlcNFAc also show line broadening upon addition of lysozyme (Figure 2b). The motion of the substrate will be restricted when bound to the active site of the enzyme, which causes an increase in the linewidth. Inhomogeneity effects, however, also contribute to line broadening and in gelatin matrices, these may dominate the linewidth. Determination of T_2-values by means of spin-echo experiments, which minimise inhomogeneity effects, is therefore preferred to direct measurement of linewidths.

As for the chemical shifts, under fast exchange conditions, the relaxation rate ($1/T_2$) will be the weighted average of the values in the bound and the free state (equation 10). The values of T_{2free} are obtained from low-concentration GlcNFAc samples without lysozyme. To determine values for T_{2bound}, the bound fraction of the substrate (P_{bound}) was calculated from the corresponding substrate concentration and K_s using equations 2 and 4, $1/T_{2obs}$ was determined and plotted against P_{bound} (Figure 4 and 5). Equation 10 was fitted to these data and T_{2bound} calculated from the gradient. For α-GlcNFAc we found T_{2free} = 340 (±10) ms and T_{2bound} = 1.10 (±0.03) ms, and for β-GlcNFAc T_{2free} = 288 (±8) ms and T_{2bound} = 1.29 (±0.03) ms. These results make clear that the mobility of the substrate when bound in the active site of the enzyme is strongly reduced compared to the substrate in free solution. The fact that the T_{2free} is bigger for the α-anomer than for the β-anomer and that the T_{2bound} appears to be smaller for the α-anomer than for the β-anomer, indicates that the two anomers experience different environments in the active site. This is in agreement with the fact that we found two different equilibrium constants for the two anomers. These figures also confirm that according to equation 6b the substrate is in fast exchange with the enzyme with regard to the relaxation parameters as it is with regard to the chemical shifts.

Using these results, it was possible to measure and analyse the relaxation parameters of the lysozyme-GlcNFAc system in gelatin matrices (Figure 2c). For the solution samples we have determined the kinetic parameters from the changes in chemical shift due to the interaction with the enzyme and used these parameters to calculate P_{bound}. In the gelatin samples, the chemical shift cannot be measured with sufficient accuracy for these calculations, since small changes in the sample preparation can induce variations in the chemical shift of the same magnitude. To circumvent this, we have considered the substrate and enzyme concentrations in the water fraction of the samples and used the equilibrium constants from the solution experiments to calculate P_{bound} (equations 2 and 4). We have performed T_2-measurements for three gelatin matrices with water contents of 85%, 75% and 65%, respectively, and analyzed the data as for the solution data. Figures 4 and 5 show the relaxation data for α- and β-GlcNFAc in the gelatin matrices together with the data for the solution state. We clearly observe a gradual increase in the gradient of the lines when going to lower water contents. This corresponds to a gradual decrease in T_{2bound} according to equation 10 and bearing in mind that $T_{2free} \gg T_{2bound}$. The values for T_{2free}, obtained from low-concentration samples of GlcNFAc without lysozyme, and calculated values for T_{2bound} are summarized in Table 1.

To ascertain that concentration effects are not distorting these results, we have measured the T_2-values of the substrate at different concentrations in the absence of lysozyme. In solution we observed about a 20% decrease in the T_2-value at 200 m*M* substrate concentration. This effect diminishes to about 5% in the 35/65 gelatin/D_2O matrix. We corrected the relaxation data for this concentration effect according to:

$$\frac{1}{T_{2obs}} = \frac{P_{bound}}{T_{2bound}} + \frac{1-P_{bound}}{T_{2free}} + \frac{d\ 1/T_2}{d\ [S]_0}[S]_0 \tag{11}$$

$$\frac{1}{T_2} = \frac{1}{T_{2obs}} - \frac{d\ 1/T_2}{d\ [S]_0}[S]_0 \tag{12}$$

Analysis of the corrected data showed that the concentration effect on the gradient of $1/T_2$ versus P_{bound}, and hence on the value for T_{2bound}, was less than 2%, which is of the same order as the experimental error.

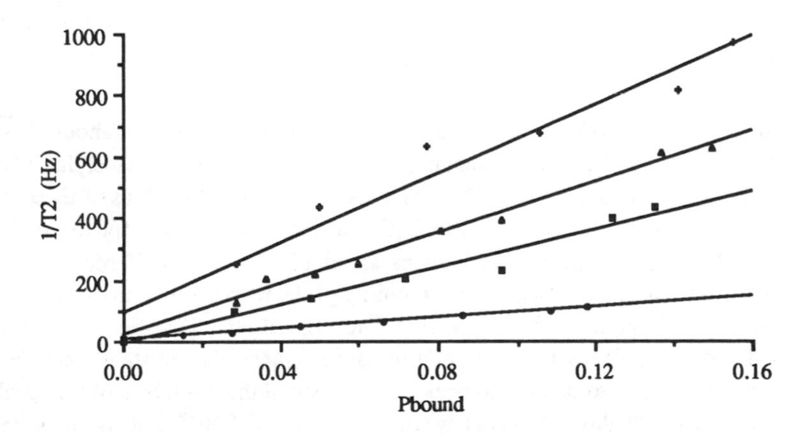

Figure 4. *Spin-spin relaxation rate constants ($1/T_{2obs}$) for α-GlcNFAc as a function of the fraction bound to lysozyme (P_{bound}) and for different matrices: D_2O (circles), gelatin/D_2O (15/85) (squares); gelatin/D_2O (25/75) (triangles); gelatin/D_2O (35/65) (plus signs); The lines are obtained by fitting equation 10 to the experimental data.*

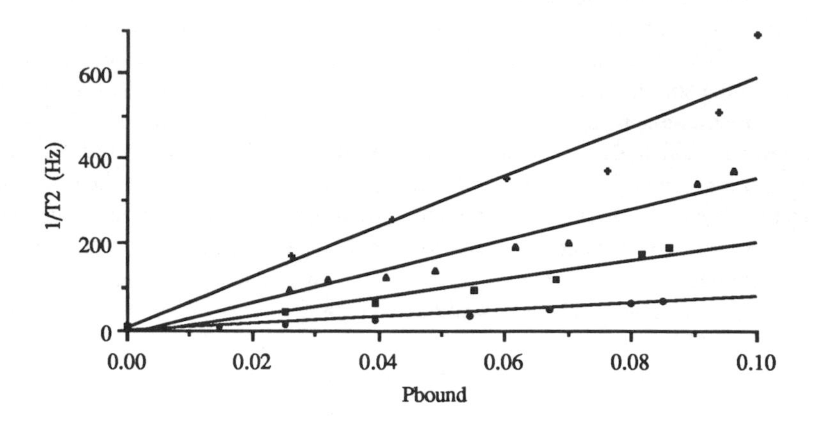

Figure 5. *Spin-spin relaxation rate constants ($1/T_{2obs}$) for β-GlcNFAc as a function of the fraction bound to lysozyme (P_{bound}) and for different matrices: D_2O (circles), gelatin/D_2O (15/85) (squares); gelatin/D_2O (25/75) (triangles); gelatin/D_2O (35/65) (plus signs); The lines are obtained by fitting equation 10 to the experimental data.*

Finally, we analyzed the data using equation 9 instead of equation 10 to verify whether there is any contribution of the exchange process to the T_2-values. We obtained the best data fits when applying values for k_d that were bigger than Δ_{bound}^2, resulting in a negligible exchange contribution and, consequently, in the same values for T_{2bound}. Even when we applied relative small numbers for k_d, corresponding to relatively slow exchange, the outcomes for T_{2bound} remained virtually unchanged. We therefore conclude that fast exchange conditions still apply to the enzyme-substrate system in the studied gelatin matrices.

3.3 Relaxation Behaviour of the Enzyme Molecule

The ^{19}F NMR spectrum of trifluoroacetylated lysozyme shows six peaks of about the same intensity in the region of -76 to -78 ppm that can be assigned to the trifluoroacetylated lysine residues according to Adriaensens *et al.*[15]. We have measured the T_2-values of these peaks in aqueous solution and in the same gelatin/D_2O matrices that we used for the substrate measurements. In solution, the six resonances all display slightly different relaxation behaviour, which indicates differences in the mobility of the respective lysine residues. The average T_2-value is 38 ms (individual T_2-values, starting with the downfield peak, are 38, 47, 21, 46, 38 and 34 (±10) ms). In the gelatin matrices these signals show extreme line broadening and converge into one broad peak. We measured the T_2-values of this peak by spin-echo experiments at three different water contents and found, consistent with the excessive line broadening, a sharp decrease at reduced water content. The resulting average T_2-values are given in Table 1 together with the substrate parameters.

Table 1. *Spin-spin relaxation time constants (T_{2free} and T_{2bound}) for both anomers of GlcNFAc and [TFA-Lys]-lysozyme in aqueous solution and in gelatin/D_2O matrices.*

	α-GlcNFAc		β-GlcNFAc		F-lysozyme
matrix	T_{2free} (ms)	T_{2bound} (ms)	T_{2free} (ms)	T_{2bound} (ms)	T_{2enz} (ms)
D_2O	340 ± 10	1.10 ± 0.03	288 ± 8	1.29 ± 0.03	38 ± 1
gelatin/D_2O (15/85)	181 ± 5	0.33 ± 0.02	173 ± 5	0.46 ± 0.02	2.9 ± 0.5
gelatin/D_2O (25/75)	115 ± 3	0.24 ± 0.02	111 ± 3	0.28 ± 0.02	2.5 ± 0.5
gelatin/D_2O (35/65)	75 ± 2	0.18 ± 0.02	66 ± 2	0.17 ± 0.02	2 ± 2

4 DISCUSSION

We developed a model enzyme-substrate system for the study of the molecular mobility of enzymes in food-like systems using NMR techniques. This model system consists of an enzyme, a competitive inhibitor as substrate and a biopolymer matrix, and allows the study of enzyme-substrate dynamics in the solution state and in the rubbery state. To be able to selectively analyse resonances of the substrate and the protein in such complex and immobilized biopolymer systems we have specifically labelled the substrate and the enzyme with ^{19}F and used ^{19}F NMR for their analysis. For the study of the dynamics of the substrate, the model system consisted of lysozyme and the competitive inhibitor *N*-monofluoroacetyl-D-glucosamine (α- and β-anomer) in gelatin gel matrices. For the study of the enzyme dynamics we used ^{19}F-labelled lysozyme, in which the lysine residues were trifluoroacetylated.

We investigated the changes in molecular motions of this enzyme-substrate system when going from aqueous solution to low-water-content gelatin matrices (100-65% water content) by measuring its ^{19}F NMR spin-spin relaxation parameters as a function of water content. For the analysis of the relaxation data we first determined the kinetic characteristics of the enzyme-substrate system and the timescale of the exchange process by chemical shift measurements. It appeared that both in solution and in the gelatin matrices the substrate is in fast exchange with the enzyme on the NMR timescale. Hence, we analyzed the data according to equations that have been derived for the fast exchange regime[16-18]. We have determined three different relaxation parameters that relate to the molecular motions of the substrate, the enzyme-substrate complex and the enzyme, respectively. We measured the relaxation time constants for both anomers of the substrate at low substrate concentration in the absence of enzyme, giving the T_2 of the free substrate (T_{2free}). The relaxation time constants for the two anomers in the presence of enzyme were also measured, enabling us to calculate the T_2 of the bound substrate (T_{2bound}). Finally, the average relaxation time constants for the trifluoroacetylated lysozyme (T_{2enz}) were determined (Table 1).

In Figure 6 these results are shown as T_2-values normalized to the value in free solution versus the water content of the gelatin matrices. A clear decrease is observed for all parameters at lower water contents, demonstrating a decrease in molecular motions under these conditions. The three relaxation parameters, however, decrease at different rates. The

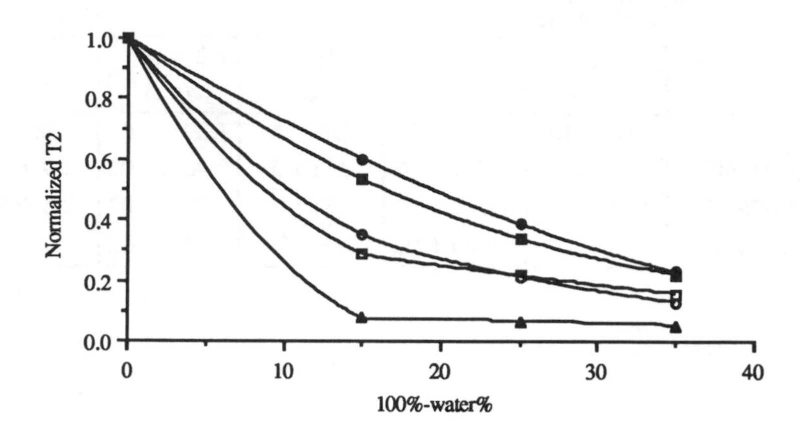

Figure 6. *Normalized spin-spin relaxation time constants as a function of water content in gelatin/D_2O matrices (T_2 in solution is 1.0) for α-GlcNFAc (T_{2free} filled squares, T_{2bound} open squares) and β-GlcNFAc (T_{2free} filled circles, T_{2bound} open circles) and [TFA-Lys]-lysozyme (triangles). The curves are obtained by interpolation.*

decrease in T_{2enz} is faster than in T_{2free}, which indicates that the enzyme is immobilized to a greater extent than the substrate in these gelatin matrices. This also follows from the faster rate of decrease of T_{2bound} compared with T_{2free}, since this must be due to interaction with the enzyme and therefore caused by a reduction of the overall mobility of the enzyme molecule.

Furthermore, the difference between the rate of decrease in T_{2bound} and T_{2enz} suggests a difference between the change in molecular motions at the active site relative to the overall dynamics of the enzyme. The trifluoroacetylated lysine residues are situated on the surface of the protein molecule and the bound substrate, to a certain extent, will be buried in the interior of the protein molecule. The results therefore indicate that the molecular motions in the active site are less restricted than at the exterior of the protein as the water content is decreased. It is likely that the decrease in T_{2enz} and T_{2bound} is mainly due to changes in the rate of the rotational tumbling of the enzyme molecule and that the additional decrease in T_{2enz} is due to the interaction of the surface lysine residues with the matrix. Consequently, it may well be that there is only a negligible change in the internal motions of the enzyme through the rubbery state region. As enzyme activity requires a certain minimum of molecular motions in the active site, this might prove an important parameter which would partially explain enzyme activity in highly immobilized systems.

Future experiments at lower water contents using the same approach will show how the relaxation behaviour of the enzyme-substrate system will develop towards the glass-rubber transition. The present results demonstrate a levelling of the T_2 parameters at lower water contents which may indicate that there is a certain minimum mobility throughout the rubbery state, which most likely will be followed by a sharp decrease at the glass-rubber transition. These insights into the molecular motions of the enzyme-substrate system under conditions ranging from the solution state to the glass-rubber transition will bring us one step closer to the understanding of the parameters that determine enzyme activity in food systems.

References

1. J.M.V. Blanshard and P.J. Lillford, 'The Glassy State in Foods', University Press, Nottingham, 1993.
2. A. Zaks and A.M. Klibanov, *Proc. Natl. Acad. Sci. U.S.A.*, 1985, **82**, 3192.
3. P.L. Poole, *J. Food Eng.*, 1994, **22**, 349.
4. J.T. Gerig, in 'Methods in Enzymology', (N.J. Oppenheimer and T.L. James, eds.), Academic Press, London, 1989, Vol. 177, p. 3.
5. T. Imoto, L.N. Johnson, A.C.T. North, D.C. Philips and J. Rupley, in 'The Enzymes', (P.D. Boyer, ed.), Academic Press, New York, 1972, 3rd ed., Vol. 7, p. 665.
6. L.J. Smith, M.J. Sutcliffe, C. Redfield and C.M. Dobson, *J. Mol. Biol.*, 1993, **229**, 930.
7. C.C.F. Blake, G.A. Mair, A.C.T. North, D.C. Philips and V.R. Sharma, *Proc. Roy. Soc. ser. B*, 1967, **167**, 365.
8. C.C.F. Blake, L.N. Johnson, G.A. Mair, A.C.T. North, D.C. Philips and V.R. Sharma, *Proc. Roy. Soc. ser. B*, 1967, **167**, 378.
9. N.C.J. Strynadka and M.N.G. James, *J. Mol. Biol.*, 1991, **220**, 401.
10. C.L. Brooks and M. Karplus, *J. Mol. Biol.*, 1989, **208**, 159.
11. T. Ichiye, B.D. Olafson, S. Swaminathan and M. Karplus, *Biopolymers*, 1986, **25**, 1909.
12. R.A. Dwek, P.W. Kent and A.V. Xavier, *Eur. J. Biochem.*, 1971, **23**, 343.
13. M. Tada, A. Oikawa, R. Iwata, T. Fujiwara, K. Kubota, T. Matsuzawa, H. Sugiyama, T. Ido, K. Ishiwata and T. Sato, *J. Labelled Compds. Radiopharm.*, 1989, **27**, 1317.
14. M. Sharma and W. Korytnyk, *Tetrah. Letters*, 1977, 573.
15. P. Adriaensens, M.E. Box, H.J. Martens, E. Onkelinx, J. Put and J. Gelan, *Eur. J. Biochem.*, 1988, **177**, 383.
16. K.L. Gammon, S.H. Smallcombe and J.H. Richards, *J. Am. Chem. Soc.*, 1972, **94**, 4573.
17. J. Feeney, J.G. Batchelor, J.P. Alband and G.C.K. Roberts, *J. Magn. Reson.*, 1979, **33**, 519.
18. O. Jardetzky and G.C.K. Roberts, 'NMR in Molecular Biology', Academic Press, New York, 1981.
19. C.G. Butchard, R.A. Dwek, P.W. Kent, R.J.P. Williams and A.V. Xavier, *Eur. J. Biochem.*, 1972, **27**, 548.

What's in an NMR Image? Characterising Structure and Transport in Heterogeneous Systems

K. J. Packer, H. J. Turner, and M. A. P. Wright

DEPARTMENT OF CHEMISTRY, UNIVERSITY OF NOTTINGHAM, UNIVERSITY PARK, NOTTINGHAM NG7 2RD, UK

1. INTRODUCTION

NMR images are representations of spatially-resolved measurements of NMR parameters[1]. These parameters can be signal intensities (proportional to the number of spins), resonance frequencies, relaxation behaviours and transport properties, such as diffusion and flow[2,3]. As with any NMR investigation the full range of NMR tricks involving multiple r.f pulses and their phases as well as gradient pulses may be used to generate these parameters in an image. It may seem obvious, but what we refer to as an image is a view, in two dimensional form, of what, in general, is a multi-dimensional data set. The data is a spatially ordered set of values of one or more NMR parameters with the image a projection, of one form or another, often chosen to illustrate some particular features of importance. Appearances of images may be altered and enhanced through use of varying mappings of NMR parameter values onto a display scale (e.g. colour, grey level etc) as well as through image processing methods designed to filter or highlight some characteristics. There is a natural tendency to regard the image as the end of the matter, partly because the eye-brain combination is particularly good at, and gratified by, viewing structured scenes and recognising patterns, and partly because a major drive for the development of NMRI has been medical examination and diagnosis. Clearly, variations from one image to another can be detected, particularly by the professionally trained and experienced eye. However, for the physical scientist, used to working with quantitative data to illustrate and test models of nature, an image is really only a means to an end. This is particularly true for systems which do not contain clearly structured and delineated objects with distinguishable NMR properties. Such conditions are to be found in many heterogeneous fluid-containing materials such as natural and fabricated foods, porous solids permeated with fluids (such as oil- and water- bearing rocks and soils) and many other disordered materials. Of course, the use of NMR imaging to study processes such as the ingress of a fluid into a solid or other fluid is an example of a quantitative use of the spatially-resolved data set and is not the focus of this paper.

The main objective of this paper is to offer some preliminary thoughts on ways in which spatially-resolved NMR data sets, arising from complex and heterogeneous materials, might be treated in order to provide quantitative measures which enable characterisation and comparisons to be achieved, for example, for related but different samples. The context of this discussion will be the transport of fluids in porous solids, although these processes will not be addressed specifically. The issues may be illustrated by Figure 1 which is one of a series of image slices for a rock containing fluid. Apart from noting that the fluid distribution is inhomogeneous there is not much else that can be said, even if the data are made quantitative by suitable calibration methods. The ideas to be outlined here were suggested briefly in a previous paper[2] in which the data sets were represented as histograms of values. It was pointed out there that comparison of such histograms from different image slices or samples would, in a simple manner, reflect aspects of the relative degrees of heterogeneity i.e. a narrow as compared to a broad histogram, for example.

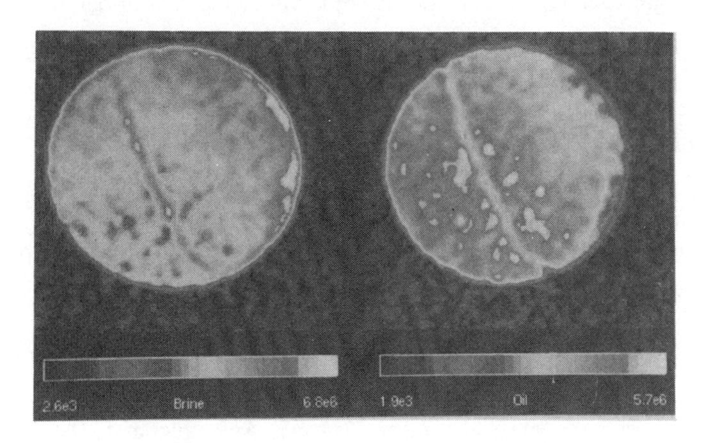

Figure 1 *A typical NMR image of an inhomogeneous object: fluid-saturated rock*

2. FLUID TRANSPORT IN HETEROGENEOUS SOLIDS

Spatially-resolved information on the ability of a solid matrix to contain and transport fluid can be obtained through both so-called k-space[1] and q-space imaging methods[4]. k-space is the term used to describe the direct imaging of the spatial distribution of NMR signal and its relaxation properties, whilst q-space has been used to designate methods based on the use of field gradient pulses to obtain structural information via the propagator for translational displacements of spins[5]. Whatever the source or nature of the NMR parameters measured the behaviour of the system with respect to fluid transport will be dependent on such aspects as value of the fluid-filled porosity, the size and connectivity of this porosity and its spatial distribution. In the next section we consider some ways in which, for a complex distribution of fluid, the spatially resolved NMR data might be treated

3. HISTOGRAMS, SPATIAL SCALING AND NMR PROPERTY AGGREGATION

Consider a 2-D spatially-resolved NMR data set comprising N^2 pixels which provides, inter-alia, the spatial distribution of some NMR parameter, x, at a spatial resolution length scale L_0. In the simplest case x could be the absolute NMR signal intensity, I. We may represent the distribution of values of x in this data set by means of a histogram, N(x), where N(x) represents the number of pixels having values within some range around x, the range determined by the number of classes chosen for N(x) over the full range of x values.

The choice of the number of classes in the histogram is not significant. It need only allow a sufficient representation of the variation of the property x. The histogram N(x) provides a view of the continuous distribution of the values of x, at the spatial scale L_0. Given such a distribution, N(x), it may be characterised, like any other distribution, in terms of a number of statistical measures[6]. Principal amongst these would be the moments of the distribution about the mean. For a discrete distribution, N(x), the moments, $M_p(n)$, relative to the mean <x>, would be defined by:

$$M_p(n) = \Sigma N_i(x_i)\{x_i(n) - \langle x \rangle\}^p \qquad (1)$$

where p = 2,3,4.... correspond to the second, third, fourth moments etc., and the label n refers to the length scale L_n. The significance of n will be made clear below.

The complexity of the distribution could be represented in two ways. The simplest is just the number of non-zero histogram bins, which can be represented as

$$\eta_n = \Sigma (q_i)_n \qquad (2)$$

where $(q_i)_n = 1$ if $N_i\{x_i(n)\} > 0$ and zero otherwise.

A more rigorous measure of complexity is the entropy of the distribution[6] given by

$$S_n = - \Sigma p_i(n) \ln\{p_i(n)\} \qquad (3)$$

where $p_i(n) = N_i\{x_i(n)\}/N_n$, with N_n the total number of pixels at spatial scale L_n.

The maximum disorder and, hence, entropy, is represented by a totally smooth histogram in which each bin contains an equal number of pixels. Any structure in the histogram then is seen as a decreased entropy compared to this state. For 100 bins in the histogram, for example,

$$S_n(\max) = - \Sigma (1/100)\ln(1/100) = 4.605 \qquad (4)$$

From the above it should be clear that the appearance of the pixel histograms and the values of the moments and complexity measures, will be a function of the spatial length scale L_n if the latter can be varied to encompass the spatial scale of variation of the property x. Clearly, a spatially resolved NMR data set obtained at a particular resolution length scale contains within it all larger length scales up to the dimensions of the field of view. It is suggested that by spanning these length scales, and aggregating the NMR property values in an appropriate manner, the behaviour of the statistical measures defined above will give quantitative characterisation of the spatial distribution and heterogeneity length scales of the system studied. In the next section these ideas are illustrated with some calculations on simple model systems and in the final section some other possible measures are considered in a general manner.

The central proposition is that the values of the NMR property determined at a given length scale L_n be aggregated together at increasing length scales. In this process the number of pixels is reduced as appropriate numbers of adjacent pixels are aggregated to give a new, reduced number, of pixels at the longer length scale. The final stage in this process is a single pixel at the length scale of the sample.

Before proceeding to illustrate these approaches the method of property aggregation must be considered. There are only two methods of aggregation. For signal intensity, which is additive, the new value at the reduced length scale is the average of the set of pixels which have been aggregated. For NMR parameters which have been derived from a fit to some model, such as relaxation behaviour, diffusion etc, the signal intensity data must first be averaged through addition and normalisation and then re-fitted to the relevant model. In what follows we restrict ourselves to the first case.

4. MODEL CALCULATIONS

A computer programme has been written which calculates the histograms and moments for a simple model "image" comprising 512x512 pixels, which is successively scaled by a factor of 2x2. Figure 2 illustrates this process whilst Table 1 defines the length scales L_n and corresponding pixel numbers.

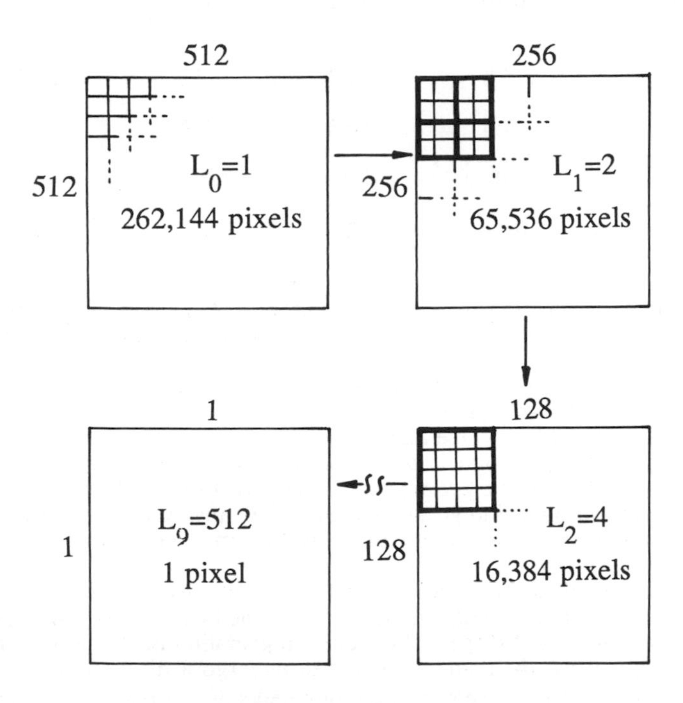

Figure 2 *The aggregation of pixels with decreasing binary length scales*

Table 1 *Binary Length Scales* L_n *in a 512x512 2-D Image*

n	0	1	2	3	4	5	6	7	8	9
L_n	1	2	4	8	16	32	64	128	256	512
N_n	2^9	2^8	2^7	2^6	2^5	2^4	2^3	2^2	2^1	2^0

The programme allows the "image" to contain a number of objects, either rectangular or circular, with a choice of the number of objects, their size and orientation(rectangles only) and their location(regular spacing or random distribution, the latter including a random degree of overlap) within the image field. Values(of the NMR parameters) may be assigned to the pixels falling within the objects with the(selected) background value(s) outside the objects. Table 2 summarises the images for which results are discussed

Table 2 *Model Images Used for the Calculations*

Objects/ No.	RegSq 256	RegC 16	RmC 50	RmSq 100	RmSq 10
Size	16x16	R=24	R=20	10x10	100x100
Value/ Backgr'nd	50/10	50/10	50/10	50/10	50/10
<x>	20	14.8	14.4	11.3	18.7
A/A_0(%)	25	11.6	10.8	2.7	21.4
% Overlap	0	0	57	29	44
$M_2(0)$	294	161	151	41	264

RegSq/C=regularly distributed squares/circles; R=radius; A/A_0 = % of image area covered by objects; % Overlap = % overlap of objects; $M_2(0)$ = second moment of histogram at L_0

The histograms corresponding to the "images" described in Table 2(length scale L_0) each comprise two non-zero bins, one at x=10, the other at x=50. The programme was used to calculate the "images" at all the binary length scales, as described in Table 1, and to generate the corresponding histograms and moments. Figure 3 shows the variation with L_n of the second moments for the two regular image structures together with their unit cells. The $M_p(n)$ are constrained to go to zero for n=9 in any "image" as this value of n corresponds to treating the image as a single pixel and thus must give the mean value <x>.

However, the transition of the moments of the histograms towards zero as n increases reflects the scale, shape and distribution of the objects in the image. This can be seen in Figure 3 where the square pixel shape and the exact registration of origin with the length scale grid allows $M_2(n)$ for the regular squares to undergo a discontinuous transition exactly at $L_5 = 2^5 = 32$, which is the size of the unit cell side. The regular circles model, however, shows a more gradual second moment transition although this reaches zero exactly at $L_6 = 2^6 = 128$ which, again, is the unit cell dimension. This is required by the regular structure. However, whereas the histograms for the regular squares model contain the same two elements for $n \leq 4$ and a single element for larger n, those for the regular circles show a considerable variation in complexity as a function of n. The values of η_n are 2,5,14,14,9,4,1,1,1 (n=0-9 respectively). This arises entirely from the mismatch in shape between the pixels and the objects. It is apparent that the complexity measure, η_n, passes through a maximum at a length scale significantly smaller than that at which the second moment reduces significantly. This would be expected as the range of pixel values will maximise when the pixel scale is of the order of a fraction of the overall object structure scale. These simple examples show the sensitivity of histogram moments and complexity to both scale of spatial variation of the NMR property and symmetry with respect to that of the pixels.

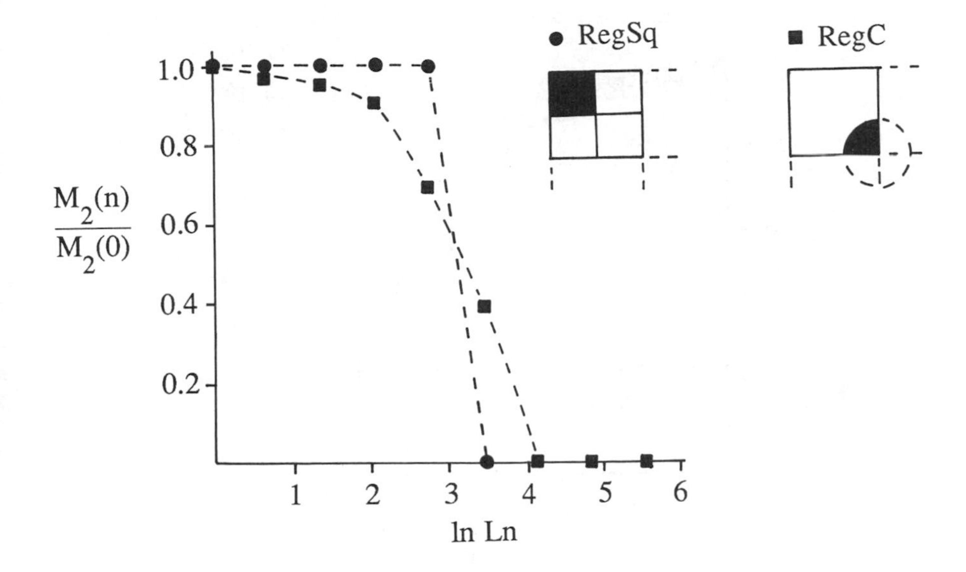

Figure 3 *Second moment variation with L_n fnd unit cells or regular square(RegSq) and circle(RegC) images*

Figure 4 shows the variation with scale L_n of $M_p(n)$ {p=2,3&4} for the random circles model. Comparison with Figure 3 shows that, despite the random circles having similar radii to the regular circles, their second moment transition with increasing n is much more gradual. All three moments show the expected transition towards zero. In order to give a quantitative interpretation to these moment transitions we propose to treat them rather like a motional narrowing transition in NMR spectroscopy by defining a half-height length scale $L_n(0.5)$ which is the value of L_n for which $[M_p(n)/M_p(0)]^{1/p} = 0.5$. For the data shown in Figure 4 the values are given in Table 3.

Table 3 *Half-height Length Scales for Moments of Random Circles Image*

p	2	3	4
$L_n(0.5)$	97	76	83

$L_n(0.5)$ values can be considered as a measure of the distance scale on which the inhomogeneous "image" appears largely homogeneous. In the case of the random circle model we may estimate an average value for this scale by noting that, given the area of a circle of radius 20 is 1316(for the implementation of a circle on our 512x512 square grid), the area of the 50 overlapped circles is equivalent to ≈ 22 separate circles of the same size. The average repeat distance for this structure can then be approximated as [40 + {512 -

$(22)^{1/2}(40)\}/(22)^{1/2}] \approx 109$ which very similar to the values in Table 3. The complexity measure η_n, however, shows a peak value at n=3, corresponding to a length scale of 8,

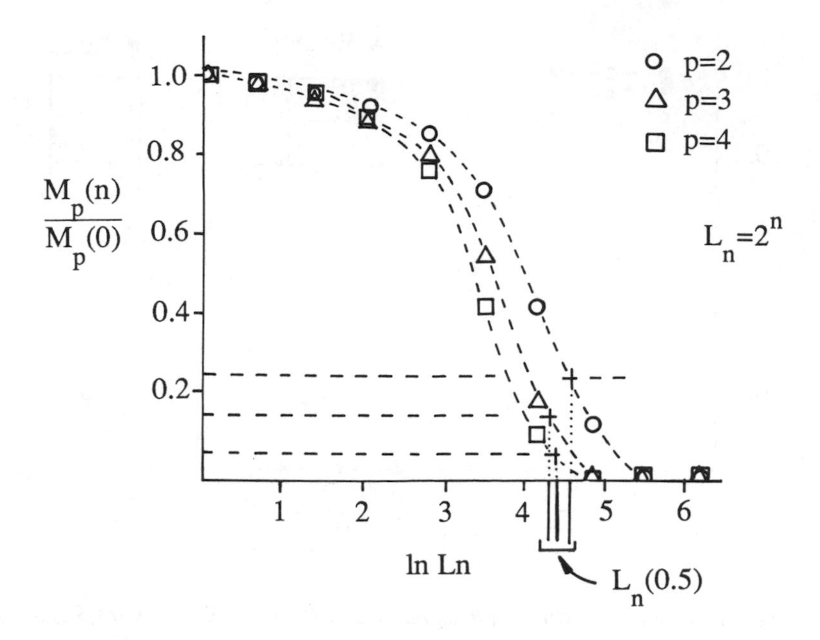

Figure 4 *Variation with L_n of the moments of pixel histograms for random circles*

with the entropy S_n also giving a maximum at this value. As for the regular structures, this scale is considerably smaller than that given by $L_n(0.5)$, the ratio being of order 10 which is very similar to the same ratio for the regular circle model($\approx 45/5.6 = 8$).

Figure 5 shows the L_n dependence of $M_2(n)$ and η_n for the two random square model images described in Table 2. It can be seen that the second moments fall off at very different distance scales and the complexity measures η_n peak at different scales and have very different widths. These differences can be related to features in the models. One model has large squares as the basic object and these, whilst overlapped to 44%, cover 21% of the image. The second has ten times as many, but smaller squares(x0.1 linear dimension), covering only 2.7% of the total image area. This sparser model, as would be expected, shows the faster decline in M_2 with $L_n(0.5) \approx 2^5 = 32$. A similar calculation to that carried out above for the random circles model gives an average repeat scale of about ≈ 60. This ratio of these two numbers of order 2 is a reflection of the sparse nature of this random squares model which leads to it appearing largely homogeneous on a scale which is considerably smaller than the average repeat. For the other model the corresponding values are $L_n(0.5) \approx 180$ with the average repeat being 125, giving a ratio of 1.44. The corresponding ratio for the random circles model is 1.1 which indicates that the use of a rectangular grid requires that a scale corresponding to around that of the average repeat must be reached before the system appears largely homogeneous.

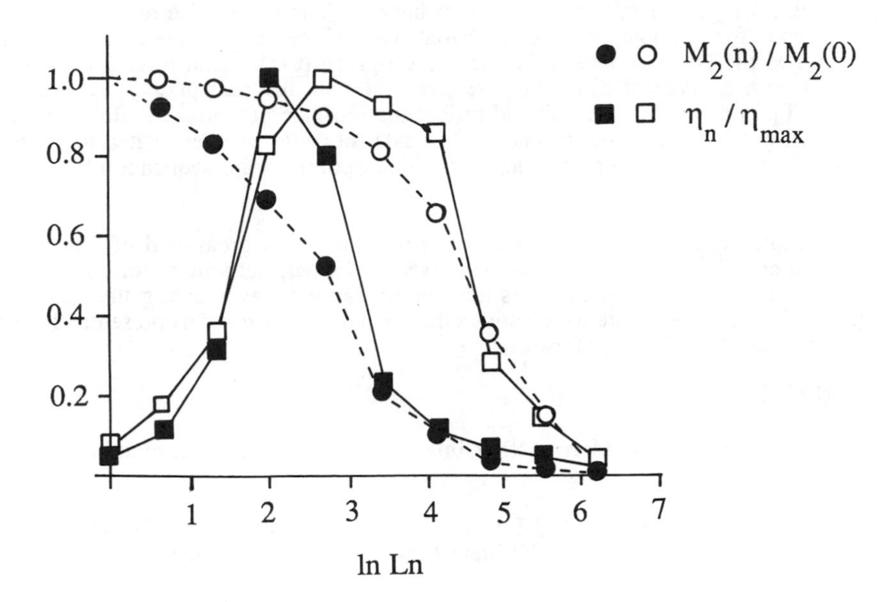

Figure 5 *Variation of $M_2(n)$ and η_n with L_n for the two random squares images*

5. DISCUSSION

The calculations described in section 4 are intended to illustrate a possible approach to the treatment of spatially-resolved NMR data sets for situations where the complexity of the data requires some means of quantitation. Other approaches may be considered in which the data is probed for the spatial relationships which exist. In the context of this it is worth pointing out that for an NxN 2-D array there are $\sim N^4/2$ binary correlations. Of course, the main interest in inhomogeneous materials such as those which are the focus of this paper will be in local correlations of properties. In this situation the radial distribution function will be an obvious quantity of interest. This would take the form

$$\Phi(d) = (1/N^2)\Sigma_{ij}\phi_{ij}(d) \qquad (5)$$

where

$$\phi_{ij}(d) = \Sigma_{\alpha}\{x_{ij}(d_{\alpha}) - \langle x \rangle\} / N_{d\alpha}$$

$x_{ij}(d_\alpha)$ is the value of the NMR property x at distance d_α from the reference pixel ij and $N_{d\alpha}$ the number of pixels at that distance.

As has been suggested elsewhere[2], when a number of NMR properties are determined for each pixel the relationship between these, both within a pixel and between different pixels, again contains information of interest. The spin-lattice relaxation time in many systems, for example, reflects the spatial scale of the porosity containing the fluid giving rise to the NMR signal. It is clear that a given pixel signal intensity I may be associated with a range of different pore size scales in different pixels. As such the product IxT_1 in different pixels should reflect the intrinsic potential for fluid transport through that pixel. In a similar manner, a 3-D data set could be interrogated to establish and rank the pathways through the data set which optimised the sequence of I or IxT_1 values.

The principal message of this paper is that there is a great deal of quantitative information contained within a spatially-resolved NMR data set which, for complex and heterogeneous systems particularly, is not readily captured by viewing the data as an image. Statistical methods are the obvious solution and the calculations presented here are meant to suggest possible ways forward.

REFERENCES

1. P.Mansfield and P.G.Morris, 'NMR Imaging in Biomedicine" Academic Press, New York, 1982

2. P.A.Osment, K.J.Packer, M.J.Taylor, J.J.Attard, T.A.Carpenter, L.D.Hall, N.J.Herrod and S.J.Doran, Phil.Trans R.Soc.Lond. A, 1990, **333**, 441

3. A. Caprihan and E.Fukushima, Phys Reports, 1990, **198**, 197

4. P.T.Callaghan, D.MacGowan, K.J.Packer and F.O.Zelaya, J.Magn.Res., 1990, **90**, 177

5. J.Karger, H.Pfeifer and W.Heink, Adv. in Magn.Res., 1988, **12**, 1

6. P.A.P.Moran, 'An Introduction to Probability Theory' Clarendon Press, Oxford, 1968, Chapter 1

750 MHz NMR Spectroscopy and Food-related LC–NMR Applications

Manfred Spraul and Martin Hofmann

BRUKER ANALYTISCHE MESSTECHNIK GMBH, SILBERSTREIFEN, D-76287 RHEINSTETTEN, GERMANY

1 INTRODUCTION

When going from 500 to 600 MHz NMR-Spectroscopy several years ago, the benefit of the higher field could clearly be shown, especially protein investigations by NMR could be improved considerably. In November 92, the first commercial 750 MHz spectrometers became available, and again the question was would the higher field fulfil the expectations of the NMR-spectroscopists. Besides the proteins a new field of applications has emerged, the investigations of mixtures by NMR. Most prominent examples in this area are the biofluids and food-related samples. The analysis of mixtures can be improved by higher field, but there also is a different approach to this problem: The on-line coupling of liquid chromatography and nuclear magnetic resonance. In this case the overlap of interesting signals is removed by having a separation step prior to NMR. Successful use of LC-NMR for biofluids and food samples also is optimally done at high field.

The aim of this contribution is to give a review, on what has been achieved in the last two years with 750 MHz and LC-NMR.

1.1 NMR at 750 MHz

The expectations of the NMR spectroscopist comparing 600 to 750 MHz can be summarised into the following features:

- higher dispersion, therefore
  - reduced peak overlap
  - more possibilities for selective excitation
  - improved integration results
  - less spin systems of higher order
- higher sensitivity, therefore
  - faster multi-dimensional experiments
  - less minimum sample amount
  - more efficient use of pulsed field gradients as due to higher sensitivity, experiments can be done with 1 or 2 scans
- at least same lineshape and solvent suppression quality as at 600 MHz
- similar magnet stability as at 600 MHz
- all experiments running at 600 MHz have to be possible at 750 MHz with at least equal performance

Higher sensitivity can only be achieved if all RF-pulses needed are short and the dynamic range of the detection system is high enough.

To clarify the importance of short pulses, one has to keep in mind a 20% increase in sweep-width when going from 600 to 750 MHz. For protons with a typical 1H-sweep-width of 15 ppm this means increasing spectral width from 9000 to 11250 Hz. To guarantee uniform excitation by the RF-pulse, the length should be less than 15 micro-seconds; this also assures correct integration values at the spectral edges.

1H spectral width is less of a problem than for example 13C. Proton-carrying 13C-atoms typically cover a sweep range of about 135 ppm, at 188.6 MHz this means ca. 25000 Hz. Full sensitivity for many heteronuclear multidimensional experiments can only be achieved if this sweep range can be uniformly excited and decoupled.

The excitation bandwidth of a RF pulse is calculated as 1/(4*P90), where P90 is the length of a 90 degree pulse; for the 13C-range of 25000 Hz, this means a pulse length of 10 microseconds. To achieve composite pulse decoupling of the same range using the GARP sequence a decoupler pulse of about 55 microseconds is needed.

In order to avoid sample heating this pulse length has to be created with as lowpower as possible.

When looking to proteins double labelled in 13C and 15N also the nitrogen sweep range is important. When covering ca 80 ppm for NH and NH2 this means ca 6000 Hz at 76 MHz (15N at 750 MHz). Uniform excitation therefore needs a pulse of ca 40 microseconds length.

To fulfil the pulse length requirements on all nuclei probes with very high quality factor are needed, which are able to stand high power RF pulses without arcing.

Experiments of the heteronuclear Hartmann Hahn type need matched spin lock fields corresponding to a 25 microsecond 90 degree pulse over several milliseconds applied simultaneously at 1H and 13C or 15N. Probeheads need to stand this power for several hours during a 2D or 3D-experiment.

The high sensitivity of a 750 MHz spectrometer can only be transferred into the final spectrum if the receiver dynamic range is high. When using the oversampling technique together with a 16 bit digitizer the dynamic range can be increased to 19 bit. This leads to better digitization of small signals in the presence of very large solvent signals, a typical situation which is found for proteins in aqueous solution or all biofluids.

When running multidimensional experiments over long time periods (can be several days) magnet drift must be small in order not to disturb the acquisition.

750 MHz magnets are available operating at 4 Kelvin or at 2 Kelvin. The super-conducting magnet wire has better stability at 2K charged with the current needed to operate at 750 MHz.

Typical drift rates observed for the magnets available are as follows:

operation at 4K >= 3 Hz/hour
operation at 2K >= 0.1 Hz/hour

Figure 1 shows the lineshape obtained on 750 MHz system operating at 2K on a sample of 1% CHCl3 in acetone-d6 nonspinning. The width of the central line is measured at the height of the carbon satellites and at 1/5 of this height. The values 4.3 and 7.3 Hz were only obtainable in the spinning mode a few years ago on smaller magnet systems.

To test the quality of water suppression a sample of 2 mmol sucrose in 90% H2O and 10% D2O is often used. Figure 2 shows the spectrum obtained with 8 scans where only the standard presaturation sequence was used. A small and symmetric residual water signal indicates good RF- and magnet homogeneity. The signal to noise ratio (S/N) of 260:1 on the anomeric proton signal (5.4 ppm) excels the values obtained routinely at 600 MHz by more than 25 %. The sucrose sample in H2O is a better sensitivity test for highfield machines compared to 0.1% ethylbenzene, as all samples typically run are supplied in aqueous solution. As a lot of proteins and biofluids are measured with salt solution the same test as in figure 2 was repeated with a 400 mmol salt and 1 molar salt solution. The S/N obtained was only reduced to 200:1 or 180:1 respectively thus indicating the large tuning and matching range needed to run biological samples.

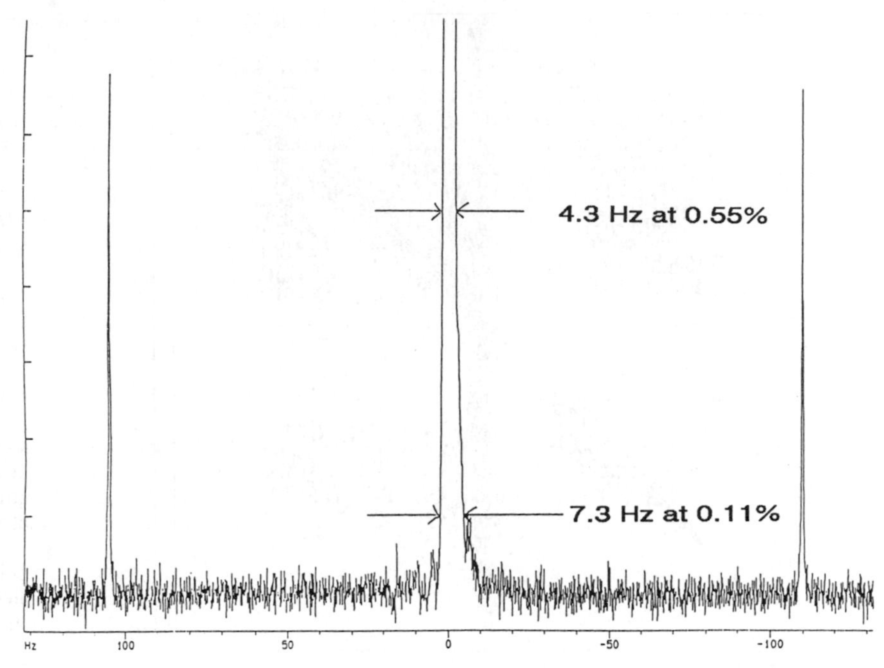

Figure 1: 1% CHCl3 in acetone-d6, nonspinning humptest, 5 mm-1H probe Bruker DMX-750

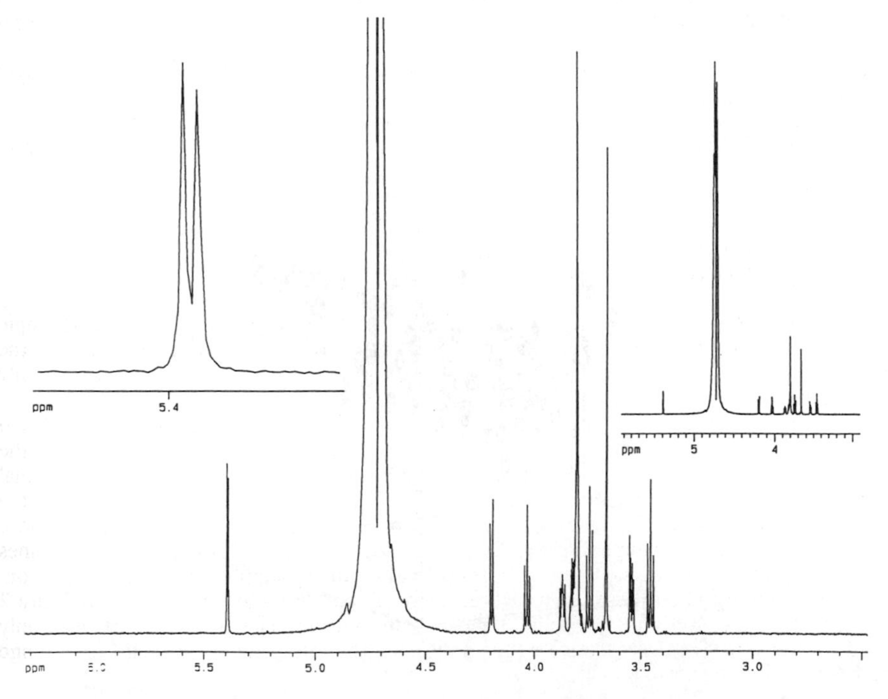

Figure 2: 2 mmol sucrose in 90% H2O/10% D2O, 8 scans standard presaturation Bruker DMX-750 5 mm inverse triple probe with Z-gradient

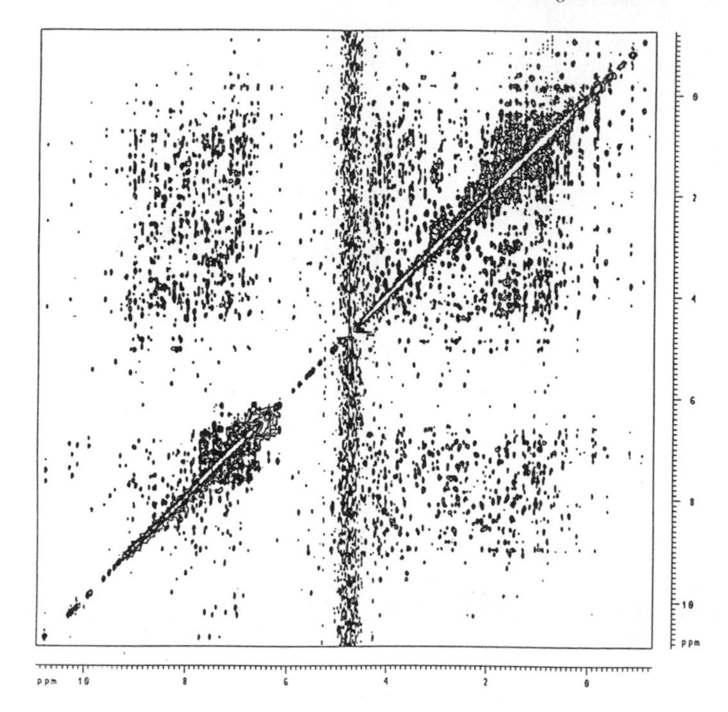

Figure 3: 2D-noesy on 0.3 mmol lysozyme in 90% H2O/10% D2O, 13.5 hours acquisition on Bruker DMX-750

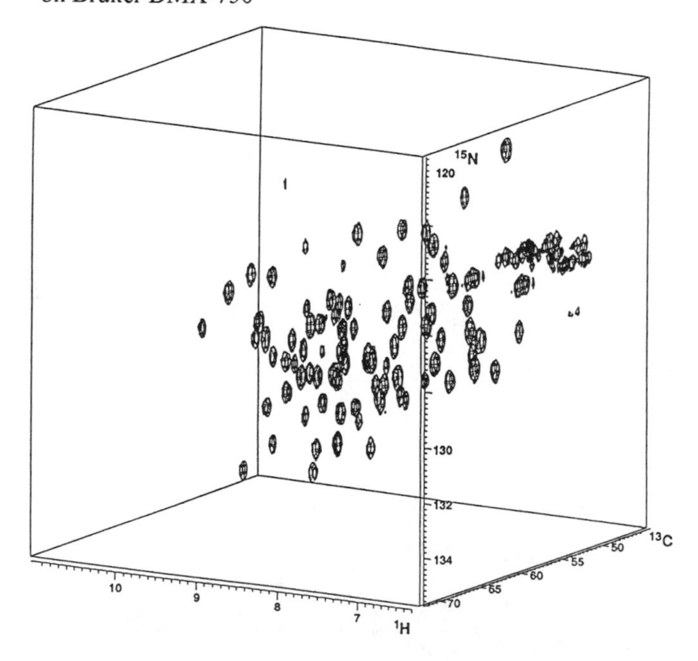

Figure 4: 3D-HNCOCA on 2 mmol double labelled ribonuclease in 90% H2O/10% D2O, matrix size 64*128*512, acquisition time 37 hours on Bruker DMX-750

Figure 3 shows a 2D-noesy experiment [1] on 0.3 mmol lysozyme solution in 90% H2O/10% D2O acquired in 13.5 hours on an 5 mm inverse triple probe. The spectrum clearly shows that homonuclear 2D`s of 0.1 to 0.5 mmol protein solutions can be measured in an overnight run at 750 MHz with more than adequate S/N on the interesting crosspeaks.

In figure 4 a HNCOCA experiment [2] on 2 mmol double labelled ribonuclease T1 in 90% H2O/10% D2O is displayed to show the long term stability of the system. The measurement was done over 34 hours and it allows to trace down the aminoacid sequencing in the protein chain. Off and on resonance shaped pulses allow to selectively excite the carbonyl and alpha C-region, by this avoiding the necessity of having a 4-channel spectrometer. The length of the high power carbon and nitrogen 90 degree pulses on the RNA sample obtained on a Bruker DMX-750 were 10 and 35 microseconds respectively. These values exactly fit the needs of pulse length necessary as mentioned above.

Figure 5 shows the use of pulsed field gradients to speed up acquisition of multi-dimensional experiments by removing the need of phase cycling as the coherence selection is done with the gradients. On the same RNA sample as in figure 4 a HSQC-experiment [3] was done in 14 minutes with 1 scan per increment without any water suppression. The aliphatic part of the spectrum is also presented although there are no 15N/1H correlations in this part, but the absence of any T1-noise demonstrates the quality of the gradient coherence selection.

Besides the proteins the biofluids urine, plasma and seminal fluid have been investigated in detail at 750 MHz. Figure 6 compares expansions of a human urine spectrum obtained at equal temperature and within 30 minutes at 600 and 750 MHz.

In the upper part the 750 MHz spectrum shows better peak separation than at 600 MHz. Mixture fluids like urine are ideal samples to show the increased dispersion as they contain hundreds of small molecules which all yield sharp NMR lines.

The better the dispersion is the higher the possibility to be able to identify individual compounds out of the mixture which for example are indicating certain diseases. Of course the same is true for 2D experiments like J-resolved or inverse correlations. Like this it was possible for the first time using 750 MHz 2D-jresolved of plasma to correctly assign the most crowded spectral region between 3 and 4.4 ppm. Figure 7 shows an expansion of this spectrum with the assignments made. J-resolved experiments [4] can be done very fast (ca 15 minutes) at high field and therefore are suitable for screening purposes.

Figure 8 shows the natural abundance gradient 1H/13C-HSQC on seminal fluid demonstrating the resolving power of the technique when coupled to 750 Mhz instruments. With this spectrum available, it was possible to assign a multitude of individual compounds automatically by comparison to database spectra. The same seminal fluid sample was used for the spectra in figure 9, comparing selective tocsy results to the overall mixture spectrum. With the increased dispersion at 750 MHz there are more possibilities to extract spin systems in a very clean manner using selective excitation. To achieve the pure subtraction of peaks not belonging to a spin system excited, the spectrometer needs to be extremely stable especially as the experiment has to be done in water. Figure 10 shows a 2D-tocsy [5] spectrum on pear-juice at 750 MHz using oversampling and digital filtering where only the region from 5.6 to 2.5 ppm was acquired with very high digital resolution. The efficiency of the digital filter is obvious by the absence of any folded peak from outside the sweep-range acquired. The spectrum was calculated with linear prediction and allows to differentiate the different sugars in the mixture like the fructose, glucose, sucrose, arabinose and mannose isomers.

1.3 Summary

A multitude of further experiments has been acquired all demonstrating the routine use of 750 MHz NMR spectrometers, which is also shown by several instruments already installed in the spectroscopists laboratories by now.

2. LC-NMR

The identification of individual compounds out of mixture spectra by 1- and 2-dimensional experiments needs methods which reduce the spectral overlap considerably. In

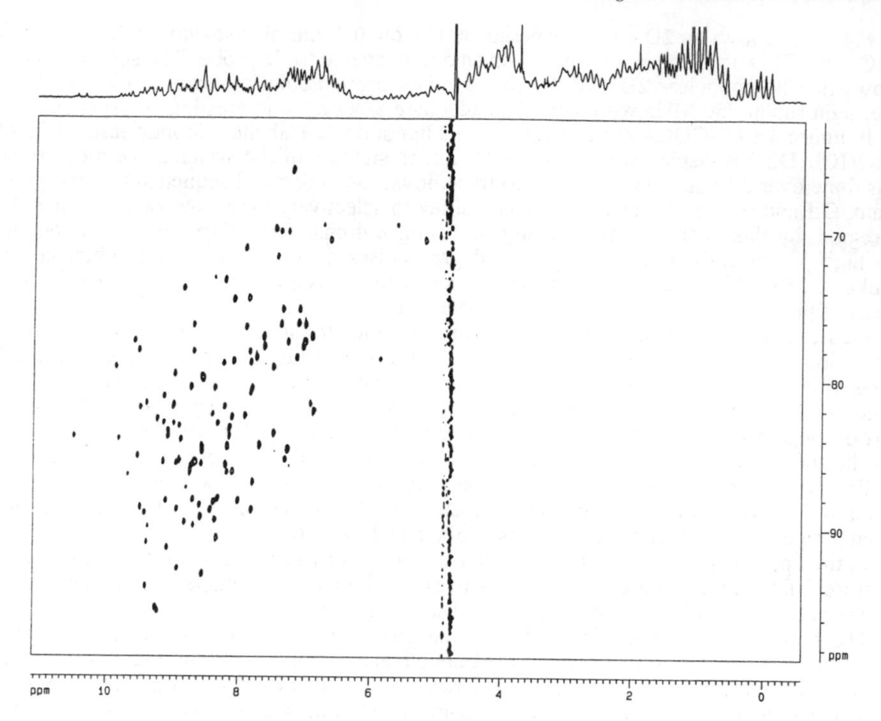

Figure 5: 2D-gradient selected 1H/15N HSQC on 2 mmol ribonuclease in 90% H2O/10% D2O, 14 minutes experiment time, scan/512 increments on Bruker DMX-750

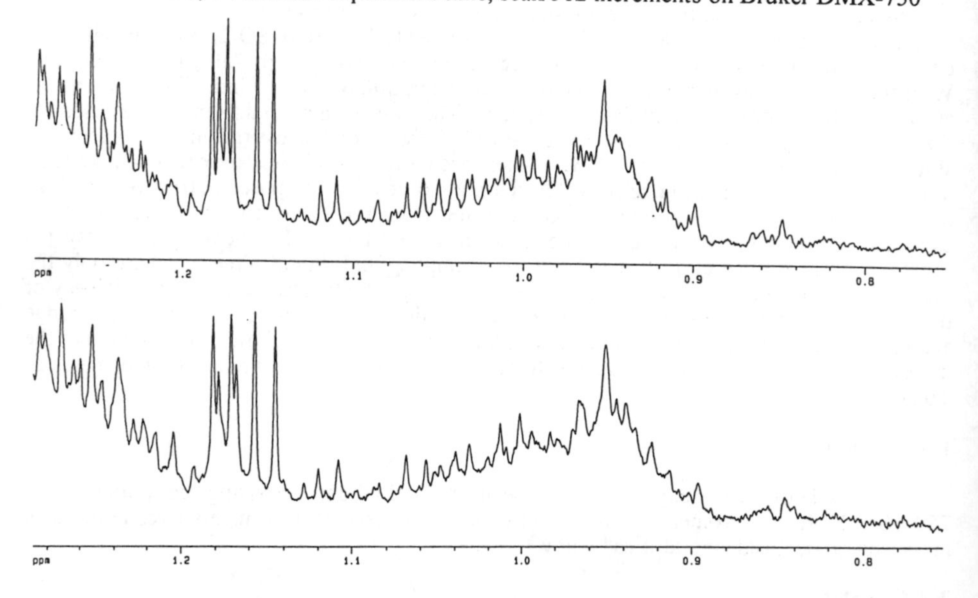

Figure 6: Comparison of 750 versus 600 MHz 1H spectrum of human urine taken after flurbiprofen dosage to demonstrate the higher dispersion, upper trace 750 MHz, lower trace 600 MHz

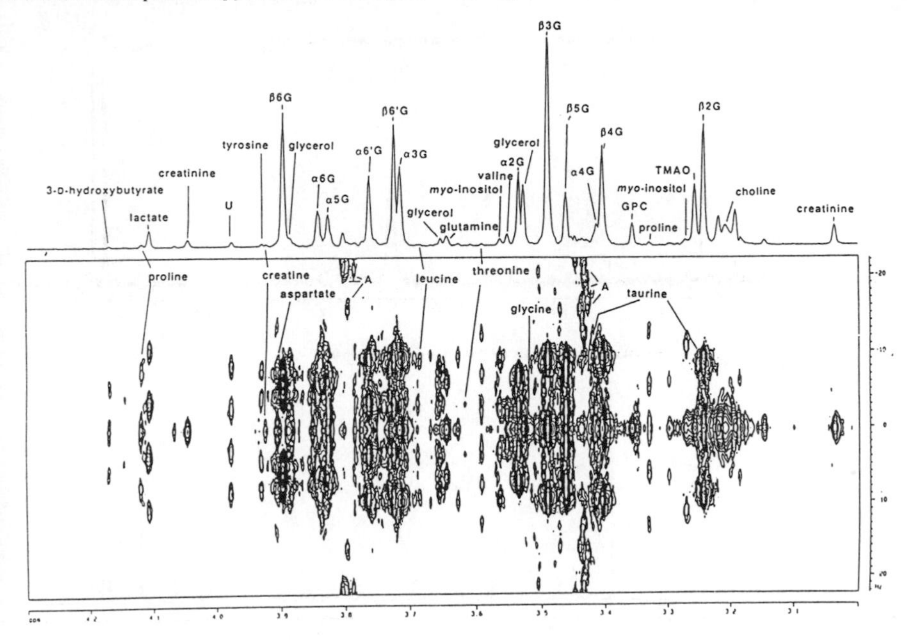

Figure 7: 2D-j-resolved experiment on human plasma, expansion of the sugar region, 16 scans/128 increments on Bruker DMX-750

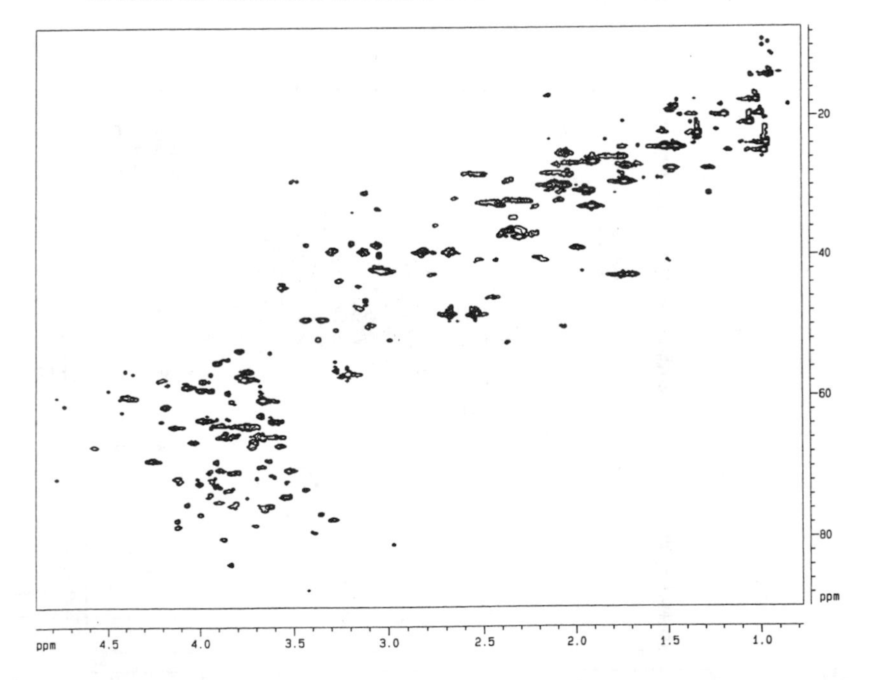

Figure 8: Natural abundance 1H/13C inverse correlation of seminal fluid, 56 scans/400 increments, linear predicted to 800 F1 points, Bruker DMX 750

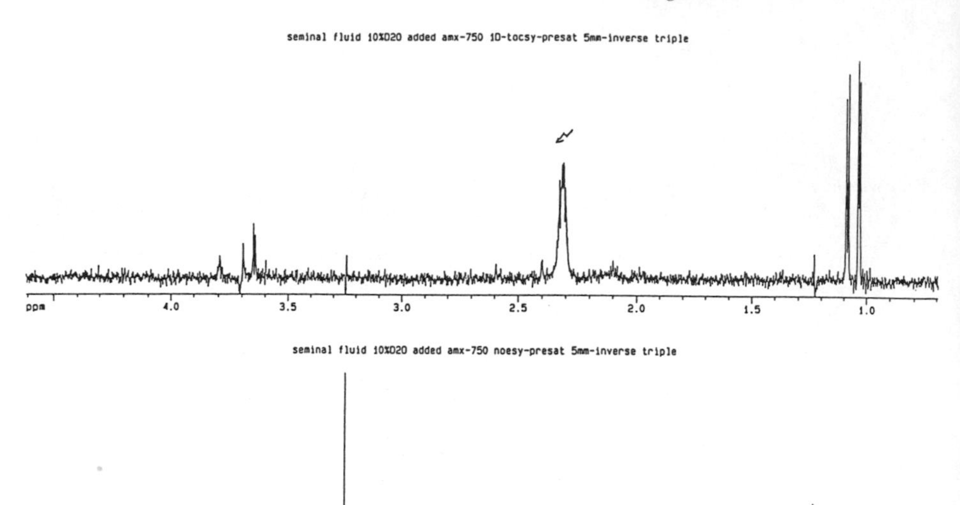

Figure 9: 1D tocsy experiment on seminal fluid showing the valine spin system selectively excited, Bruker DMX-750

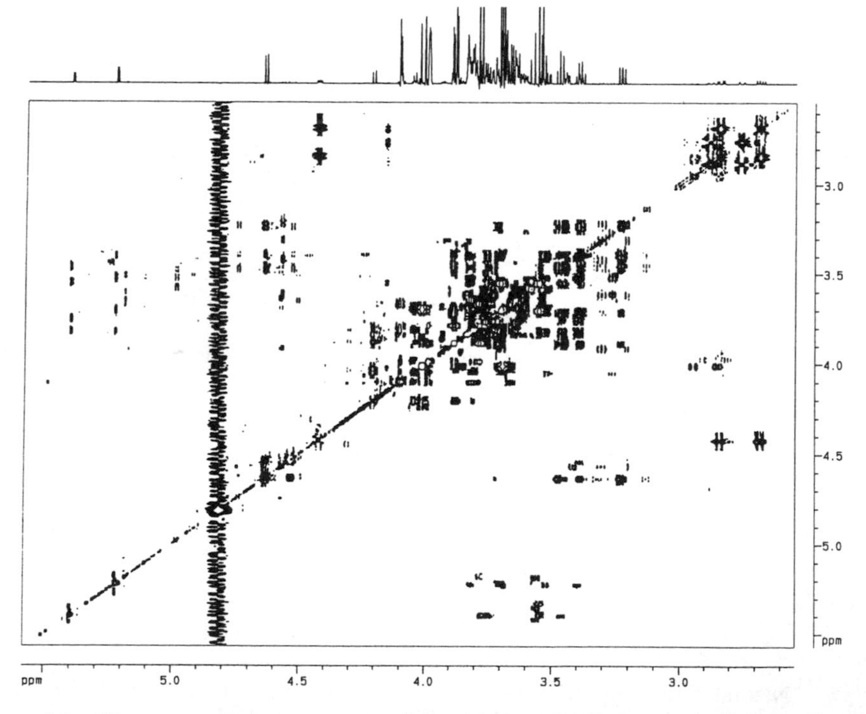

Figure 10: 2D-tocsy experiment on pear juice 16 scans/400increments digitally filtered to 3.3 ppm sweep width, Bruker DMX-750

several cases this is not possible especially when compounds of similar chemical structure are investigated. By implementing a non NMR-related separation step before NMR measurement such cases can also be solved. The most promising way to go was to couple liquid chromatography on line to NMR spectroscopy. The hardware requirements and some application examples are described, which prove the efficiency of LC-NMR.

While HPLC-MS has proven its usefulness and sensitivity in many cases, LC-NMR has been perceived to be less developed although several publications demonstrated its possibilities [6-12]. White spread use of HPLC-NMR was prevented in the past by insufficient sensitivity, the need to use deuterated solvents, lacking multiple solvent suppression routines and restricted dynamics of the receiving system. This situation has changed drastically in the last two years. The development of dedicated flow probes, where the detection coil is directly mounted to the flow cell in the NMR-probe, guaranteed optimal filling factor and better sensitivity. Modifications to the NMR-detection coil have improved solvent suppression. The introduction of pulse sequences enabling the further reduction of solvent NMR-signals [13,14] also allowed to obtain better detection limits and the use of non deuterated eluents. By adding the oversampling technique to the NMR-acquisition system the dynamic range was improved, for typical 1H-observations the digitizer resolution increased to 18 or 19 bit instead of 16 bit. Digital filtering reduces the noise in the NMR-experiment further.

Having the tools in hand described above, LC-NMR was able to generate useful results in the following areas:

- detection of drug metabolites in body-fluids like urine or bile [15-18]
- investigations of plant extracts
- sorting of peptide mixtures for biologically active compounds
- investigations of wines and fruit-juices for rare sugars and phenolic compounds
- structure analysis of polymers, also by GPC-NMR coupling

Solvent signal suppression is obsolete in 19F-LC-NMR. As more and more newly introduced drugs contain fluorine, the method allows a rapid screening of body fluids for fluorine containing metabolites [19].

2.1 Description of the hardware

Figure 11 shows the LC-NMR probe. The best compromise between NMR-sensitivity and retaining the chromatographic resolution was achieved by using a flow cell of 3 mm inner diameter, resulting in ca 120 microliters volume. If a LC-peak is assumed with a half width of 20 seconds, about 50% of the peak is in the flow-cell, when the peak maximum is at the flow cell centre. A increased flow cell volume would allow a larger part of a LC-peak to be trapped, but the chromatographic separation would be degraded. A typical case for a larger flow cell volume is the GPC-NMR coupling, where LC-peak half width is substantially increased.

The flow-probe is handled like a normal NMR-probe, it is connected to the chromatographic system by a PEEK-capillary whose length depends on the field strength of the NMR-magnet which is used. The LC-pump has to be located outside the stray-field of the magnet in order to avoid operational disturbances. For a 500 MHz NMR-system a distance of 2.5 meters is appropriate, transfer times to the flow-cell cover a range of 15 to 40 seconds depending on the flow rate. Tests with a second UV-detector after the transfer-capillary showed no recognisable degradation of the chromatographic separation.

Figure 12 shows a LC-NMR interface which was developed to improve the handling of the LC-NMR coupling and allows to implement automatic measurements. As with regard to the NMR sensitivity, it is often necessary to acquire NMR-spectra when the flow is stopped. A LC-peak is normally detected first by UV or refraction index, but before the flow can be stopped for NMR measurement the transfer time to the NMR flow cell has to be waited for. Software has been developed to allow autodetection of the LC-peak maximum and stop the

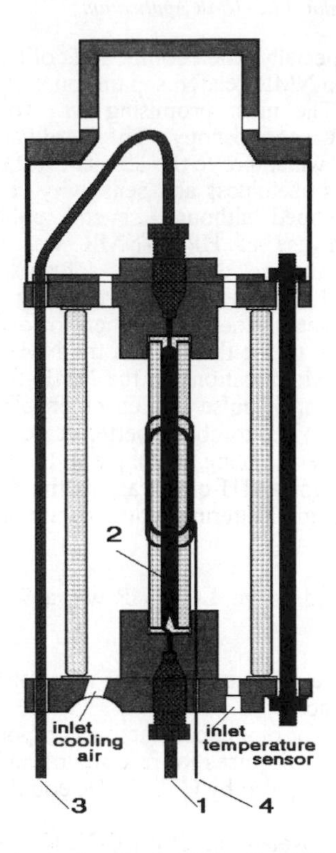

Figure 11 Schematic drawing of a flow probe for LC-NMR coupling 1 inlet capillary, 2 flow cell, 3 outlet capillary, 4 detection coil

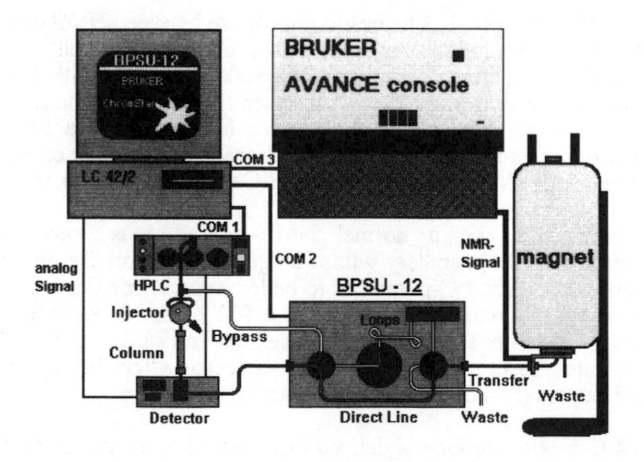

Figure 12: LC-NMR interface with 12 loops for system internal intermediate storage of 12 LC-peaks also allowing on-flow and direct stop-flow measurements.

flow exactly when the peak maximum is in the NMR-cell. In addition chromatographic conditions can be frozen when the flow is stopped and resumed after the NMR-measurement is finished. This is vital especially for gradient reversed phase runs.

The LC-NMR interface also allows the intermediate storage of up to 12 LC-peaks in loops. The loop volume corresponds to the volume of the flow cell. The software allows to transfer the loop-contents automatically to the flow cell and start the NMR measurement. The loop concept was developed to avoid degradation of the chromatographic resolution by stopping the flow repeatedly to do NMR acquisition. As could be shown for reversed phase separations even longer stops did not deteriorate the separation; however in GPC-NMR runs, especially when CH2Cl2 was used, the opposite was true.

In the separations performed up to now, Bruker LC-22 and Merck-Hitachi 6200 pump systems have been used. For both pumps interfaces have been developed to the LC-NMR system. When on-flow LC-NMR experiments are acquired, it is vital for a good quality solvent suppression, that the pumps have minimum pulsation.

The NMR-machines which were used where Bruker AMX 400, DRX 400, AMX500 and DMX600.

2.1 The solvent requirements for LC-NMR

Normal HPLC grade solvents are rarely usable for LC-NMR. Tests of HPLC-grade acetonitril, methanol, methylenechloride and chloroform of various suppliers showed a low UV absorbance but considerable NMR-background. Naturally this background renders the measurement of low concentrated elutes difficult, especially when in the nanogram range. Riedel de Haen's Pestanal series of solvents has been used in all the separations shown because of the best NMR purity. For LC-NMR minimal UV absorption is not critical as the UV sensitivity is higher than the NMR sensitivity, so a slight constant increase in UV absorbance can be tolerated.

In reversed phase separations acetonitril is preferred to methanol as it only gives rise to one proton signal and therefore covers a minimum area of the spectral window. As D2O is relatively cheap, it is used instead of H2O for all reversed phase or ion exchange separations. A minimum amount of deuterated solvent is needed as the spectrometer is referenced to the deuterium frequency. To bypass this problem, a flow probe has been developed which has a second small-diameter-flow-cell inserted centrally into the main flow cell. This second flow cell can be filled with deuterated solvent, then the eluent mixture can be used without deuterated addition.

2.3 Results obtained with the LC-NMR coupling

Figure 13 shows an on-flow detected reversed phase gradient run on a lyophilysed urine sample obtained from a person which was dosed with 400 mg Ibuprofen, a constituent of several headache and influenza tablets. The structure of ibuprofen and its major metabolites in human urine is shown in figure 14. The urine was concentrated by a factor 6 through lyophilysation and redissolved in eluent mixture.

In figure 13 the horizontal axis shows the 500 MHz-1H-NMR spectral window, while the vertical axis represents the chromatographic retention time. At about 2.5 minutes all the polar compounds like sugars and several amino acids are eluted together, then hippuric acid can be readily identified. Most of the signals of the later eluting compounds are due to the ibuprofen metabolites. The strength of NMR is the structural information obtained even with the restricted acquisition time possible in a LC-NMR run, time resolution in this experiment was 12 seconds per NMR-trace. The whole run consisted of 300 traces acquired consequently with 16 transients accumulated per trace. Three different signals around 5.5 ppm along the retention time axis can immediately be identified as belonging to glucuronides of ibuprofen metabolites. A horizontal slice through those signals at 5.5 ppm shows the residual signals belonging to the metabolites. By inspection of the region around 1 ppm in an expansion the multiplicity of the signals defines, whether the metabolite is an oxidised species or whether the alkyl chain is still intact.

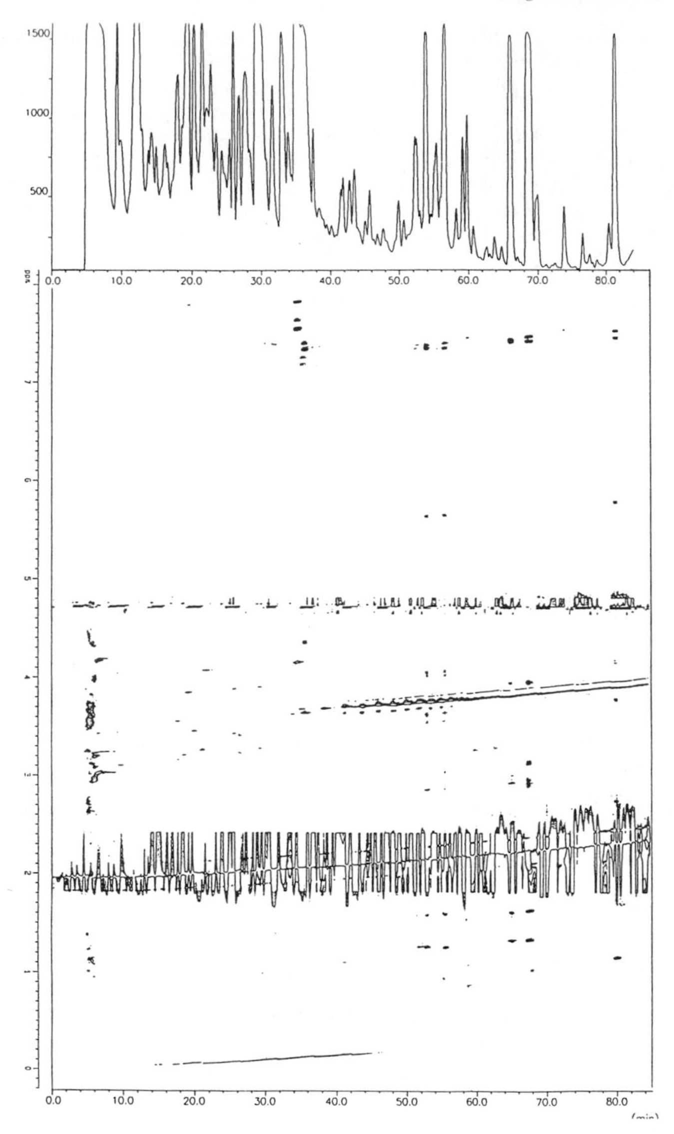

Figure 13: On-flow gradient 500 MHz-LC-NMR run of a human urine, obtained after dosage of 500 mg ibuprofen, sixfold concentrated by freeze drying and redissolved in CH3CN/D2O

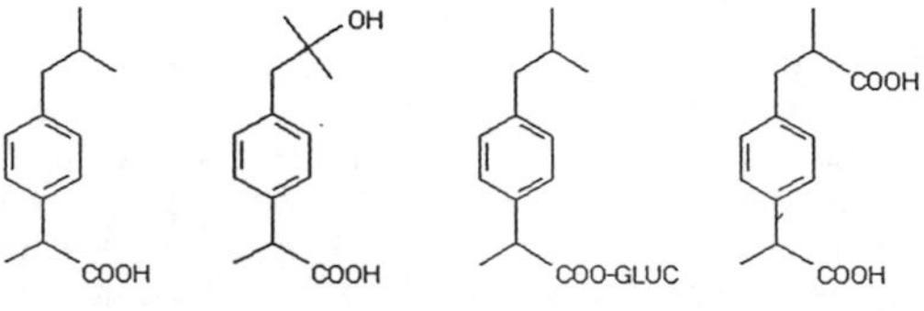

Figure 14: Structures of the main ibuprofen metabolites in human urine

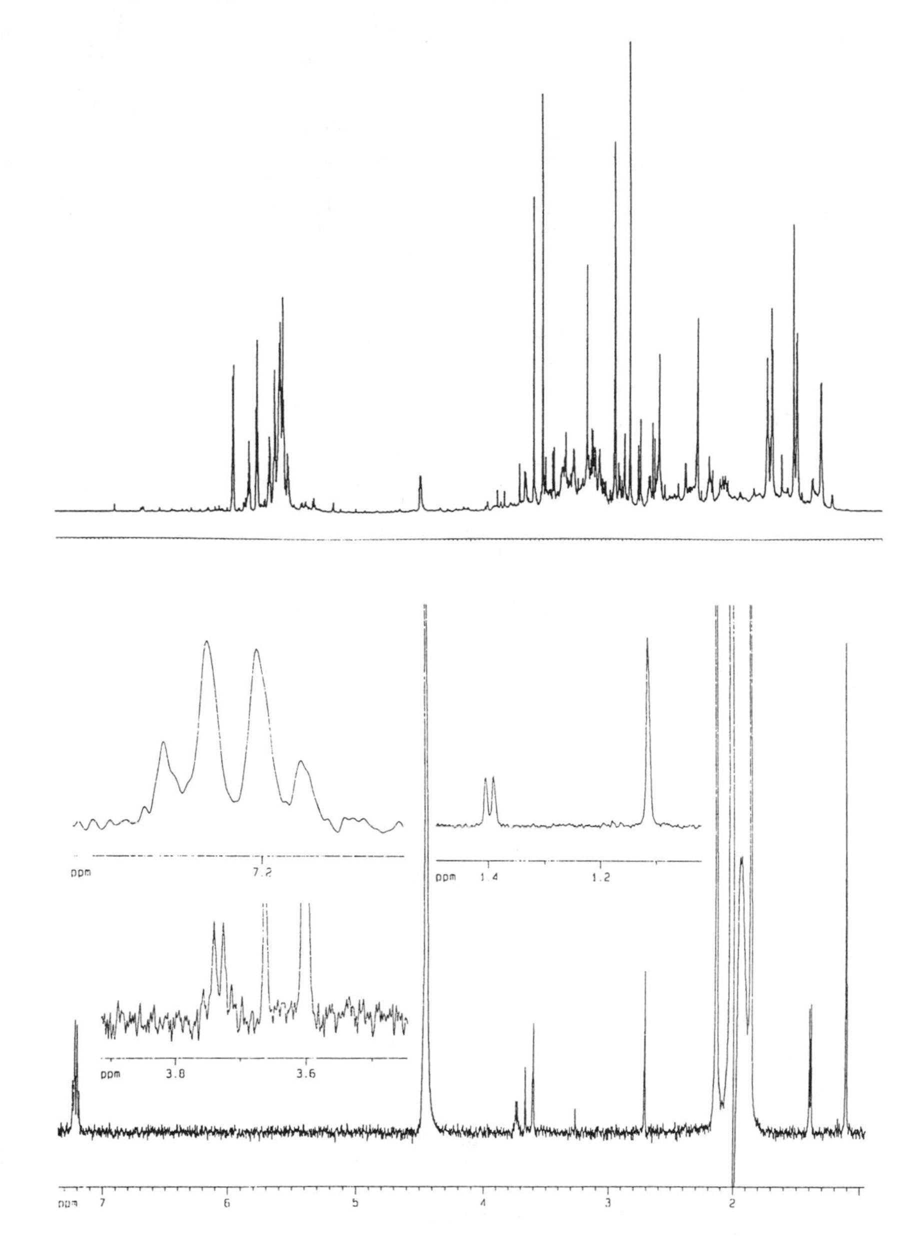

Figure 15: The stop-flow LC-NMR spectrum of metabolite 2 (b) at 500 MHz

Figure 15 shows in the upper trace a spectrum of the complete urine and in the lower trace a glucuronated metabolite spectrum as obtained by stop-flow in less than 5 minutes.

To obtain an on-flow detected reverse phase gradient run is not straightforward for the NMR-part. As the resonance frequency of acetonitril is changing with increasing percentage of acetonitril, the solvent suppression scheme has to follow that considerable change in frequency. When looking to the path of the residual acetonitril signal in figure 3 starting at 2 ppm, the moving resonance position with increasing run time is visible. The way to solve the problem is to acquire first an on-flow detected run without solvent suppression. A peak picking routine picks the two largest peaks and writes the values found in each trace into two frequency lists. The real experiment is performed with solvent suppression and for each trace the corresponding values are taken from the solvent frequency lists.

In figure 16 a tripeptide mixture starting from the amino acids tyrosine, methionine and alanine is injected onto a reversed phase column using a D2O/CH3CN gradient. The UV-spectrum is superimposed onto the on-flow LC-NMR run at 600 MHz. It clearly reveals how misleading the UV trace can be. At early retention times the tripeptides consisting of methionine and alanine only are eluting. Their UV response is very weak, in this case the NMR-detection shows the real amount of this peptides is much larger as UV would suggest. It becomes immediately obvious which are the tripeptides with two alanines in the sequence, because they show two doublets in one trace. From the chemical shift of the two signals one can immediately conclude whether its ala-ala-met, ala-met-ala, ala-tyr-ala or ala-ala-tyr. This example again shows the high structure resolving power of LC-NMR, it also demonstrates that NMR gives a real representation of the relative amounts of compounds in the mixture.

Figure 17 shows a wine concentrate injected onto a sugar calcium column using D2O with addition of 0.1% sulphuric acid as the eluent, detected at 500 MHz on flow. When counting the anomeric proton signals around 5.2 ppm along the retention time axis, at least 15 different sugars can be detected. With the aid of NMR it is easy to differentiate for example alpha and beta glucose, which are also slightly separated by retention time, as the chemical shift of the anomeric protons considerable deviates (5.22 ppm alpha, 4.64 ppm beta). Special interest for future investigations will go into the oligosaccharides, which elute first in this separation.

2.4 Summary

With the examples shown, it could be demonstrated, that LC-NMR has reached the sensitivity level to work successfully on analytical HPLC-columns. Further improvements in sensitivity would be possible by going to smaller diameter columns, to have a LC-peak completely in the measurement cell. Smaller diameter columns also will provide the possibility to use fully deuterated solvents as the flow rate is reduced drastically. Also sensitivity would improve in this case.

References

[1] A.Kumar, G.Wagner, R.Ernst a. K.Wuthrich J.Am.Chem.Soc. **103**, (1981) 3654
[2] A.Bax and M.Ikura, J.Biomolec.NMR, **1**, (1991) 99.
[3] A.Davis, J.Keeler, E.D.Laue a. D.Moskau, J.Magn.Res.,**98**, (1992) 207
[4] W.P.Aue, J.Karhan a R.R.Ernst, J.Chem.Phys., **64**, (1976) 4226
[5] A.Bax and D.Davis, J.Magn.Res., **65**, (1985) 355
[6] E.Bayer, K.Albert, M.Nieder, E.Grom, G.Wolff, M.Rindlisbacher, Anal.Chem., **54**, (1982) 1747.
[7] H.C.Dorn, Anal.Chem., **56**, (1984) 747A
[8] D.A.Laude, Jr., C.L.Wilkins, Trends Anal.Chem., **5**, (1986) 230.
[9] K.Albert, E.Bayer, Trends Anal.Chem., **7**, (1988) 288.
[10] K.Albert, M.Kunst, E.Bayer, M.Spraul, W.Bermel, Chromatogr., **463**, (1989) 355.
[11] K.Albert, M.Kunst, E.Bayer, J.J. de Jong, P.Genissel, M.Spraul, W.Bermel, Anal.Chem., **61**, (1989) 772.

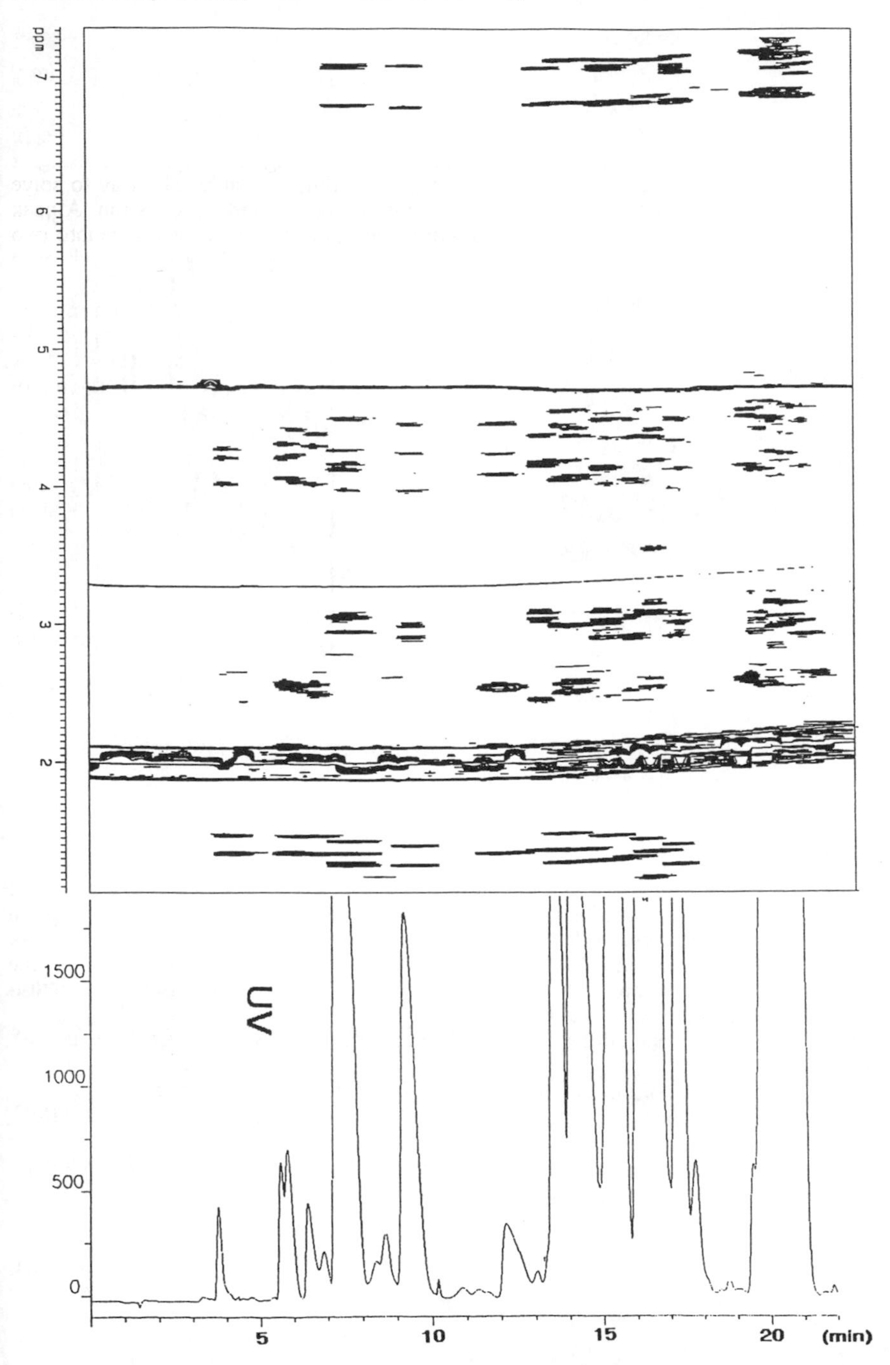

Figure 16: On-flow gradient 600 MHz-LC-NMR run of a tripeptide mixture based on the amino acids alanine, methionine and tyrosene with superimposed UV-chromatogram.

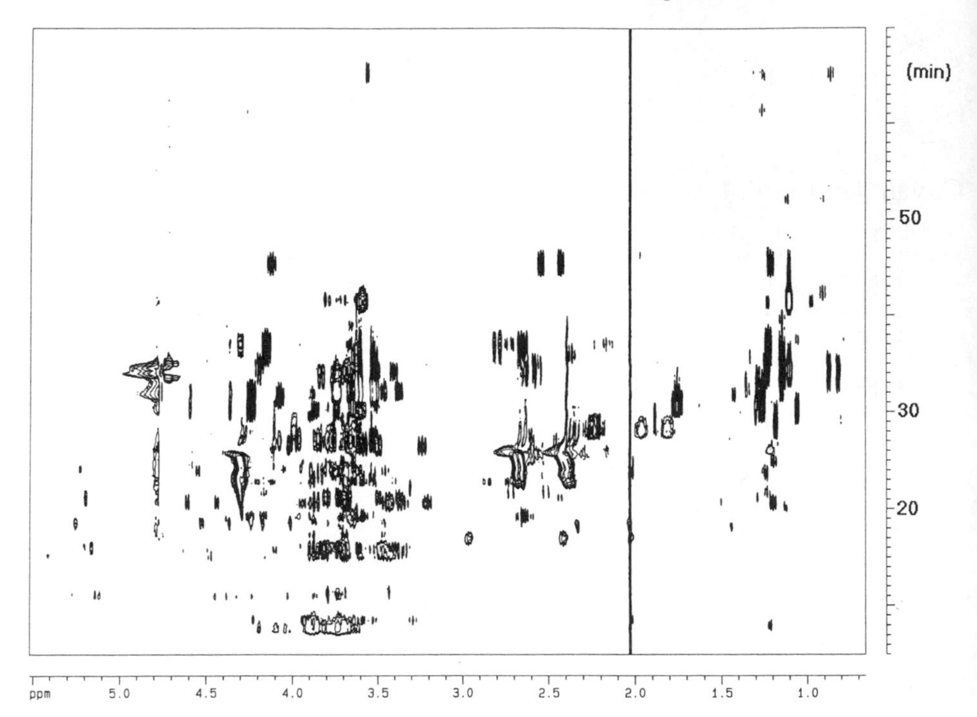

Figure 17: On-flow 500 MHz-LC-NMR run of a wine concentrate (100 ml distilled into 5 ml) on a sugar-Ca-column

[12] L.A.Allen, T.E.Glass, H.C.Dorn, Anal.Chem., **60**, (1988) 390.
[13] E.Bodenhauser, H.Kogler, R.R.Ernst, J.Magn. Res., **58**, (1984)370
[14] A.Bax, J.Magn.Reson., **65**, (1985) 142
[15] M.Spraul, M.Hofmann, P.Dvortsak, J.K.Nicholson, I.D.Wilson, J.Lindon, Pharm.Biomed.Anal., **8**, (1992) 601.
[16] M.Spraul, M.Hofmann, P.Dvortsak, J.K.Nicholson, I.D.Wilson Anal.Chem., 65, (1993) 327.
[17] I.D.Wilson, J.K.Nicholson, M.Hofmann, M.Spraul, J.Chromatogr., in press
[18] M.Spraul, M.Hofmann, J.C.Lindon, J.K.Nicholson, I.D.Wilson, Anal.Proc.,**30**, (1993) 390 -392
[19] M.Spraul, M.Hofmann, I.D.Wilson, E.Lenz, J.K.Nicholson and J.C.Lindon, J. Pharmaceut. Biomed.Anal., **11**, (1993) 1009-1015

Molecular Mobilities around the Glass Transition in Sugar–Water Systems

C. van den Berg[1], I. van den Dries[1], and M. A. Hemminga[2]

[1] FOOD PHYSICS GROUP, DEPARTMENT OF FOOD TECHNOLOGY, AND [2] DEPARTMENT OF MOLECULAR PHYSICS, WAGENINGEN AGRICULTURAL UNIVERSITY, PO BOX 8129, 6700 EV WAGENINGEN, THE NETHERLANDS

1 INTRODUCTION

Upon cooling and/or drying many amorphous components in foods undergo a rubber-glass transition, which affects their physical properties and stability. This transition, which is predominant with many carbohydrates and some proteins, represents for pure low molecular weight carbohydrates a transition from viscous liquid-like to vitreous solid-like behaviour. For carbohydrate polymers it signifies a transition from elastic rubbery to a brittle glassy behaviour. Water, as a universal plasticizer of the carbohydrate structure, strongly influences the temperature at which the glass transition occurs.

Generally, a glass is an amorphous solid in which the molecules are forming a nonperiodic and nonsymmetric network. In theory, the molecular mobility in glassy state systems is many orders of magnitude lower than in liquid systems (1). Below the glass transition temperature (T_g), the very low molecular mobility would predict a good stability of the food product under consideration. The glass transition will affect food processing operations such as drying, humidification, extrusion and freezing, and also quality attributes such as hardness and fluidity, stability, hygroscopicity, ageing behaviour and reactivity in the solid state. The importance of this effect for practice, especially with respect to an improved understanding of food handling and preservation processes, is realized only fairly recently, and using this approach, during the past few years numerous studies have been mounted to elucidate the glassy state behaviour of foods (2, 3, 4). In this respect not all aspects are well understood yet, e.g. a discrepancy in practice is that amorphous food matrices below the glass transition can be dried out easily showing that water and some other small molecules have higher mobilities than theoretically expected.

In order to obtain an improved understanding of the molecular dynamics in the glassy and liquid states in comparison to macroscopic properties, a study was started to investigate in more detail the mobility of sugar-water systems around the glass transition using spin probe electron spin resonance spectroscopy (ESR) and differential scanning calorimetry (DSC). ESR spectroscopy has been shown to be a good method to study molecular motion in sugar-water systems but has not been much applied yet (5-8) . It yields direct information about the average rotation correlation times (τ_C) of paramagnetic spin probe molecules in the systems under study, whereas DSC measures the glass transition temperatures (T_g) macroscopically by monitoring the difference in specific heat before and after the glass transition.

This work continues earlier ESR investigations into molecular mobilities around the glass transition of aqueous sucrose, maltodextrin and some other carbohydrate solutions, where the nitroxide radicals Tempol and maleimide spin label have been used as spin probes. All investigated sugar-water systems exhibited a characteristic sudden increase in rotation correlation times near the presumed glass transition temperature (5-9),. However, these earlier measurements had not been backed up by simultaneous independent determinations of T_g. The latter is required in view of the scatter in values for T_g' in the literature (up to ten degrees) (3, 5-7, 9- 11), due to differences in pretreatment, equipment, scan speed and interpretation.

This paper reports the ESR and DSC results obtained with four sugars, being the monosaccharides glucose and fructose, and the disaccharides sucrose and maltose, at low temperatures and different concentrations: 20, 40, 55, and (if solubility at elevated temperatures allowed it) 70 wt%.

Figure 1 gives the state diagram of these four sugars, showing the glass transition temperatures as a function of composition together with the liquidus and solidus lines, which are connected at the eutectic temperature. More important for this study however, is T_g', the characteristic glass transition temperature of the maximally frozen aqueous saccharide solutions. Because the four saccharides do not easily crystallize in the cold, upon freeze-concentration the eutectic point (T_e) is passed easily, bringing the system in a non-equilibrium state. A high viscosity near the eutectic composition generally favours glass formation. At further lowering the temperature freeze-concentration continues and the remaining saccharide solution attains its glassy state viscosity, where ice crystallization has become so slow that it is not observable any more on practical time scales. Here the remaining saccharide-water mixture outside the ice crystals is maximally freeze concentrated at the (specific) glass temperature T_g' and composition C_g' (weight percent of solute in the glassy phase). At temperatures and/or compositions above T_g' (in the rubbery state) the sample becomes unstable in its supersaturated state and risks crystallization of the solute being dictated by nucleation and crystal growth kinetics.

In general, when a maximally frozen aqueous saccharide solution is warmed from below its T_g there is a sudden increase in molecular mobility due to passing the glass-rubber transition. This mobility increase is immediately followed by and cannot be separated from a further strong increase in mobility due to the subsequent melting of ice and accompanying dilution of the dissolved solids. During melting, the mixture reverses the liquidus curve (Figure 1). This melting process is also illustrated in Figure 2 showing the warming up trace of an aqueous 25 w% sucrose solution from 225

K up to ambient temperature. The glass transition is signified by the change in specific heat of the sample (jump in baseline) at the onset of ice melting.

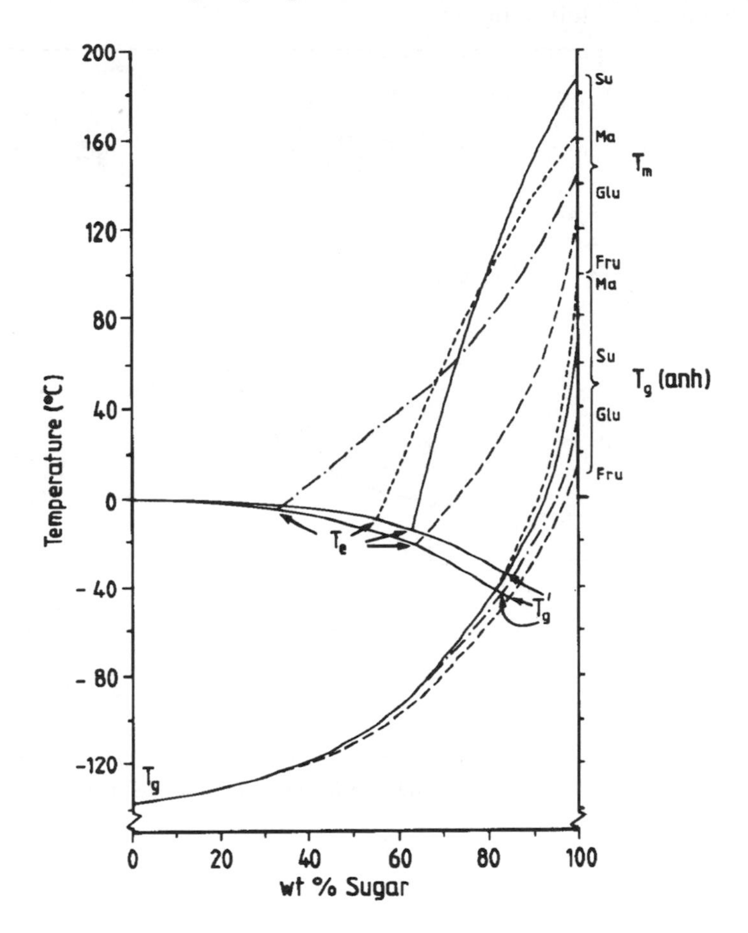

Figure 1. State diagram with liquidus, solidus and glass transition temperatures of the glucose (glu), fructose, (fru), sucrose (su) and maltose-water (mal) systems as a function of composition. Adapted from references (9, 12-15).

2 MATERIALS AND METHODS

Preparation of solutions

D-glucose (Janssen Chimica, p.a.), D-fructose, sucrose and D-maltose (Merck, p.a.) were mixed with the spin probe 4-hydroxy-2,2,6,6-tetramethylpiperidine N-oxyl (Tempol) and water. The final concentration spin probe was 0.2-0.5 mg/ml. Anhydrous glycerol was mixed with the spin probe and dried over phosphorus pentaoxide under vacuum for a few weeks. The water content afterwards was smaller than 0.5 w% (Karl Fisher method). The ESR samples were sealed in 100 µl tubes (Brand).

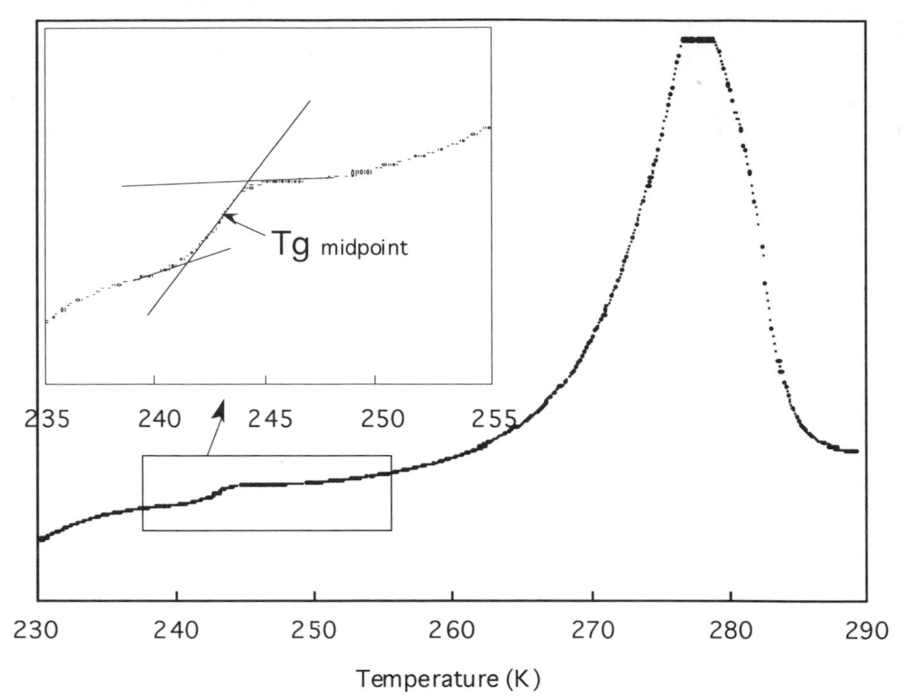

Figure 2. Differential scanning calorimetry trace of 20 w% sucrose solution. (scan speed 20 K/min). Enlarged part shows interpretation of $Tg_{midpoint}$.

Spectroscopy

ESR spectra were recorded on a Bruker ESR spectrometer ESP 330E with nitrogen temperature control. For conventional ESR the microwave power was 2 mW. The scan range, scan rate, time constant and modulation amplitude were adjusted so that distortion of the spectra was avoided. The rotational correlation time (τ_c) of weakly immobilized probes was estimated from the relation (16):

$$\tau_c = 6.5 * 10^{-10} \Delta H_0 \left\{ \left({h_c}/{h_h} \right)^{0.5} - 1 \right\} \quad (1)$$

where h_h and h_c are the heights of the high field and central lines, respectively. ΔH is the line width of the central line in tesla. The rotational motion of the spin probe is assumed to be isotropic.

In the slow motional region of the spin probes (τ_c between 10^{-6}-10^{-8}), the rotational correlation time was obtained by using the method of Goldman (17):

$$\tau_c = a\left(1 - A_z'/A_z\right)^b \tag{2}$$

A'_z is the separation of the outer hyperfine extrema in the ESR spectra, and A_z is the rigid limiting value for the same quantity. Both a and b depend on the nature of the motion of the probe and of the intrinsic line width of the spectra. The parameters a and b are determined resulting in $a = 3.34*10^{-10}$s, $b = -1.92$.

Saturation transfer ESR was applied in the very slow motional region (10^{-7} -10^{-3} s). Spectra were recorded under the following conditions: microwave power 100 mW, a modulation amplitude of 0.5 mT, modulation frequency of 50 kHz. The ESR signal was recorded 90° out-of-phase with respect to the modulation signal. τ_c was estimated by comparing the recorded spectra with standard spectra (18), recorded in anhydrous glycerol under similar conditions as for the sugar solutions. The rotational correlation time of the standard spectrum is determined by extrapolation of the correlation times of the spin probe in anhydrous glycerol between 0 and 100°C which is assumed to follow the Stokes-Einstein equation.

The sugar-water mixtures were rapidly cooled to -90°C and rewarmed to the temperature at which the spectrum was recorded. This procedure ensured not too highly concentrated samples to be maximally frozen initially. The sought T_g was found at the crossing of tangents before and after the steep decrease in τ_c (fig. 3).

Differential scanning calorimetry

The glass transition temperatures of the samples have been determined with a Perkin-Elmer DSC-7, equipped with a liquid nitrogen refrigeration unit. Aluminium sample cups, Perkin Elmer 15 µl, were scanned against an empty blank. Scan speeds of 20 K/min for cooling and, after equilibration, of 2.5, 5, 10 and 15 K/min for heating were used, enabling the extrapolation to zero scan speed. We used the generally accepted $T_{gmidpoint}$, extrapolated to zero scan speed, for comparison with the T_g obtained from the τ_c-diagrams for the same samples. It may be assumed that the $T_{gmidpoint}$ temperature indicates the midst of the strong decrease in viscosity occurring at the glass transition. The here not recorded T_{gonset} was found to be one to nine degrees lower than $T_{gmidpoint}$.

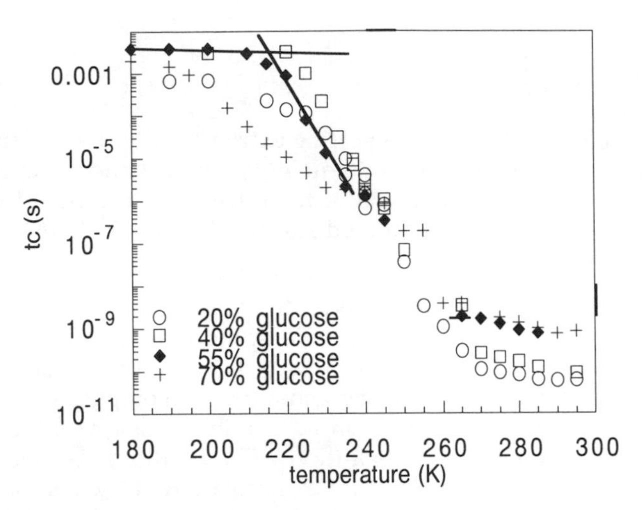

Figure 3. Rotational correlation times (τ_c) of Tempol at low temperatures in glucose-water system at various glucose. concentrations.

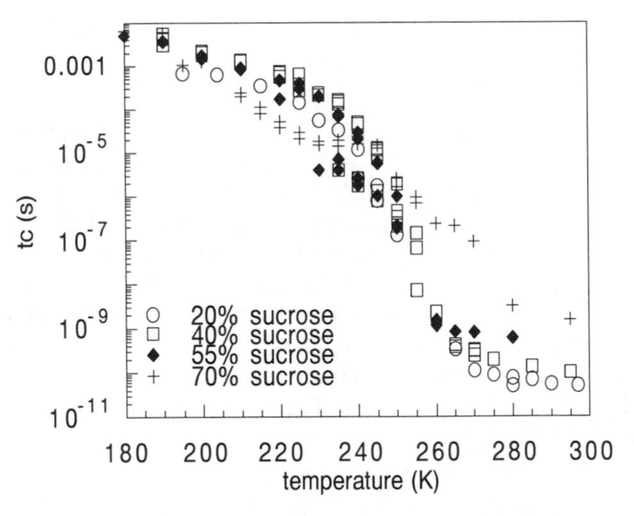

Figure 4. Rotational correlation times (τ_c) of Tempol at low temperatures in sucrose-water systems at various sucrose concentrations.

3 RESULTS AND DISCUSSION

Figures 3 and 4 show the rotational correlation times (τ_c) of Tempol, obtained from ESR measurements, as a function of temperature, at four different concentrations, for the glucose-water system and the sucrose-water system respectively. The results for fructose and maltose basically exhibit the same trend and are not shown.

Table 1 compares and summarizes the various T_g values obtained with ESR and DSC. These values for T_g obtained with DSC agree within five degrees with comparable literature values (3, 19).

concentration	glucose		fructose		sucrose		maltose	
(w%)	ESR	DSC	ESR	DSC	ESR	DSC	ESR	DSC
20	227	228	228	229	234	240	244	243
40	222	228	224	226	233	239	244	243
55	217	228	227	228	234		244	243
70	193	201	205	210	200	197		

Table 1. Temperatures (K) obtained for T_g of four saccharides as determined by ESR and DSC.

In general there is a good agreement in T_g's observed with ESR and with DSC. This observation confirms for these saccharide-water systems the direct relation between molecular mobility and change in specific heat. For 20% glucose and for 20, 40 and 55% fructose and maltose the agreement in found values for Tg is almost perfect, i.e. two or less degrees difference. Mobility during warming up is earlier detected by ESR than by DSC, in 40 and 55% glucose with four and nine degrees respectively, and for 20 and 40% sucrose with six degrees. This may be related to a difference in local structure of the glassy state or an interaction between Tempol and the saccharide. The values obtained at the highest concentration (70 w%) show a scatter of several degrees. In that case, however, no maximally equilibrium freezing can be expected under the given experimental conditions (20).

At temperatures below the glass transition Tempol in all samples was found to be rather immobile with τ_c roughly 10^{-3} s, slightly raising with temperature in logarithmic fashion until Tg is attained. Except for the highest concentration, around the glass transition a steep increase in mobility is observed, enabling the construction of T_g from the tangents. Due to dilution caused by ice melting τ_c decreases to the order of 10^{-9} s in the solution when all ice has been melted. In this high temperature area the effect of higher viscosity due to differences in saccharide concentration is visible.

During cooling of the highest concentration (70 w%), some ice is formed due to strongly delayed ice nucleation and crystal growth. Here beyond T_g in the rubbery state, τ_c increases more gradually with a slope deviating from the slope before the glass transition is reached, indicating that no dilution from ice melting occurs. A shoulder in τ_c is visible in the area of T_g' of the diluted solution, that is probably due to some recrystallization of ice made possible by the increased mobility. At higher temperatures these ice crystals melt again as is seen by a steeper increase in τ_C.

The mobility in the glassy state is affected significantly by the initial saccharide concentration. In all cases the glassy system made from dilute 20% concentration exhibits a higher mobility than the 40 and 55% concentrations. This effect of concentration is somewhat unexpected since, theoretically, one would expect equal mobilities in maximally frozen systems produced with initially different concentrations. During ice crystallisation the spin probes are concentrated in the amorphous solution (6, 21). This effect may cause an inhomogeneous spin probe distribution depending on the amount and size of ice crystals.

4 REFERENCES

1. S. R. Elliot, *Physics of amorphous materials.* (Longman Scientific, Harlow, 1990).
2. L. Slade, H. Levine, *Food science and Nutrition* **30**, 115 (1991).
3. H. Levine, L. Slade, *Cryo letters* **9**, 21-63 (1988).
4. J. M. V. Blanshard, P. J. Lillford, *The glassy state in foods.* (Nottingham University Press, Nottingham, 1993).
5. M. A. Hemminga, M. J. G. W. Roozen, P. Walstra, *Molecular motions and the glassy state.* J. M. V. Blanshard, P. J. Lillford, Eds., The glassy state in foods (Nottingham University Press, Nottingham, 1993).
6. M. J. G. W. Roozen, M. A. Hemminga, *J. Phys. Chem.* **94**, 7326-7329 (1990).
7. M. J. G. W. Roozen, M. A. Hemminga, P. Walstra, *Carbohydrate research* **215**, 229-237 (1991).
8. D. Simatos, G. Blond, M. Le Meste, *Cryo-Letters* **10**, 77-84 (1989).
9. C. Van den Berg, B. Nagy, M. A. Hemminga, *Unpublished Results.*
10. Y. Roos, M. Karel, *Food Technology* , 66-71 (1991).
11. D. Simatos, G. Blond, *Some aspects of the glass transition in frozen foods systems.* J. M. V. Blanshard, P. J. Lillford, Eds., The glassy state in foods (Nottingham University Press, Nottingham, 1993).
12. W. Schaertl, H. Sillescu, *Journal of Statistical Physics* **74**, 687-703 (1994).
13. R. H. M. Hatley, C. Van den Berg, F. Franks, *Cryo-Letters* **12**, 113-124 (1991).
14. R. C. e. Weast, *Handbook of chemistry and physics.* (CRC Press, Cleveland, 1975).
15. G. G. Birch, K. J. Parker, *Sugar: science and technology.* (Applied Science Publishers, London, 1979).

16. P. F. Knowles, D. Marsh, H. W. E. Rattle, *Magnetic resonance of biomolecules*. (John Wiley & Sons, London, 1976).

17. J. H. Freed, *Theory of slow tumbling ESR spectra for nitroxides*. L. J. Berliner, Ed., Spin Labeling (Academic: New York, N. Y., 1976).

18. J. S. Hyde, *Methods Enzymol.* **49**, 480-511 (1978).

19. P. D. Orford, R. Parker, S. G. Ring, A. C. Smith, *Int. J. Biol. Macromol.* **11**, 91-96 (1989).

20. S. Ablett, M. J. Izzard, P. J. Lillford, *J. Chem. Soc. Far. Transactions* **88**, 789-794 (1992).

21. H. Yoshioka, *Chem. Lett.* , 1153-1154 (1977).

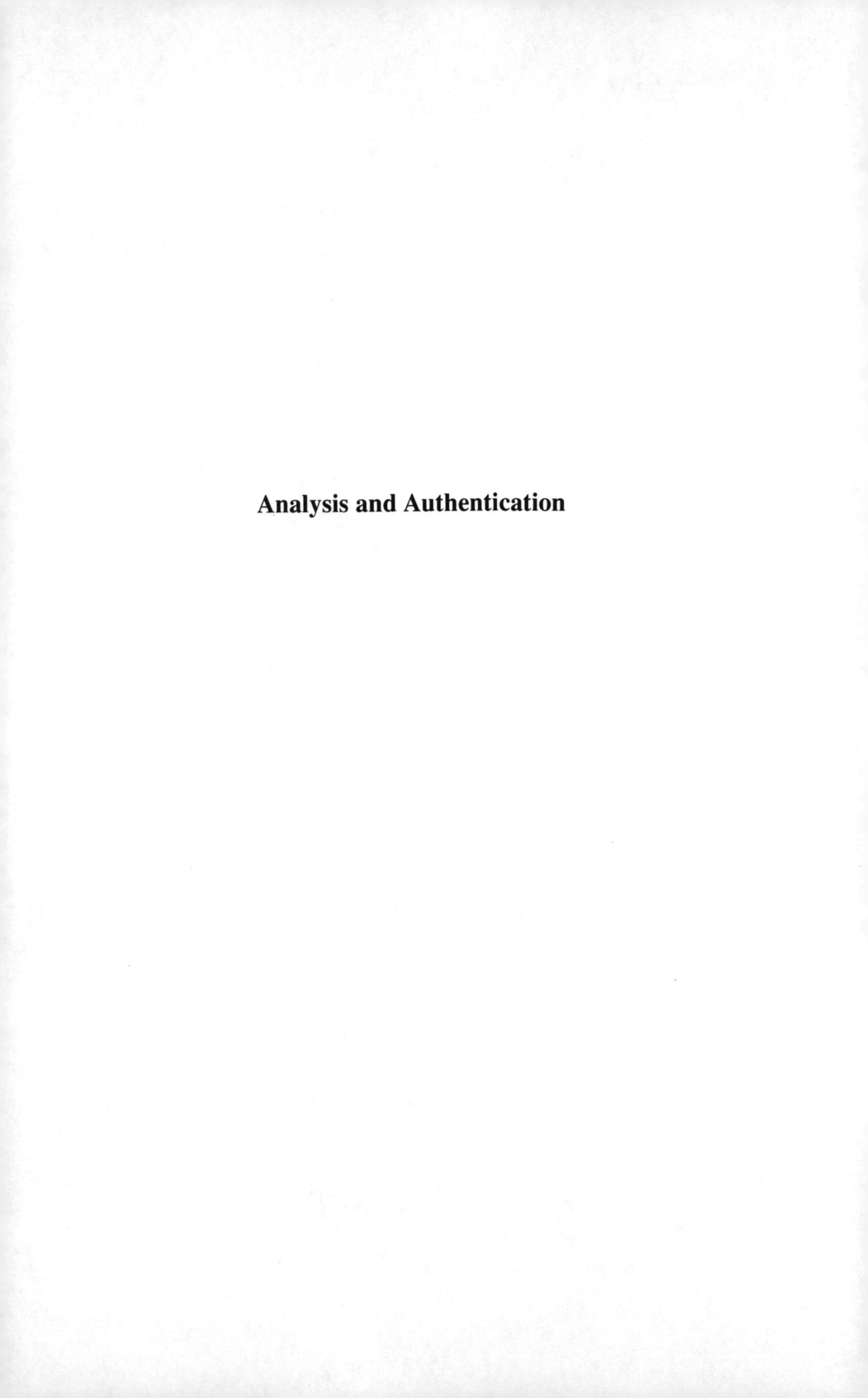

Analysis and Authentication

Analytical Performance of High Resolution NMR

Gérard J. Martin

LABORATORY OF NMR AND CHEMICAL REACTIVITY, UNIVERSITY OF NANTES, CNRS URA 472, FRANCE

1 INTRODUCTION

The first applications of HR-NMR spectroscopy to Food Science were described in the papers of Bloch and Purcell in 1946 (1) which used bee wax to investigate the resonance behaviour of solids and by Arnold et al (2) in 1951 which gave the first proton spectrum of ethanol ! However it is only at the beginning of the sixties that the true analytical applications of HR-NMR to Food Science began to appear in the literature. Johnson and Shoolery (3) demonstrated the ability of ^{1}H-NMR to characterise fats and oils, and at the end of that decade Shoolery and Smithson (4) published the first ^{13}C study of these products. Thirty years after this pioneering work, NMR literature now contains a number of applications to Food Science since the advances in magnet, electronic and computer technologies have enormously increased the performances of this technique. The existence of the International Conference on the Applications of NMR to Food Science has firmly established this new branch of NMR spectroscopy. Indeed, foods and beverages have a special status in the kingdom of organic compounds : they are often engaged in complex matrices and have to obey severe nutritional, toxicological and regulatory requirements. The specificity of food products has generated a number of new analytical methodologies in HR-NMR spectroscopy and the purpose of this contribution is to discuss the nature and the importance of the different criteria which govern the performances of NMR spectroscopy in Food Science.

2 THE PERFORMANCES OF ANALYTICAL HR-NMR

First of all, the different criteria which can be used to describe the performances of an analytical methodology should be clearly defined. Generally speaking, two kind of criteria i.e. qualitative and quantitative criteria - are considered.

Qualitative performance criteria are usually the *linearity*, the *selectivity* and the *ruggedness* of the method. From the point of view of *linearity*, NMR spectroscopy is intrinsically the best analytical method since the intensity I of the resonance signal is strictly proportional to the number N of nuclei which resonate at a given frequency for a given magnetic field in an unlimited range of concentrations :

$$I = k\,N \qquad \text{(eq 1)}$$

Obviously, electronic artefacts or bad spectrometer settings may obscure this very favourable situation as compared to most of the other analytical techniques and this point will be discussed latter.

The *selectivity* of NMR spectroscopy is also unique in analytical chemistry since the method differentiates by nature all the isotopes of the elements which have a nuclear spin, and even, for a given isotope, NMR spectroscopy is able to show up minute differences in chemical environment. This property directly arises from the nature of the energy range ΔE in which NMR transitions occur.

The *ruggedness* of a method may be understood as its resistance to induce bias when the environmental conditions change more or less significantly. As far as HR-NMR spectroscopy is concerned, it may be stated that this technique is relatively robust when frequencies are measured and less so in quantitative determinations. However, it should be strongly emphasised that the performances of an analytical technique depend not only on the instrument itself, but also on the different treatments that the sample has undergone before the measurement. This is also true for quantitative HR-NMR and great care must be exercised in the sample preparation step. When discussing the quantitative performance criteria the intricate dependence of sample preparation and instrumental determination must be permanently kept in mind.

2.1 Quantitative performance criteria

Sensitivity, *precision* and *accuracy* are the most representative quantitative performance criteria for HR-NMR spectroscopy. The probabilistic nature of these criteria should be made clear and any quantitative evaluation requires that a confidence level is chosen : usually, a 95 % confidence level is assumed which means that no more than 5% of the occurrences of the event to be evaluated will fall outside the range.

2.1.1 Sensitivity and detection limit. The sensitivity is the extent of change in signal intensity with the concentration of the analyte. IUPAC has given new definitions for old and somewhat misunderstood concepts of sensitivity (5) First of all, the critical level Lc is defined as a function of the standard deviation of the blank which can be assimilated in the case of HR-NMR to the standard deviation of the noise for a given spectrometer setting σ_n

$$Lc = 1.645\,\sigma_n \qquad \text{(eq 2)}$$

Then the detection limit Ld and the quantification limit Lq, are given by :

$$Ld = 2\,Lc = 3.3\,\sigma_n \qquad \text{(eq 3)}$$

$$Lq = 3\,Ld = 10\,\sigma_n \qquad \text{(eq 4)}$$

When applied to a typical NMR situation, equations (2-4) imply that a signal-to-noise equal to 10 is required to quantitatively detect a signal in the noise.

Indeed, the signal-to-noise S/N in NMR spectroscopy is strictly defined at the 95% confidence level as

$$S/N = 2H/h \qquad \text{(eq 5)}$$

where H and h are respectively the heights of the signal and of the noise. If the noise is N=10 in arbitrary units (au) and its standard deviation equals 5, the quantification limit should be equal to 50. NMR is known for being an insensitive analytical technique and the constant k in eq 1 is far from being equal to unity at room temperature ! This fact is unfortunately inherent to hertzian spectroscopies and only hardware clues such as very high fields and well designed coils are able to partly overcome this drawback.

2.1.2 Precision. Precision is a measure of the random errors which are associated to the analytical process considered. The replicate measurement of a given analyte is the only way to estimate precision which is related to the standard deviation S_i of the mean $\bar{x}_i$ of the n replications of the measurement of analyte i :

$$S_i = \sqrt{\frac{(x_i - \bar{x}_i)^2}{n-1}} \qquad \text{(eq 6)}$$

For a normal distribution of a series of n measurements, it is apparent that the random errors decrease when the number of replicates increases. In fact, like in most of the analytical techniques, precision is directly proportional to the signal-to-noise ratio, the higher S/N, the lower S_i. From a practical point of view, it is important to distinguish clearly two faces of precision, i.e. repeatability, r, and reproducibility, R, which are sometimes improperly invoked. These concepts are defined in terms of instrumental methodology, operator and time intervals between measurements. Repeatability is the internal precision obtained by the operator with a given instrument according to a well defined protocol in a short period of time. For example, running several times consecutively the same NMR spectrum is the easiest way to estimate repeatability. It should be noted that when spectrum averaging is carried out, the product of the number of scans acquired, NS, by the number of replicates, NE, may be a good criterion for judging the shortness of the period of time. On the other hand, when the same analyte is measured with different spectrometers, by different operators for a long period of time (say weeks or months), the precision which is computed is referred to as the reproducibility, R, of the measurement. Thus, the reproducibility involves two sources of variation : the within laboratory $S^2(r)$ and the between laboratory $S^2(L)$ variances. The first term is known as the repeatability variance and is the mean internal precision of the p different laboratories or sets of measurements involved. According to the ISO norm (6),r and R are given by the following equations :

$$S^2(r) = \frac{1}{p}\sum_{j}^{p} S_j^2 \qquad \text{(eq 7)}$$

$$S^2(L) = \frac{p\sum_j^p x_j^2 - \left(\sum_j^p x_j\right)^2}{p(p-1)} - \frac{S^2(r)}{n} \quad \text{(eq 8)}$$

$$S^2(R) = S^2(r) + S^2(L) \quad \text{(eq 9)}$$

r and R are obtained by multiplying the corresponding standard deviations by $f_{(0.95,\,p)}\sqrt{2}$, which is nearly equal to 2.8. It should be noted that the reproducibility decreases greatly with the amount of matter to be analysed. The so-called Horwitz trumpet (7) (Fig 1) illustrates the fact that the inter laboratory coefficient of variation changes from 5 to 50 when the analyte concentration goes from 0.1% to 1ppb. As compared to other analytical techniques, not as many inter-laboratory comparison studies have been carried out with NMR spectroscopy. Recently, two ring tests were organised in Europe to determine the reproducibility of ^2H measurements. These comparison studies involved the sample treatment and the NMR determination steps and the results were gratifying if we consider the former bad reputation of NMR in terms of precision. Indeed, the mean standard deviation of quantitative ^2H-NMR is on the order of 0.2% (standard deviation of repeatability). At the 95 % confidence level, the mean of ten replicated measurements of a given sample should be included in the ± 0.15 % confidence interval. It is also interesting to note in Table 1 that the NMR measurement does not depend very significantly on the spectrometer nor on the operator, providing that the right experimental protocol is followed : the standard deviation of reproducibility is equal to 0.3 %. It should be emphasised that a well-trained operator and a carefully described experimental procedure are prerequisites for obtaining precise results. For example, in five years the standard deviations of repeatability and reproducibility have significantly decreased despite the fact that the number of steps of the whole analytical sequence considered has increased :

-One-step determinations (1989) CRM 123 - BCR

(ethanol samples : ^2H-NMR measurements)

S(r) = 0.53 % S(R) = 0.30 %

-Two-steps determinations (1991) JRC Ispra

(Wine samples : distillation + ^2H-NMR measurements)

S(r) = 0.20 % S(R) = 0.35 %

-Three-steps determinations (1994) CEN-AOAC

(Fruit juices samples : fermentation + distillation + ^2H-NMR measurements)

S(r) = 0.21 % S(R) = 0.29 %

2.1.3 Accuracy. The accuracy is the closeness of agreement between the true value and the mean of a series of replicates. Since it may be difficult to determine the true value of a statistical variable, accuracy is considered as the limit of the precision when the number of assays is infinitely great. However, when a limited number of experiments is considered, inaccuracy is related to the concept

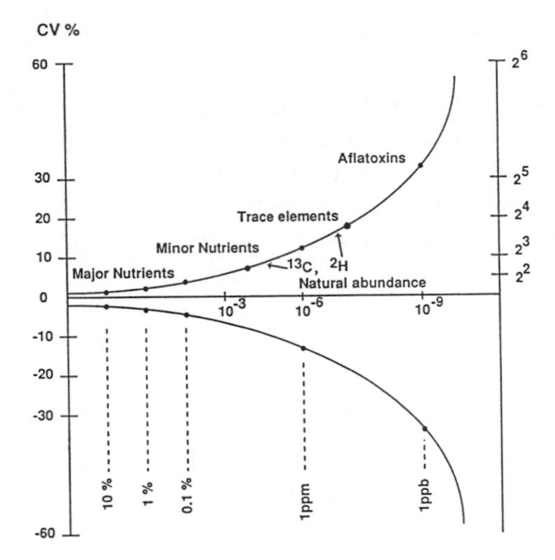

Figure 1 *The HORWITZ trumpet : Interlaboratory coefficients of variation as a function of analyte concentrations*

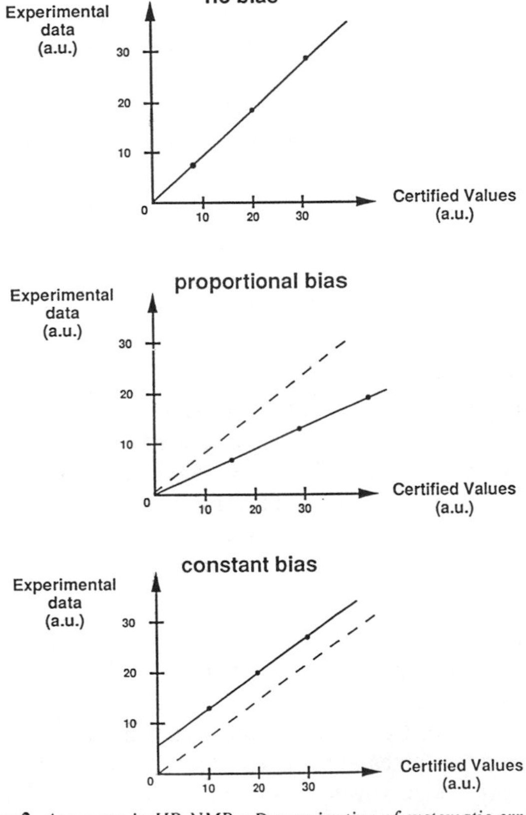

Figure 2 *Accuracy in HR-NMR : Determination of systematic errors by use of certified reference materials*

of systematic error or bias of the analytical process. Different methods may be used for detecting bias :

-using reference materials having certified values (i.e. BCR, NIST.).This procedure is illustrated in Fig 2. The bias of the instrument is easily corrected for an interpolation procedure. The reference materials M and L, having the certified values M_o and L_o are measured by a laboratory i which obtains the values M_i and L_i. The corrected value X_{cor} of an unknown X measured in lab I as X_i, is given by equation 10 :

$$X_{cor} = L_o + \left(\frac{(M_o - L_o).(X_i - L_i)}{(M_i - L_i)} \right) \qquad \text{(eq 10)}$$

-adding to the analyte to be determined well known quantities of a standard (isotopic dilution method)

-comparing the results obtained for the same analyte by two independent methods, such as NMR and mass spectrometry for instance. Thus the deuterium isotope ratios of three different analytes, ethanol, anethol and vanillin, having different levels of enrichment, were determined by ^{2}H-NMR and by Isotope Ratio Mass Spectrometry (IRMS) (8) (table 2). The mean standard deviation (MSD) either expressed in isotope ratio units or in relative percentages, between IRMS and HR-NMR data for the three couples of products studied is of the order of 0.40 ppm. The MSD is of the same order of magnitude as the standard deviation of reproducibility of the HR-NMR measurements (Table 1) and it may be safely assumed that MS and NMR give very consistent and true values.

To conclude this section, it is interesting to compare the performances of quantitative HR-NMR, to those of a more conventional technique such as gas chromatography (GC) which is largely used in analytical laboratories. These two techniques were applied to the quantitative analysis of triglycerides from olive oils in the framework of a joint research program sponsored by the BCR in Brussels and some results will be discussed now (9, 10).

If we consider linearity and selectivity both techniques seem to have, at first glance, quite different performances. NMR spectroscopy is very selective in energy and able to observe very tiny different states of matter even at a low field and with unsophisticated spectrometers. The NMR response is strictly linear in a very large range of concentrations. On the other hand, it is well known that GC needs frequent and careful calibration to give reproducible and accurate results and its selectivity depends to a rather large extent on the materials used and on the skill of the operator.

The sensitivity of both techniques are comparable, but to compete with GC NMR requires high fields and expensive spectrometers. Obviously, the sensitivity of NMR depends also on the nature of the nucleus studied and from this point of view the two stable isotopes of hydrogen are drastically different since with the same system, deuterium is 106 times less sensitive than proton for a given number of nuclei. However, the very high selectivity of NMR helps it to compete with GC successfully. For example, capillary GC can detect and quantify nanograms of a product but is unable, even with sophisticated detectors (such as Atomic Emission

Table 1 *Reproducibility of the determination of isotope ratios by ^{2}H-NMR (Participating laboratories (15), CEN-AOAC Study on Fruit Juices)*

PRODUCT	TEST LEVEL (ppm)	REPEATABILITY (ppm)		REPRODUCIBILITY (ppm)	
	m	S(r)	r	S(R)	R
Orange juice A	104.51	0.22	0.61	0.28	0.79
Orange juice B	102.34	0.19	0.53	0.24	0.67
Orange concentrate	101.27	0.23	0.64	0.35	0.99
Grapefruit	102.51	0.20	0.56	0.32	0.91
Grape	102.18	0.21	0.59	0.21	0.59
Apple	99.00	0.20	0.57	0.29	0.80
Mean values	/	0.21	0.58	0.29	0.80

Table 2 *Accuracy in HR-NMR : Comparison of two methods*

PRODUCT		MS		NMR	
		(D/H)	n	(D/H)	n
ETHANOL	Sugar beet	113.7	5	113.9	72
	Vine	123.2	16	121.8	385
	Sugar cane	124.4	8	122.2	76
	Synthesis	135.2	4	134.7	4
ANETHOL	Star anise	142	17	143.2	10
	Fennel	143.5	11	142.7	11
	Synthesis	150	15	150.4	4
VANILLIN	V. planifolia	144.8	6	143.7	10
	Lignine Ç	126.8	6	127.1	9
	Synthesis			162.5	8

Detectors), to detect the different ^{2}H, ^{13}C, ^{15}N natural abundance isotopomers of an organic compound.

A comparable degree of precision may be reached by GC and NMR but the long term repeatability (same product, same operator, same apparatus, including the column) of a given GC system is far below that of NMR.

In both techniques systematic deviations are likely to occur due to improper calibration for GC or to differential NOE effects for NMR but other sources of bias exist (low hydrolysis and derivatisation yields, weighing errors ...). The behaviour of GC and NMR in the determination of the fatty acid composition of olive oils are compared in Fig. 3. It should also be noted that ^{13}C-NMR spectroscopy is able to observe the branching position of fatty acids on the glycerol molecule without any treatment of the sample whereas GC cannot, but conversely it is difficult to evaluate the relative percentage of stearic and palmitic acids by ^{13}C-NMR and GC can !

3 THE METHODOLOGY OF ANALYTICAL HR-NMR

In order to achieve the top analytical performances described in the previous section, optimum acquisition parameters and FID treatments must be selected. A number of books and papers deal with the theoretical and practical aspects of quantitative NMR (11, 12) and some have focussed on the case of specific nuclei, ^{2}H (13), ^{13}C (14), ^{15}N (15,16). As far as sensitivity is concerned, the magnetic field strength Bo and the coil geometry are fixed by the proprietary NMR system and the highest sensitivity will be obviously obtained with the highest field available for a given coil geometry. However, the acquisition parameters (pulse width, PW, acquisition time, AT, pulse delay, PD,...) should be carefully selected and the number of scans NS fixed in order to have a compromise between the minimum overall experiment time and the maximum signal-to-noise. From this point of view, note also that, in order to recover 99.9% of the true magnetisation the delay between two pulses must be equal to about 9 or 6 times the maximum T_1 value according to whether NOE is present or not. Moreover several replicates, NE, are required to increase the precision. The price to be paid for obtaining a given precision will depend on the product (AT+PD)*NS*NE. AT being the acquisition time and PD the pulse delay, for example, the determination of the isotope ratio of an ethanol sample with a precision of $\pm$ 0.2 % at 7.05, 9.4 and 11.4 T requires 6, 3 and 1 hours respectively.

On the other hand, it should be noted that when the signal in the time domain is sampled during a very long time (5 T_2) the noise increases severely (12). Under these circumstances, it could be useful to add a delay between pulses in order to match the 6-9 T_1 condition of 99.9% of magnetisation recovery. Note also from this point of view that numerical integration requires a rather high digital resolution and acquisition time. The zero filling procedure (17) may improve the performance of numerical integration by carrying out an interpolation between two adjacent data points.

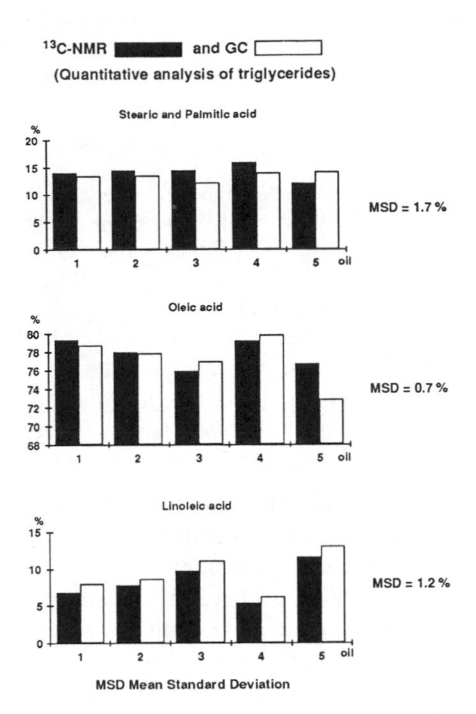

Figure 3a *Precision :Comparison of the Analytical Performances of Two Methods*

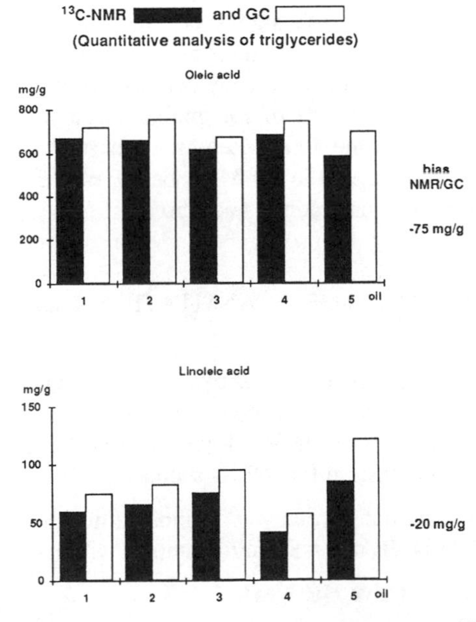

Figure 3b *Accuracy : Comparison of the Analytical Performances of Two Methods*

In a general sense, quantitative work is carried out using the ninety degree pulse width but the random errors associated with signal sampling increase. A smaller Pw value would improve the precision, but frequently it induces bias and decreases the accuracy. Then, sensitivity compete with precision and accuracy and a compromise should be found for any specific problem.

Random errors arise frequently from a poor digital resolution associated to the use of improper algorithms of numerical integration. It should be noted that the so-called "zero-filling" method (17) does not add any supplementary information to the signal and does not improve the quality of the result of a good curve fitting algorithm. Moreover, when series of NE spectra are recorded with a view to increasing the precision, great care should be exercised when phasing the spectra since even small phasing inconsistencies may induce significant random errors. However, as stated previously, the precision of a quantitative NMR measurement depends mainly on the signal-to-noise ratio available and the HR-NMR determinations are more frequently spoiled by systematic errors.

Indeed, the accuracy of quantitative HR-NMR is mainly governed by the quality of the excitation (B_1) and decoupling (B_2) electromagnetic fields : small irradiation windows and unstable decoupling powers for instance induce baseline and NOE artefacts which lead to bias in the signal intensities. Appropriate algorithms for curve fitting may obviate most consequences of the experimental defects. From this point of view it has been demonstrated theoretically that whatever the investigated domain, time or frequency, a complex least-squares curve fitting treatment is the best quantitative solution over any other treatment when the NMR signal has a reasonable signal-to-noise ratio (18). The Interliss algorithm developed in Nantes (18) considers the phasis, the frequency height and the half-height line width of each signal as independent parameters to optimise during the iterative curve fitting in the complex plane. Each lorentzian curve is thus integrated over the whole frequency window investigated. This procedure meets the theoretical requirements of integration limits which should be at least equal to a thousand times the line width in order to recover 99.99 % of the lorentzian area. With an Intel 486-66 MHz based PC, the computation of a 10 signals spectrum with 16 K data points takes 50 s.

4 TYPICAL SITUATIONS OF QUANTITATIVE ANALYSIS BY HR-NMR

The quantification concepts developed above apply to a great number of situations and it is far beyond the scope of this report to produce an exhaustive presentation of the analytical applications of NMR. However typical examples will be selected in different fields of quantitative HR-NMR such as :

- standard 1-D quantitative analysis : most situations discussed previously concerned in fact 1-D NMR and no other examples will be presented :

- hetero-atomic quantitative HR-NMR

- in-situ quantitative analysis

- quantitative solid state HR-NMR analysis
- hyphenated HR-NMR techniques (flow, LC, ...NMR)

4.1 Hetero-atomic quantitative analysis

HR phosphorus NMR is a very valuable tool for compositional determinations and in particular for the detection of trace products. In the field of food science, quantitative ^{31}P NMR has been successfully applied to the analysis of pesticides, either at the manufacture level or in the form of residues in crops. Gurley and Ritchey demonstrated in 1976 that 10 ppm of phosphorous containing compounds could be observed with a 100 MHz instrument (19). However the signal-to-noise quoted (S/N=2) is five times lower that the quantification limit recommended by IUPAC (see §2). More recently, Mortimer and Dawson (20) studied the feasibility to determine, in vegetables, traces of P-derivatives in the range 0.1 - 10 ppm using three spectrometers working at 5.85 T (100.8 MHz), 9.4 T (161.9 MHz) and 11.7 T (201.6 MHz). The detection limits were of the order of 3 for a 30 min acquisition time and the signals of malathion, parathion, phosmet and phosalone were easily observable. The minimum quantities of phosphorous pesticides detected were 1 ppm (5.85 T), 0.3 ppm (9.4 T) and 0.12 ppm (11.7 T). The quantification of the pesticides was carried out by comparing the signal height of the pesticide to that of a carefully weighed internal reference. The standard deviations of the ratios of the signal heights (pesticide and reference) computed from 5 replications were of the order of 3% and 5% in the range of the concentrations investigated. The same authors also observed residues of fluorine pesticides in wines and peas in the 0.1 to 1 ppm range.

High performances of quantitative analysis are also required by natural abundance isotopic determinations. The SNIF-NMR method has now found a wide range of applications. For instance the molecule of caffeine has been thoroughly studied by Stefaniak et al (21) and after careful assignment of the different resonances of $^{1}H(^{2}H)$, ^{13}C, ^{15}N and ^{17}O the site-specific isotope ratios : $^{2}H/^{1}H$, $^{13}C/^{12}C$, $^{15}N/^{14}N$ and $^{17}O/^{16}O$ could be obtained. Until now, the results are good for ^{2}H, encouraging for ^{13}C and ^{15}N and deceptive for ^{17}O. Work is in progress to determine the all-isotopes fingerprint of different caffeine samples.

4.2 In-situ quantitative solid-state HR-NMR

Since most of the molecules of interest in food science are contained in more or less complex matrices, a growing need appears for in-situ quantitative measurements. Promising results were obtained with conventional HR equipment (^{1}H or ^{13}C spectroscopies). One of the most interesting approach was the determination in 1976 of anethol in intact star anise seeds (22). Some years later, the direct study of seeds and other solid food products was greatly improved by the development of HR solid-state NMR which involves rotation of the sample at the magic angle 54° 44 ' (MASS) and cross polarisation (CP). In 1991, Wollenberg (24) developed a quantitative method to determine the triglycerides content of whole vegetable seeds by ^{1}H and ^{13}C MASS NMR. When the analyte exhibits a noticeable degree of mobility, cross polarisation is not required and good ^{1}H or ^{13}C spectra may be obtained by resorting only to Magic Angle Sample Spinning. Rutar et al (23) observed in this way very good ^{13}C spectra of fir seeds containing terpenes or triglycerides.

It is interesting to mention that even at a moderate spinning rate, ^{1}H spectroscopy is able to quantify water, simple carbohydrates and organic acids in intact fruit tissues and to follow the change in liquid and solid phases compositions during the ripening of fruits such as bananas (25, 26).

4.3 Quantitative analysis with hyphenated NMR techniques

A hyphenated technique involves the on-line coupling of two instruments, such as GC and MS for example. In the case of HR-NMR, it has been recognised a long time ago that flowing liquids through the NMR coil could improve the sensitivity of this spectroscopy and permit the study of relatively fast processes. A liquid chromatograph, LC, or a chemical reactor, CR, can be coupled to a NMR spectrometer and an optimization of the experimental parameters must be carried out in order to take advantage of the magnetic properties of flowing liquids. The intensity I depends on the pre-magnetisation M_o' and on the effective longitudinal relaxation time $T_{1eff.}$, which in turn, depend on the flow rate, V, and on the residence time τ, of the product in the coil :

$$1 = M_o'\left[1 - \exp\left(-\frac{T}{T_{1eff.}}\right)\right] \qquad \text{(eq 11)}$$

$$\text{with } M_o' = \left[1 - \exp\left(-\frac{k}{V\,T_1}\right)\right] \qquad \text{(eq 12)}$$

$$\frac{1}{T_{1eff.}} = \frac{1}{T_1} + \frac{1}{\tau} \qquad \text{(eq 13)}$$

For long relaxation times, flowing a liquid even at a moderate rate such as 5 ml/min may decrease the pulse repetition time and therefore the overall experiment time by a factor of 30 for the same S/N value (27).

Flow HR-NMR is mainly applied in biological chemistry and for in-vivo studies. In food science, a method was developed for the on-line determination of proteins, lipids and water in thinly-minced meat for instance (28). An HR-NMR detector was preferred to a low resolution NMR system in order to optimise the chemical shift separation between the lipids and water signals. The precision of the method is of the order of 1% but flow NMR induces a bias with respect to off-line conventional methods. The system worked conveniently and the performances met the specifications of the industrial line. However, the heterogeneity of the meat was much larger than the precision of the method.

Satisfactory performances have also been obtained in compositional determinations by exploiting a coupling between liquid chromatography and ^{1}H or ^{19}F-NMR in continuous flow or stopped flow experiments (27,29).

5 CONCLUSION

To conclude, it should be emphasized that at the end of the analytical sequence, the raw results of a given technique must be handled in the most appropriate way in order to deliver the right diagnosis. From this point of view, it is worth recalling that all the techniques are not equivalent. As emphasised by Booksh (30), analytical instruments may be classified, *inter alia,* according to the nature of the data set obtained. When a machine generates only one datum per sample, i.e. pH, red-ox potential, ..., it is called a zero-order instrument ; GC and HR-NMR for example, are first-order instruments. They generate an array of data (retention values, frequencies/intensities) but it should be kept in mind that GC and hetero-atomic NMR may be considered as first order methods on condition that a mixture of products or isotopomers is studied, whereas ^{1}H-HR-NMR is a true first order technique which gives an array of data even for only one compound. 2-D NMR, GC-MS and other hyphenated techniques are second order instruments since a matrix of data is obtained. Note that when a given spectrum is recorded NE times in order to increase the precision, the whole set of data is consistent with that of a second-order instrument. From the point of view of accuracy, high order instruments are very efficient because the calibration procedure always involves a greater number of data than low order techniques :

zero-order $r = f(c) + e$ (eq 14)

first-order $\mathbf{r} = f(c_1,c_2,...c_n) + \mathbf{q} + e$ (eq 15)

second-order $R_{ij} = g_j[f_i(c)] + Q_{ij}$ (eq 16)

In these equations eq 16, $(c_1,c_2,...c_n)$ are for example the concentration (or the intensities) of the n compounds present in the mixture, an element $\mathbf{q}$, is the baseline response of the sensor $\mathbf{j}$ considered (polarity index, resonance frequency) and $\mathbf{e}$ is the residual error. Similarly R_{ij}, Q_{ij} are the matrix elements of the multidimensional instrument and baseline responses. Multivariate statistics applied to the results of high order instruments enables the error associated to the measurement to be, at least in a part, eliminated. Eigenvalues decomposition of raw data with reconstruction of a pure data set and least squares analysis of over determined data sets are frequently carried out and are indeed very effective calibration procedures. However, it should be reminded that these multi-variate treatments assume that the variables which describe the phenomenon observed behave linearly. This is not really the case in a number of situations where the apparent linearity is only a consequence of the short range of variation of the variables considered. The determination of the composition of a mixture by a least squares approach or the classification of an unknown in pre-defined groups by means of discriminant analysis incorporates more or less severe biases when the matrix effects and the non-linearity of the variable are non negligible. In such situations, it is preferable to apply a Monte-Carlo analysis of the data set. This unbiased method is based on a great number of estimations of mixtures compositions from a random balloting of pure reference compounds. The greater size of the reference groups and the number of random selections, the better Monte-Carlo analysis is. Usually 2000 random tests lead to a confidence interval at the level of 99% of certitude and with the new generation of PC microprocessors a Monte-Carlo simulation needs only a few seconds.

References

1a. F. Bloch, W.W. Hansen and M.E. Packard, *Phys. Rev.*, 1946, **69**, 127

1b. E.M. Purcell, H.C. Torrey and R.V. Pound, *Phys. Rev.*, 1946, **69**, 37

2. J.T. Arnold, S.S Dharmatti and M.E. Packard, *J. Chem. Phys.*, 1951,**19**, 507

3. L.F. Johnson and J.N. Shoolery, *Anal. Chem.*, 1962, **34**, 1136

4. J.N. Shoolery and L.H. Smithson, *J. Amer. Oil Chem.*, 1969, **47**, 153

5. IUPAC Recommandations fort the presentation of results of Chemical Analysis Ed. Currie L.A., Svehla G., *Pure Appl. Chem.*, 1994

6. ISO Norm 5785, Precision of measurement and testing,1986

7. N. Horwitz, R.K. Laverne and W.K. Boyer, *J. Assoc. Off. Anal. Chem.*, 1980, **63**, 1344

8. G.J. Martin and N. Naulet, *Fresenius Z. Anal. Chem.*, 1988, **332**, 648

9. A. Royer, DEA Report, 1994

10. C. Cluzelle, DEA Report, 1994

11. M.L. Martin, J.J. Delpuech and G.J. Martin, Practical NMR Spectroscopy, ed. Heyden-Wiley, 1990

12. D.L. Rabenstein and D.A. Keire, in Modern NMR techniques and their application in chemistry, Eds. A.I. Popov, K. Hallenga and M. Dekker, New-York, 1991, Vol. 11, p. 323

13. M.L. Martin and G.J. Martin, ^{2}H-NMR in the study of Site-specific Natural Isotope Fractionation in NMR Basic Principles and Progress, Eds. P. Diehl, E. Fluck, H. Günther, R. Kosfeld, J. Seelig, Springer-Verlag, 1990, Vol. 23, p. 1

14. C.H. Sotak, C.L. Dumoulin and G.C. Levy, in Topics in carbon-13 NMR Spectroscopy, Ed. Levy, G.C. Wiley, New-York, 1984, Vol. 4, p. 91

15. M. Witanowski and G.A. Webb, Nitrogen NMR, Plenum Press, 1973

16. G.J. Martin, M.L. Martin and J.P. Gouesnard, ^{15}N-NMR Spectroscopy in NMR Basic Principles and progress, Eds P. Diehl, E. Fluck and R. Kosfeld, Springer-Verlag, 1981, Vol. 18

17. E.D. Becker, J.A. Ferretti and P.N. Ghambir, *Anal. Chem.*, 1979, **51**, 1413

18. Y.L. Martin, *J. Magn. Res. A110*, 1994, (in press)

19. T.W. Gurley and W.M. Ritchey, *Anal. Chem.*, 1976, **40**, 1137

20a. R.D. Mortimer and B.A. Dawson, *J. Agric. Food Chem.*, 1991, **39**, 911

20b. R.D. Mortimer and B.A. Dawson, *J. Agric. Food Chem.*, 1991, **39**,1781

21. L. Stefaniak, J. Sitkowski, L. Nicol, M.L. Martin, G.J. Martin and G.A. Webb, 2° International Conference on the applications of NMR in food science Aveiro,1994

22. M. Kainosho and H. Konishi, *Tetrahedron Letters,* 1976, **51**, 4757

23. V. Rutar, M. Kovac and G. Lahajnar, *J. Magn. Res.*, 1988, **80**, 1233

24. K. Wollenberg, *J. Amer. Oil Chem. Soc.*, 1991, **68**, 391

25. Q.W. Ni and T.M. Eads, *J. Agric. Food Chem.,*1993, **41**, 1026

26. Q.W. Ni and T.M. Eads, *J. Agric. Food Chem.*, 1993, **41**, 1035

27. D.A. Laude, R.W.K. Lee and C.L. Wilkins, *Anal. Chem.*, 1985, **57**, 1286 and 1464

28. C. Tellier, M. Trierweiler, J. lejot and G.J. Martin, *Analusis,* 1990, **18**, 67

29. P. Lancelin, P. Cleon, *Spectra 2000,* 1991, **33**, 161

30. K.S. Booksh and B.R. Kowalski, *Anal. Chem.*, 1994, **66**, 782A

Application of SNIF-NMR and Other Isotopic Methods for Testing the Authenticity of Food Products

Gilles G. Martin

EUROFINS LABORATORIES, SITE DE LA GÉRAUDIÈRE, CP 4001, 44073 NANTES CEDEX 03, FRANCE

Introduction

Food authenticity has always been a challenging problem for the analyst as the sophistication of frauds regularly increases and analytical method developers are thus having to constantly invent efficient procedures to detect new forms of adulteration.

Food adulteration must be controlled as it represents a severe danger for the consumer and the industry. Through several examples of adulterations which occurred in the past ten years, it is observed that food adulteration can often lead to :

- food poisoning due to lack of control of adulterants by the perpetrators

- monetary damage to the consumer who is sold an inferior product

- lack of confidence in food and company brands by the consumer

- elimination of honest producers from the market as their products can not compete with cheaper adulterated ones

- deterioration of overall quality of products on the market as most producers must adjust to the lower standards in order to survive.

In the past ten years new analytical methods based on isotope analysis by Nuclear Magnetic Resonance and Mass Spectrometry have enabled a quantum leap of the possibilities to detect adulterations of various food products. Application of these methods to solve large adulteration problems in various countries is discussed.

The need for guaranteeing food authenticity

Food authentication encompasses several questions and is useful at several levels of the food chain from the vegetable or animal to the consumers plate. Methods for authentication are of interest to the consumer, represented by Consumer Organizations and Regulatory Authorities but also to the food industry to describe quantitatively and document well the intermediate product (food ingredients, fruit concentrates, etc...) that are utilized. Finally, methods to check products authenticity are also crucial to producers of agricultural products and raw materials in order to guarantee equal competition and avoid market disruption by adulterated goods produced at a far cheaper price.

As discussed in a recent review concerning, more specifically the authentication of fermented beverages (1), food adulteration is as old as trade itself. However it has constantly evolved, with the progress of analytical detection methods, towards more and more sophisticated practices. Food adulteration or contamination involving the addition of molecular components not normally present in the considered product is usually easily detected, at least when the exogenous substance is in noticeable amounts. A large number of methods, and in particular most spectroscopic techniques, are able to identify and quantify added molecules structurally different from those normally expected. The problem is much more difficult to solve when it consists in detecting the presence of species with the same molecular formula but originating from different sources. For instance two mixtures constituted from the same molecular species present at exactly the same concentrations may have different prices according to whether have been extracted from plants or result from chemical synthesis. Here we shall associate the notion of authenticity to that of conformity to a claimed origin of the product. This notion of origin may itself involve stringent requirements since it must be understood not only in terms of chemical or biochemical pathway, for natural products, in terms of botanical geographical or temporal origin of the precursors and possibly in terms of the technological procedure of preparation or treatment.

Various forms of frauds with respect to authenticity have been elaborated, going from mixing with more or less important quantities of components chemically identical but from a different (less expensive) origin to complete substitution. Although such kinds of fraud usually do not create health risks, they are detrimental to the consumer who buys for example a fruit juice originating only partly, or sometimes not at all from the fruit or who pays in fact for synthetic vanillin the price (about hundred times higher) of natural aroma that he is supposed to receive. Obviously such practices when discovered may greatly affect the reputation of the firm involved in the distribution of the product and lead to disastrous economical consequences. This situation has been illustrated recently by known examples in the field of fermented beverages (1) and table 1 summarizes different kinds of adulteration of wines or alcohols.

Isotopic methods in authenticity insurance

Many attempts to identify the origin of a product have been based on the detection of very minor accompanying components, the presence or amount of which are supposed to be typical of a given natural or synthetic source. However these procedures, which involve determining the molecular contents, often at trace levels, are severely limited due to large natural compositional variations and to improvements in the purification of industrial products.

Consequently the most efficient methods now resort to the determination of the natural abundance isotope contents of the main constituent(s) of a given substance.

Thus the overall carbon isotope content of a pure molecular species or of a mixture can be precisely determined by Isotope Ratio Mass Spectrometry (IRMS). The product is burnt into CO_2 and H_2O and the ratio of the numbers of ^{13}C and ^{12}C atom is measured on the carbon dioxide. The isotopic parameter thus obtained is expressed either in absolute values of the $^{13}C/^{12}C$ ratio, in (%), or on the relative δ scale (in ‰), which refers the $^{13}C/^{12}C$ ratio of the sample, S, to that of an international standard PDB (2).

$$\delta^{13}C(‰) = 1\,000\,[(^{13}C/^{12}C)_S - (^{13}C/^{12}C)_{PDB}]\,/\,(^{13}C/^{12}C)_{PDB}$$

It has been shown for instance that the $^{13}C/^{12}C$ ratio is about 1.101 % (δ = - 20‰) for vanillin extracted from vanilla beans whereas it is about 1.090 % (δ = - 30 ‰) for vanillin synthetized from guaiacol and 1.093 % (δ = - 27 ‰) for ex-lignin vanillin (3 - 5). More

Table 1 - Various forms of adulteration of fermented beverages (1)

	Adulteration	Products where it applies
- 1 -	Dilution with water	wines ; ciders.
- 2 -	Addition of sugar	wines (chaptalisation or sweetening) : grape musts or concentrated grape musts used for wine making ; apple concentrates used for cider making ; kirsch and fruit musts used to make spirits, etc...
- 3 -	Geographical mislabelling	wines ; cognacs ; Scotch whiskies ; etc...
- 4 -	Alcohol, botanical origin not corresponding to the product definition	use of synthetic alcohols in any spririt ; rum made with beet alcohol ; geniever made with alcohol from beet molasses ; pure single malt whisky containing corn alcohol ; beet alcohol in wine brandies or fruit brandies, etc...
- 5 -	Variety mislabelling	wine (use of grape varieties of higher yield, or not corresponding to the rules for a given origin, or of higher colouring power) ; cider.
- 6 -	Undeclared addition or organic acids	wine ; cider, etc...
- 7 -	Addition of colouring subtances	wine ; brandies ; liqueurs (blackcurrant, etc...)
- 8 -	Addition of synthetic flavours	all beverages except where allowed and correctly labelled
- 9 -	Mislabelling of age (year of production)	wine ; whiskies ; brandies ; etc...

generally, a given molecular species exhibits a higher ^{13}C content when it has been elaborated by a plant with a C_4 photosynthetic metabolism than when it is issued from a C_3 plant (6). This property enables for instance the adulteration of citrus juices (C_3 plants) by cane sugar or corn syrup (C_4 plants) to be detected. Thus IRMS is recognized as an official method for guaranteeing the purely natural status of different kinds of food products : honey, maple syrup, vanillin, etc...

However since the product is burnt prior to the mass spectrometry measurement, only an average value of the isotope content of the pure molecular species or of the mixture is accessible. The situation is the same for the hydrogen gas resulting from the reduction of water and also for the $^{18}O/^{16}O$ ratio. Although these overall parameters frequently provide very reliable proofs of the origin of a product ; the IRMS method has been increasingly circumvented by isotopic adjustment. For instance since the average ^{13}C content of vanillin extracted from beans is lower than that of its synthetic conterpart, a "naturalization" is sometimes carried out by appropriate ^{13}C enrichment of the synthetic product. The same practice applies to hydrogen since deuterium is easily introduced to mimic an overall D/H ratio of natural components.

Low activity liquid scintillation counting is also efficient for characterizing the natural or synthetic origin of a product. As the radioactive decay period of carbon-14 is relatively short (5 730 years) products synthetized from fossil raw materials are devoid of radioactivity whereas the equivalent molecular species originating from living plants, which incorporate slightly radioactive carbon dioxide from the atmosphere, contain a small amount of ^{14}C atoms. In addition ^{14}C counting performed on natural products may apply to short scale dating by exploiting the known variations of atmospheric radioactivity due to nuclear explosions carried out in the recent years. However once again appropriate adjustments of the radioactivity parameters to the "natural" value have been carried in some circumstances.

With respect to these methods, which usually measure a single average parameter, the isotopic NMR method (7) offers the advantage of characterizing the product by a multi-isotopic fingerprint which is much more difficult to mimic. Although no analytical method is ensured of never being circumvented, fraud may be expected to stop when the effort required exceeds the expected profit !

It should be mentioned that introducing an exogeneous compound is not always illegal. For instance addition of beet or cane sugar to must before fermentation is allowed in certain regions of production. However since this permission is controlled by strict regulation as regards the country considered and the amount of sugar added it is desirable to dispose of analytical methods capable of verifying whether or not sugar has been added and in what proportion.

Such a challenging problem, as well similar ones in the field of fruit juices, vinegars, aromas etc... can only be solved by resorting to relatively sophisticated isotopic methods.

Site-specific natural isotope fractionation studied by NMR in the characterization of food authenticity

Nuclear Magnetic Resonance enables in principle, most isotopomers mono-labelled in ^{2}H, or ^{13}C to be simultaneously observed and the study of Site-specific Natural Isotope Fractionation by NMR (SNIF-NMR) has already found numerous applications in the authentication domain (8 - 9).

The analytical methodology developed for exploiting the SNIF-NMR method has now been described in several publications (8 - 9). However work is in progress to constantly improve the performance of the method and in particular to increase its precision and accuracy. A noticeable improvement has been recently introduced thanks to a new data processing algorithm grounded on complex least squares curve fitting (10). The method which has been theoretically proved to provide the most accurate quantitative results for spectra with a reasonable signal to noise ratio (> 20) has also the advantage of permitting full automatization of the measure. Moreover the quality of the curve-fitting convergence can be checked for individual peaks as illustrated in figure 1 which compares the automatically obtained experimental spectrum (no need for phase adjustment) to the simulated one.

Strong deviations with respect to a statistical distribution of deuterium being exhibited by most natural or synthetic products the NMR isotopic fingerprint often constitutes an identification criterion which is highly resistant to falsification.

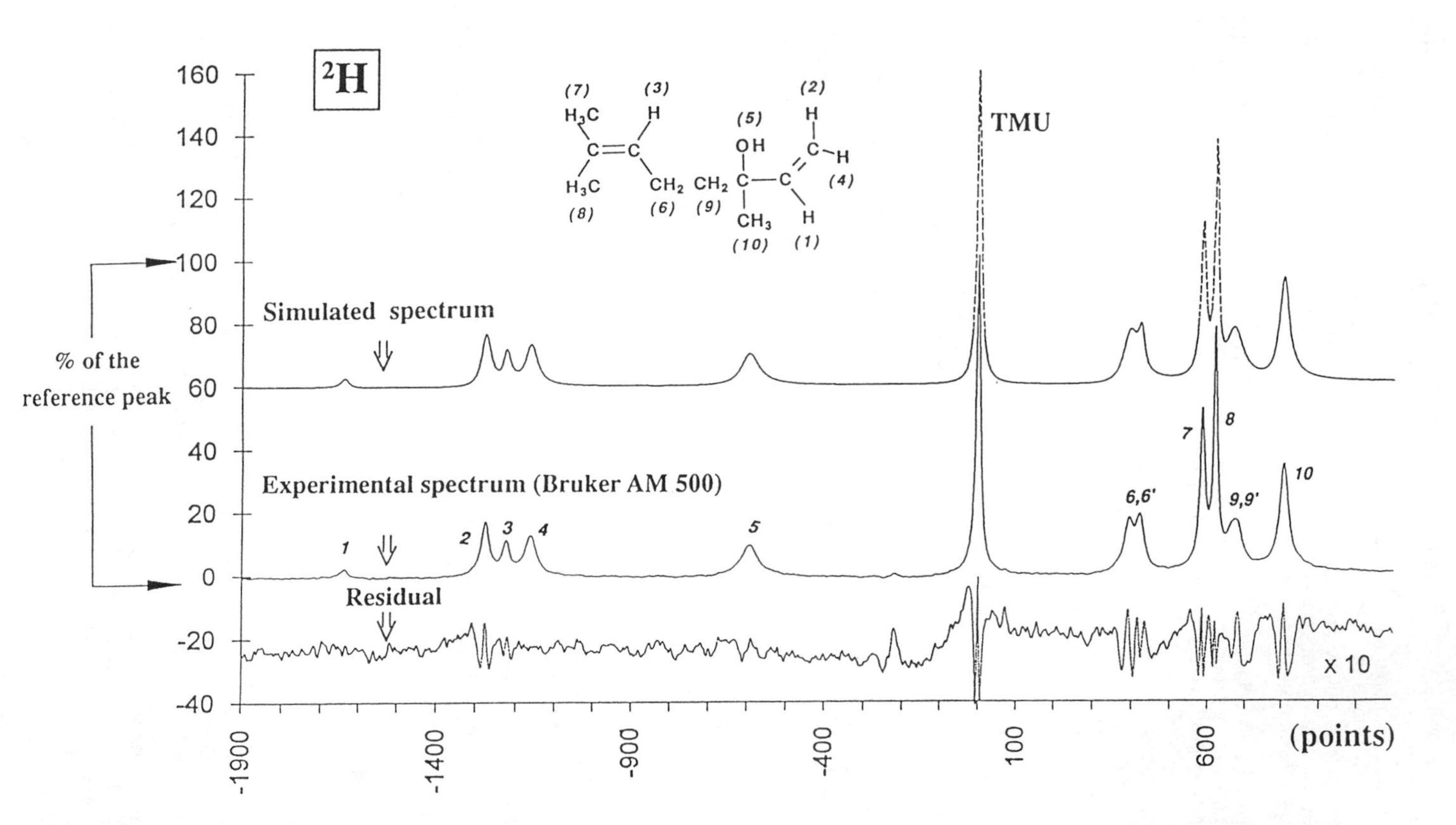

Complex least-squares curve fitting of the natural abundance deuterium NMR spectrum of linalool (93040040). The algorithm, which works without approximation, involves an integrated management of all the experimental parameters including the phases of the individual resonance signals. It is fully automatized. The residual which has been enhanced 10-fold illustrates the quality of the Lorentzian model.

Different kinds of authentication problems are amenable to SNIF-NMR :

- identification of the natural synthetic or hemi-synthetic status : for instance identification of anethole extracted from fennol, synthetized from aminole or resulting from isomerization of estragole

- characterization of the photosynthetic metabolism of the plant precursor (C_3, C_4, CAM...) : for instance characterization of glucose from potato (C_3) from corn (C_4) or from pineapple (CAM)

- characterization of the botanical species within a given metabolic family : thus the distinction of many C_3 plants (cereals, fruits...) is possible

- determination of the geographical area of production : continent, country and sometimes even region (wines and spirits in particular)

- information about the year of production (vintage of a cognac)

- adulteration by addition of the same molecular species from a different origin (beet sugar in fruit juices...)

- identification of the reaction pathway used to produce a synthetic component : for instance recognition of benzaldehyde synthetized by toluene oxidation or by hydrolysis of α,α-dichlorotoluene.

To conclude it should be emphazised that in spite of the very high efficiency of each isotopic method and in particular of SNIF-NMR the increasing complexity of frauds sometimes calls for the combined use of complementary isotopic methods (NMR and IRMS...). Moreover authentication in terms of geographical area of production for instance is significantly improved by resorting to the joint determination of natural isotopes and trace elements (11).

MAIN PRINCIPLES OF SNIF-NMR® ANALYSIS

PRODUCT TYPE	TYPICAL RANGES OBSERVED FOR	
	Delta Carbon 13	(D/H)I by SNIF-NMR
ORANGE JUICE	-25 - -27	103 - 105
CANE/CORN SUGAR	-10 - -12	109 - 111
BEET SUGAR	-25 - -27	91 - 93
ORANGE + 25 % BEET	-25 - -27	100 - 102

REFERENCES

(1) G.G. Martin, P. Symonds, M. Lees and M.L. Martin
in Fermented beverage production
Eds A. Lea and J.R. Piggott. Blackie Academic & Professional 1994

(2) H. Craig
Goechim. Cosmochim. Acta 12, 133, 1957

(3) J. Bricout and J. Koziet
Ann. Fals. Exp. Chim. 76, 845, 1976

(4) P.G. Hoffman and M. Salb
J. Agric. Food Chem. 27, 352, 1979

(5) C. Mauber, C. Guerin, F. Mabon and G.J. Martin
Analusis 16, 434, 1988

(6) M.H. O'Leary
Bioscience 38, 328, 1988

(7) G.J. Martin and M.L. Martin
Tetrahedron Lett. 22, 3525, 1981

(8) G.J. Martin and G.J. Martin
in Modern methods to detect fruit beverage adulteration
Eds S. Nagy and R. Wade 1994

(9) G.G. Martin, G. Remaud and G.J. Martin
Flavours and Fragance Journal 8, 97, 1993

(10) Y.L. Martin
J. Magn. Res. A 110, 1994

(11) M.P. Day, B.L. Zhang and G. J. Martin
Am. J. Enol. Vitic. 45, 79, 1994

NMR Studies of Lactic Acid Bacteria Involved in Wine Fermentation

Helena Santos

INSTITUTO DE TECNOLOGIA QUÍMICA E BIOLÓGICA, UNIVERSIDADE NOVA DE LISBOA, APARTADO 127, 2780 OEIRAS, PORTUGAL

1 INTRODUCTION

Lactic Acid Bacteria (LAB) have a long history in the food and feed industries. They are part of normal human and animal flora and, in addition, play key roles in the production of fermented food and beverages; thus they underpin important sectors in the dairy, meat, bread, vegetable, wine, and animal feed industries. Not surprisingly they have been the target of intensive research which attempts to extend fundamental knowledge of these bacteria in order to achieve a more scientific control of their activities and applications[1]. However, most of this research effort has been invested in LAB involved in dairy fermentations and, in general, the degree of knowledge available for the other types of fermentations is rather limited[2,3].

The usage of bacterial starter cultures to induce malolactic fermentation in wines is becoming widespread[4]. Among the strains able to perform the conversion of malic acid to lactic acid in a single step, *Leuconostoc oenos* is often favoured for the preparation of successful commercial starters. In the natural wine fermentation, the lactic acid bacteria surviving the alcoholic fermentation commence multiplication and perform malolactic fermentation. Almost invariably, *L. oenos* is the main species that develops since it is best adapted to the conditions of low pH (3 to 3.5) and high ethanol concentration that are found in wine[5].

Malolactic fermentation is an important step in the vinification process since it reduces acidity, enhances organoleptic characteristics due to the production of flavour compounds, such as acetic acid or diacetyl, and improves the microbiological stability of wine[5]. But, in order to be able to control malolactic fermentation in wine, detailed knowledge about the metabolic and regulatory processes used by these bacteria is essential.

Nuclear Magnetic Resonance has proved to be an useful technique to study cellular metabolism[6]; in particular, ^{13}C-NMR provides information on metabolite levels and

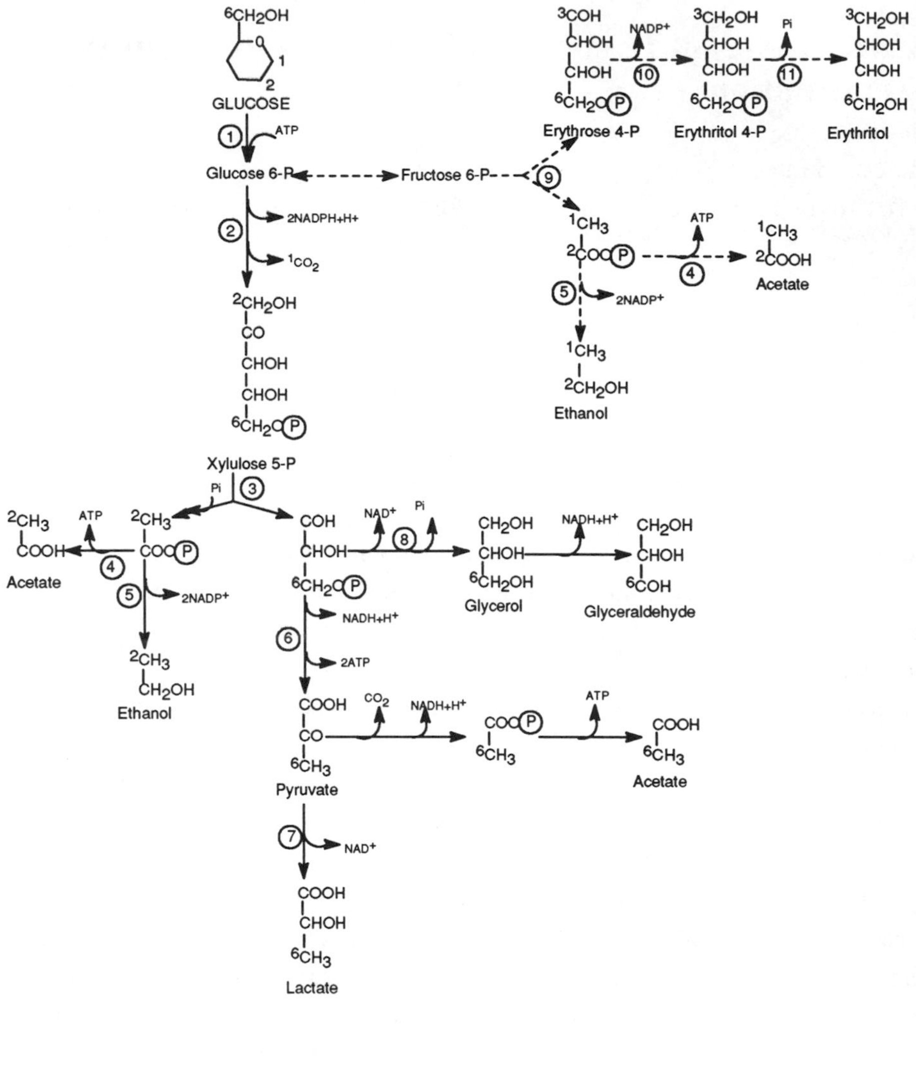

Figure 1 *Pathway of glucose metabolism in* L. oenos. *Broken lines represent the pathway for erythritol production as established in this work*[10,11]. *1, hexokinase; 2, glucose 6-phosphate dehydrogenase, 6-phosphogluconate dehydrogenase, and ribulose 5-phosphate 3-epimerase; 3, phosphoketolase; 4, acetate kinase; 5, phosphotranscetylase, acetaldehyde dehydrogenase, and alcohol dehydrogenase; 6, enzymes of the Embden-Meyerhof-Parnas pathway; 7, lactate dehydrogenase; 8, glycerol dehydrogenase, glycerol 3-phosphate phosphotransferase; 9, phosphoketolase; 10, erythritol dehydrogenase; 11, erythrose 4-phosphate phosphotransferase.*

fluxes through biochemical pathways[7-9] Here we summarize our results on the application of Nuclear Magnetic Resonance techniques to elucidate the pathway and regulation of erythritol formation from glucose in *L. oenos*[10,11]. Erythritol is a strong edulcorant and therefore is expected to influence wine flavour.

2. PATHWAY OF ERYTHRITOL PRODUCTION BY *L. OENOS*

Together with malic acid, *L. oenos* uses other carbon sources, e.g., residual sugars and citrate. Glucose (0.6-4.5 mM) and fructose (5.5-8.0 mM) are the major sugar components after ethanolic fermentation, and also important carbon sources for growth of wine lactic acid bacteria.

The metabolism of glucose in heterolactic acid bacteria is described to occur via the so called heterofermentative phosphoketolase pathway or the 6-phosphogluconate pathway[12]. The process is initiated by the oxidation of glucose to gluconate 6-phosphate; this is oxidatively decarboxylated to ribulose 5-phosphate which is subsequently cleaved into acetyl phosphate and glyceraldehyde 3-phosphate by a pentose-phosphate phosphoketolase present in all heterolactic acid bacteria. Then, acetylphosphate is converted to ethanol and/or acetate, and glyceraldehyde is metabolized to lactate via pyruvate (Figure 1).

2.1 Effect of aeration conditions on glucose metabolism

The end-products from glucose metabolism by *L. oenos* GM (Microlite Techniques, Sarasota, Florida), were monitored by *in vivo* ^{13}C-NMR using an air-lift system in the NMR tube to supply different gases to the cell suspensions[13]. Under anaerobic conditions (N_2 or CO_2), in addition to the expected acetate, lactate and ethanol, *L. oenos* was found to produce erythritol as a major end-product (37 μmol produced per 100 μmol glucose) and glycerol (12 μmol per 100 μmol glucose catabolized). Under aerobic conditions (O_2 or air) erythritol was replaced by glycerol and glyceraldehyde was also found (11 μmol per 100 μmol glucose used). In the absence of O_2, the concentration of erythritol formed was three times higher than that of glycerol and six times more acetate than ethanol was detected. When O_2 was present, the production of glycerol increased threefold and erythritol was not detected. There were less ethanol and lactate formed and the already high level of acetate produced anaerobically was 1.3-fold higher.

Interestingly, when *L. oenos* fermented either fructose or ribose, neither erythritol or glycerol was produced.

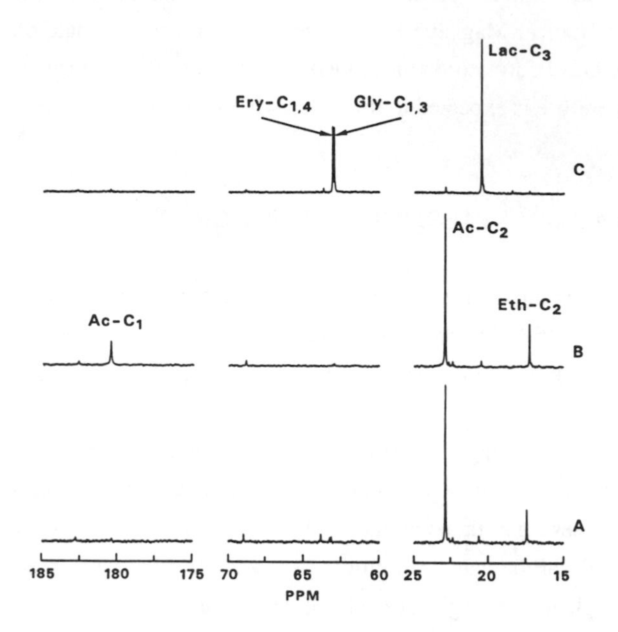

Figure 2 *^{13}C-NMR spectra of the products formed anaerobically from [1-^{13}C]glucose (A), [2-^{13}C]glucose (B) and [6-^{13}C]glucose (C).*
Ac-C_1 and Ac-C_2, C_1 and C_2 of acetate; Eth-C_2, C_2 of ethanol; Lac-C_3, C_3 of lactate; Gly-$C_{1,3}$, C_1 and C_3 of glycerol; Ery-$C_{1,4}$, C_1 and C_4 of erythritol.

2.2 Elucidation of the pathway of erythritol production by using ^{13}C-labelled glucose

In order to elucidate the pathway of erythritol production, cells were allowed to metabolize [1-^{13}C]glucose, [2-^{13}C]glucose, or [6-^{13}C]glucose and the labelling pattern of the end-products was analysed by ^{1}H-NMR and ^{13}C-NMR. All the spectra were run in Bruker AMX500 or AMX300 spectrometers.

According to the heterofermentation pathway proposed for the catabolism of hexoses by heterolactic bacteria (Figure 1, reactions indicated by full lines), the isotopically enriched carbon of [2-^{13}C]glucose should be recovered in the methyl groups of acetate and/or ethanol. This was not the case and instead the ^{13}C-NMR spectra of the cell supernatant solution after the metabolism of [2-^{13}C]glucose revealed the presence of label not only on the CH_3 groups of acetate and ethanol, but also in the COOH group of acetate and in the CH_2OH group of ethanol. The hypothesis that not all of the glucose

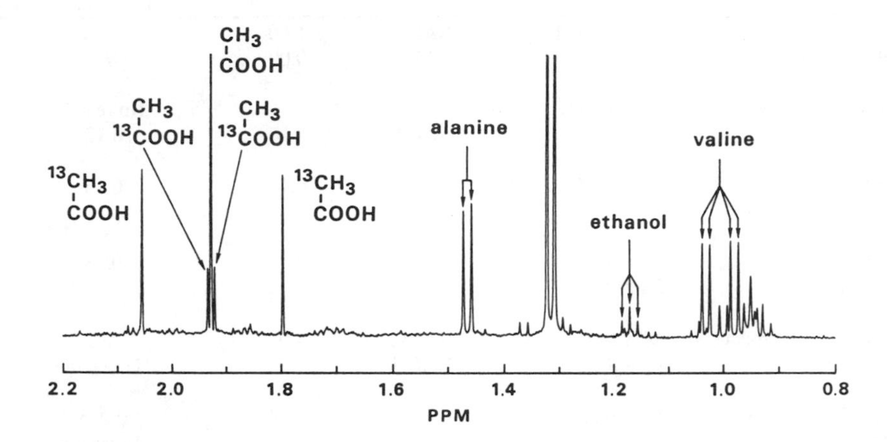

Figure 3 *500 MHz ^{1}H-NMR spectrum of the products formed anaerobically from [2-^{13}C]glucose by non-growing cells of* L. oenos *GM.*

was metabolized heterofermentatively was confirmed when the experiment was repeated using [1-^{13}C]glucose. In this case (Figure 2A) the label was not lost as $^{13}CO_2$, and instead the methyl groups of acetate and ethanol were found labelled. When [6-^{13}C]glucose was used under anaerobic conditions the label ended in the CH_3 of lactate and in the CH_2OH groups of erythritol and glycerol (Figure 2).

These results show that part of the glucose fermented was not decarboxylated and instead was diverted to an alternative novel acetate-ethanol-erythritol-forming pathway.

The ^{1}H-NMR analysis of the supernatant solutions (Figure 3) confirmed this conclusion and provided a straightforward method to determine isotopic enrichments of the carbon atoms in acetate and lactate, since the several isotopomers are readily identified. The ratio of the three isotopomers of acetate was found to be 10:8:3 (CH_3COOH, $^{13}CH_3COOH$, $CH_3{}^{13}COOH$). Therefore, about 75% of glucose was metabolized heterofermentatively and the remaining 25% was channelled to the production of erythritol. The non-labelled acetate, accounting for approximately 50% of the total acetate produced, results from the utilization of endogenous carbon reserves.

The information obtained with specifically labelled glucose led to the proposal for the pathway of erythritol production as indicated by the broken lines in Figure 1.

Table 1 *Activities of several enzymes involved in sugar metabolism by* L. oenos

Enzyme	*Substrate*	*Cofactor*	K_m *for substrate (mM)*	V_{max} *(μmol · min⁻¹ . mg⁻¹)*
Glucose 6-phosphate dehydrogenase	Glucose 6-phosphate	NAD^+	1.60	1.27
		$NADP^+$	0.09	0.53
6-Phosphogluconate dehydrogenase	6-Phosphogluconate	NAD^+		No reaction
		$NADP^+$	0.16	0.13
Phosphoglucose isomerase	Fructose 6-phosphate		0.10	0.46
Triose phosphate isomerase	Glyceraldehyde 3-phosphate		0.56	0.33
Glycerol 3-phosphate dehydrogenase	Dihydroxyacetone phosphate	NADPH	4.0	0.42
		NADH		No reaction
Erythritrol 4-phosphate dehydrogenase	Erythrose 4-phosphate	NADPH	0.56	0.025
		NADH		No reaction
Lactate dehydrogenase	Pyruvate	NADPH		No reaction
		NADH	1.25	10.9
Mannitol dehydrogenase	Fructose	NADPH	20	0.010
		NADH	21	0.012
Alcohol dehydrogenase	Acetaldehyde	NADPH	<0.01	0.82
		NADH	0.077	0.23
Acetaldehyde dehydrogenase	Acetyl coenzyme A	NADPH	2.5	0.007
		NADH		No reaction

2.3 Regulation of erythritol production

The above mentioned proposed pathway was confirmed by measurement of all the enzymatic activities involved (Table 1). $NADP^+$ is the preferred cofactor in all the redox reactions except in those catalyzed by glyceraldehyde 3-phosphate dehydrogenase and lactate dehydrogenase which are NAD^+ specific. Furthermore, both NADH and NADPH oxidase activities are present in cell extracts of *L. oenos* which provide an additional way for disposal of reducing equivalents under aerobic conditions.

In order to understand the observed metabolic shift from glycerol production under aerobic conditions, to erythritol production under anaerobic conditions, phosphorylated intermediate metabolites in the catabolism of glucose were identified and quantified by ^{31}P-NMR in perchloric acid extracts of bacteria metabolizing glucose in the presence or absence of oxygen (Figure 4). The concentrations of glucose 6-phosphate, fructose 6-

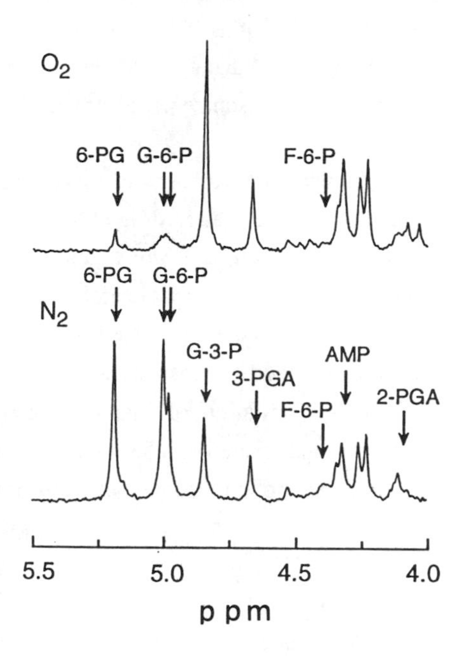

Figure 4 *^{31}P-NMR sprecta of perchloric acid extracts of cell suspensions of* L. oenos *ATCC 23277, metabolizing glucose under an oxygen (upper spectrum) or an N2 (lower spectrum) atmosphere.*
Resonance assignments: 6-phosphogluconate (6-PG); glucose 6-phosphate (G-6-P); glycerol 3-phosphate (G-3-P); 3-phosphoglycerate (3-PGA); fructose 6-phosphate (F-6-P); 2-phosphoglycerate (2-PGA).

phosphate and 6-phosphogluconate increased significantly under anaerobic conditions whereas the concentrations of later intermediates such as xylulose 5-phosphate, decreased. A 10-fold increase was observed in the concentration of glucose 6-phosphate and the concentration of fructose 6-phosphate increased by a factor of 8.5. These results are accounted for by the low activity found for acetaldehyde dehydrogenase (Table 1) which is expected to cause slowing down of NADPH oxidation under anaerobic conditions; the accumulation of NADPH will inhibit the two first reactions in the catabolism of glucose. Therefore, accumulation of glucose 6-phosphate and 6-phosphogluconate will occur in the absence of O_2, but not when O_2 is present, since then the presence of NADPH oxidases provides an alternative way for regeneration of $NADP^+$. Fructose 6-phosphate is produced upon accumulation of glucose 6-phosphate in a reaction catalyzed by phosphoglucose isomerase and is subsequently cleaved by phosphoketolase leading to the production of erythrose 4-phosphate and ultimately to erythritol biosynthesis. We also demonstrated that a single phosphoketolase enzyme uses both xylulose 5-phosphate and fructose 6-phosphate as substrates although it has a lower affinity for the latter one[12].

This explanation is consistent with the observation that the erythritol-producing strains were the only ones in which the concentration of hexose 6-phosphate increased in the absence of O_2. Glucose metabolism by four *L. oenos* strains and three *Lactobacillus* strains was examined for this purpose both under aerobic and anaerobic conditions. This interpretation of results also accounts for the fact that erythritol is not formed when either fructose or ribose are supplied; in fact, fructose can serve as an effective electron acceptor for NADPH leading to the formation of mannitol, and ribose is directly converted to xylulose 5-phosphate and therefore the two first NADPH-producing reactions in the catabolism of glucose are not part of the metabolism of this pentose.

3 CONCLUDING REMARKS

These studies revealed erythritol as a major product of glucose metabolism by wine lactic acid bacteria under anaerobic conditions, and allowed to elucidate the pathway for erythritol production and the corresponding regulatory process. The erythritol produced by growing cells (230 mg l^{-1}) after 25 days of incubation is in agreement with the range of values found for the concentration of this edulcorant in table wines (90 to 700 mg l^{-1})[14]. Until now, the presence of erythritol in table wines has been associated with mold-infected grapes[14]; however, on the basis of our results it is likely that at least part of the erythritol found in wine results from the activity of lactic acid bacteria. It is also interesting the observation that only the wine strains isolated from wine (*L. oenos*) were able to produce erythritol.

Acknowledgments

This work was done with the collaboration of M. Veiga-da-Cunha, Paula Firme, Vitória San-Romão and E. van Schaftingen.

References

1. M. J. Gasson, *FEMS Mcrobiol. Rev.*, 1993, **12**, 3.
2. N. F. Olson, *FEMS Mcrobiol. Rev.*, 1990, **87**, 131.
3 W. P. Hammes, *Food Biotechnol.*, 1991, **5**, 293.
4 R. E. Kunkee, *FEMS Mcrobiol. Rev.*, 1991, **88**, 55.
5 R. E. Wibowo, C. R. Davis, G. H. Fleet and T. H. Lee, *Am. J. Enol. Vitic.*, 1985, **36**, 302

6 P. Lundberg, E. Harmsen and H. J. Vogel, *Anal. Biochem.*, 1990, **191**, 193.
7 R. E. London, *Prog. Nucl. Magn. Reson. Spectrosc.*, 1985, **17**, 241.
8 H. Santos, P. Fareleira, J. LeGall and A. V. Xavier, *FEMS Mcrobiol. Rev.*, 1990, **87**, 361.
9. A. Ramos, K. Jordan, T. Cogan and H. Santos, *Appl. Env. Microbiol.*, 1994, **60**, 1739.
10 M. Veiga-da-Cunha, P. Firme, M. V. San-Romão and H. Santos, *Appl. Env. Microbiol.*, 1992, **58**, 2271.
11 M. Veiga-da-Cunha, H. Santos and E. van Schaftingen, *J. Bacteriol.*, 1993, **175**, 3941.
12 O. Kandler, *Antonie van Leeuwenhoek*, 1983, **49**, 209.
13 H. Santos and D. L. Turner, *J. Magn. Reson.*, 1986, **68**, 345.
14 W. R. Sponholtz, "Wine Analysis", Modern Methods of Plant Analysis, New Series, vol 6, Springer-Verlag, Berlin, 1988.

The Use of Amino Acids as a Fingerprint for the Monitoring of European Wines

M. V. Holland[1], A. Bernreuther[1], and F. Reniero[2]

[1] EUROPEAN COMMISSION, JOINT RESEARCH CENTRE – ISPRA SITE, ENVIRONMENT INSTITUTE, ISPRA (VA), I-21020, ITALY

[2] INSTITUTO AGRARIO DI SAN MICHELE ALL'ADIGE, SAN MICHELE ALL'ADIGE, TRENTO, I-38010, ITALY

1 INTRODUCTION

The authentication of foodstuffs has presented a considerable challenge to food chemists in the field of consumer protection for many years. As the surveillance agencies become armed with more sophisticated methods of fraud detection, the producer of fraudulent foodstuffs are applying more refined methods, forcing the surveillance agencies to look for other techniques for food control. Wine, being a product which receives a great deal of interest especially concerning quality and origin control, merits sophisticated methods in order to avoid adulteration.

There are several methods already in use to ascertain the quality of a wine; for example IRMS[1] (isotope ratio mass spectrometry), which measures the overall isotope ratio of $^{18}O/^{16}O$, $^{13}C/^{12}C$ and D/H, and SNIF-NMR (site specific natural isotope fractionation-nuclear magnetic resonance) which is used to measure the D/H ratio at specific sites in the wine alcohol[2]. Amino acids may also be used to provide a fingerprint of the wine thus providing another method for wine control. Little is known on this subject at present. Experience to date, however, seems to imply that the quantitative measurement of amino acids can be used to indicate possible tampering with the wine.

Amino acids belong to the minor compounds present in wines in which they represent 15 to 30% of all nitrogen compounds. The amount of amino acids in grapes is influenced by N-fertilization, soil type, degree of ripeness, grape variety and its state of health (e.g. infection with *Botrytis cinerea*). The average concentration of amino acids in musts is about 2 to 3 g/l; in wines it is about 1 to 2 g/l due to fermentation. During yeast or malolactic fermentation some amino acids are assimilated (e.g. threonine, phenylalanine, arginine, etc.) whereas others remain almost unchanged (e.g. proline, alanine, lysine, etc.).[3-5] Despite these changes amino acids can, together with other compounds present in wines, be used to classify wines of different origin.[6] Generally it can be said that red wines contain a higher amount of amino acids than white and that wines from cooler regions (northern Europe) often show higher amounts than those from warmer regions (southern Europe).[3]

In this work we further explore, by means of high resolution ^{13}C NMR, the applicability of an amino acid fingerprint of various European wines. One of the advantages of this method is that it can be applied to wine concentrates without any previous separation or derivatization. Hence, the possibilities already available for use in consumer protection may be enhanced.

2 MATERIALS AND METHODS

2.1 Wine Samples

Authentic wine samples were obtained from Italy, France, Spain, Portugal, Greece and Germany. Various grape varieties were studied from each country; these and the alcoholic grade are given in Table 1. The vintages span from 1991 to 1993.

2.2 Sample Preparation

Following a previous work[7], an aliquot of 100 ml of each wine was concentrated to 10 ml using a rotary evaporation unit with a water bath set at 30 °C. The wine concentrate was adjusted to pH 2 using a 2N DCl in D_2O solution. Then 0.5 ml of the concentrate was filtered into a 5 mm o.d. NMR tube and 20 μl of 1,3-propanediol (corresponding to 18.993 mg) was added as internal standard, 100 μl of D_2O was added as the lock substance.

Table 1 *Characteristics of the wine samples analysed.*

Country of Origin	Grape Type	Alcoholic Grade(%)	Sample Code
Italy	Chardonnay	11.03	I1 (W)
	Aglianico	12.06	I2 (R)
	Aglianico	9.27	I3 (R)
Germany	Müller Thurgau	11.92	D1 (W)
	Riesling	12.79	D2 (W)
Greece	Agiorgitiko	11.97	G1 (R)
	Assyrtiko	16.99	G2 (W)
France	Riesling	9.7	F1 (W)
	Cinsault	11.98	F2 (W)
Portugal	Trajadura	11.90	P1 (W)
	Valdigiuie	10.98	P2 (W)
Spain	Tempranillo(V)	11.32	E1 (R)
	Tempranillo(N)	12.47	E2 (R)

W=white wine, R=red wine, V=Valladolid region and N=Navarra region

2.3 NMR Parameters

All spectra were recorded on a BRUKER AMX 500 NMR spectrometer operating at a carbon frequency of 125.7 MHz (proton frequency of 500 MHz) using a 5 mm o.d. dual probe and a spinning rate of 17 Hz. A typical ^{13}C spectrum consists of 22000 transients using data points over a 27777.78 Hz band width and a 30° (2.67 μs) radiofrequency (rf) pulse. Total acquisition time was about 13.5 hours. The signal-to-noise ratio of the spectra was improved by multiplying each free induction decay (FID) with an additional exponential factor, corresponding to 1 Hz in the Fourier transformed spectrum. The spectra were proton-decoupled with the help of composite pulse decoupling (CPD). The measurement temperature was maintained at 302 K and a relaxation delay of 1 s was used.

2.4 Quantitative Determination

To evaluate the concentration of the different amino acids in the wine concentrate by using their signal intensities the following points should be considered. The linewidth at half-height of all peaks must be similar and as narrow as possible, i.e. their spin-spin relaxation time (T_2) should be less than 0.5 s. The viscosity of the sample must not be too high as this can cause the peaks to broaden. The peaks used for the evaluation should have gone through their spin-lattice relaxation time (T_1) and thus be totally relaxed. The carbon atom of the carboxylic group in amino acids, as in all acids, has a longer relaxation time, (T_1), as compared to the relatively short one of the other signals; therefore the NMR signal of the carboxyl carbon atoms have not been used for the present evaluation. The α-C atom (C2) should not be used either as those carbons which are directly bond to nitrogen which leads also to broader signals.

If one uses only one or two signals to evaluate the quantity of a compound in the sample it is necessary to calculate the proportion these signals make up of the total number of signals. Using this factor (f) one can calculate the actual concentration using only one or two signal intensities. In order to calculate the concentration of a given amino acid the following relation between the concentration of the amino acid, the concentration of the standard and their corresponding peak intensities are used:

$$C_u = \frac{\sum I_u}{\sum I_{st}} \cdot C_{st} \cdot \frac{CF_u}{CF_{st}} \cdot f \qquad (1)$$

where C_u is the concentration of the amino acid to be determined in g/l, I_u and I_{st} are the relevant peak intensities of the amino acid to be determined and of the added standard (i.e. 1,3-propanediol); C_{st} is the concentration in g/l of the added standard; CF_u and CF_{st} are the carbon factors of the amino acid to be determined and that of the added standard (2.11 for the 1,3-propanediol), and f is the factor which accounts for the fact that not all peaks of one amino acid are used for the calculation of its concentration.

The carbon factor, CF, is the ratio of the molecular weight of the compound to the molecular weight of the number of carbon atoms in the compound.

3 RESULTS AND DISCUSSION

To identify the amino acids in the wines, their chemical shifts with respect to 1,3-propanediol were obtained by enriching a D_2O solution with each amino acid of interest plus the internal standard. The pH of the solution was adjusted to pH 2. The choice of the pH value assured the best signal-to-noise ratio[7]. In Table 2 the chemical shifts of some of the principle amino acids in wine are presented.

The same solution was used to determine the calculation factor (f) for the amino acids of interest (Table 3).

A wine concentrate was enriched with varying amounts of the amino acids considered in this study ranging from 0.2 g/l to 1 g/l. This concentrate was used to control the quantitative ability of the experimental method. It is evident from Figure 1 that agreement obtained between the measured concentration and that of the weighed amount of the amino acid in the enriched sample is very good. This clearly demonstrates the applicability of this method in the quantitative evaluation of amino acids in wines.

Table 2 *Chemical shift of amino acids in D_2O at pH 2, 302 K, compared with 1,3-propanediol.*

	C1	C2	C3	C4	C5	C6	C7
Proline	174.78	62.43	31.01	26.05	48.95	-	-
Arginine	174.87	55.56	29.71	26.48	43.06	159.52	-
Alanine	175.42	51.48	17.91	-	-	-	-
Histidine	173.49	55.03	27.88	120.79	136.87	129.36	-
Serine	172.95	57.49	62.05	-	-	-	-
Glutamic Acid	174.39	55.14	28.17	33.19	179.82	-	-
Isoleucine	174.31	60.09	38.47	27.41	13.56	16.83	-
Leucine	175.47	54.25	41.63	26.70	23.66	24.31	-
Lysine	174.91	55.60	31.98	24.08	28.93	41.74	-
Phenylalanine	174.24	56.95	38.32	136.76	131.91	132.09	130.69
Threonine	173.32	61.29	67.98	21.67	-	-	-
Tyrosine	174.20	56.94	37.46	128.40	133.53	118.66	157.87

Table 3 *Parameters for the quantitative determination of amino acids in wine.*

Amino acids	Chemical Shift	Carbon Factor	Calculation Factor
Proline	C3, C4	1.93	1.83
Arginine	C3	2.42	4.24
Alanine	C3	2.50	1.94
Glutamic Acid	C3, C4	2.47	1.88
Serine	C3	2.92	2.03

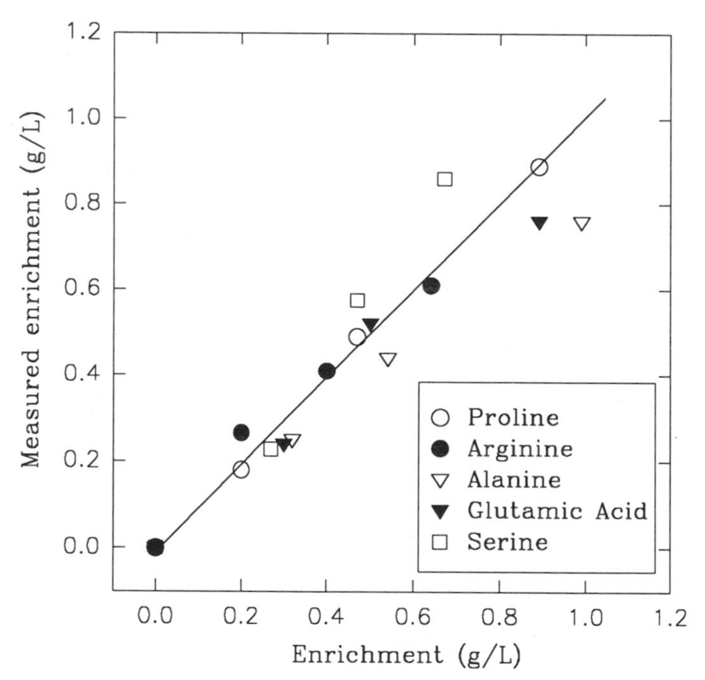

Figure 1 *Good agreement is obtained between the measured concentration and that of the weighed amount in the enriched sample.*

Figure 2 shows a typical ^{13}C NMR spectrum of one of the studied wines. The presence of a large quantity of amino acids is clearly visible in the region surrounding the added internal standard (1,3-propanediol).

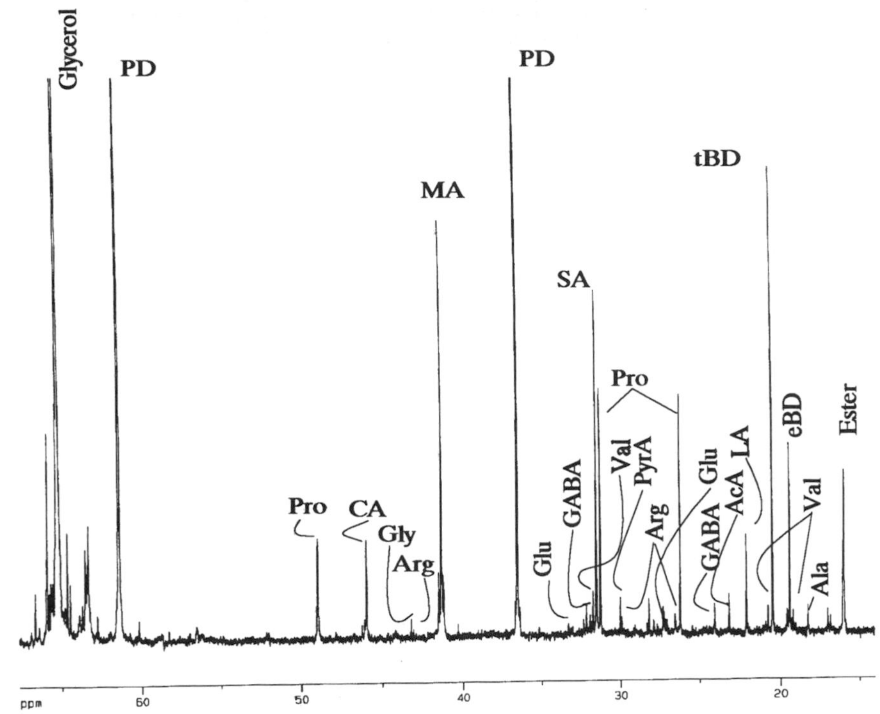

Figure 2 *Typical* ^{13}C *spectrum of the studied wines (AcA = acetic acid; Ala = alanine; Arg = arginine; CA = citric acid; eBD = erythro-2,3-butanediol; GABA = γ-aminobutyric acid; Glu = glutamic acid; Gly = glycine; LA = lactic acid; MA = malic acid; PD = 1,3-propanediol; Pro = proline; PyrA = pyruvic acid; SA = succinic acid; tBD = threo-2,3-butanediol; Val = valine).*

The spectra of all the wines under study were processed to obtain the peak intensities of the corresponding amino acids. These intensities were treated using equation (1) together with the parameters from Table 3 to obtain the corresponding concentrations. Noticeable differences were found between wines from different countries and also between wines from within the same country. Several peaks from the spectra could be unambiguously assigned due to their intensities and their chemical shifts; other peaks, which were less evident due to their lower content in the wine, could be assigned after enrichment of that particular amino acid.

The concentration of proline in red wines was found to be higher than that of white wines and in particular the Italian red wines (grape type Aglianico, vintage 1993) presented the highest value. This is apparent in Figure 3 where the concentration of proline is depicted for all the wines investigated in this study. The proline concentration of the Italian white variety is in good agreement with the values found by other authors using HPLC[8] for the same grape variety coming from a different country. On the other hand, the concentration of arginine in the wines under investigation does not seem to present a clear trend as can be seen in Figure 4, where the concentration levels vary without a clear difference between red and white wines. This might be due to its assimilation during yeast or malolactic fermentation. However the Spanish wines (grape variety Tempranillo, vintage 1992) presents the highest concentration (approximately 0.3 g/l).

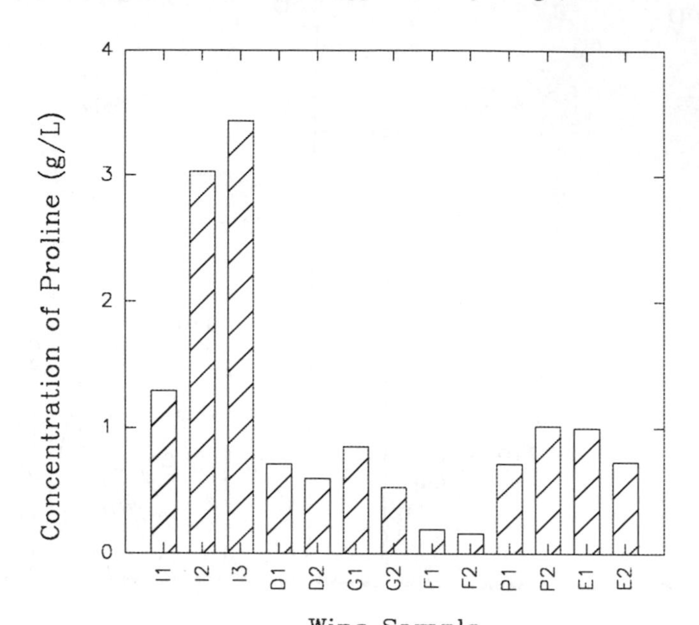

Figure 3 *Concentration of proline (g/l) of the different wines in this study.*

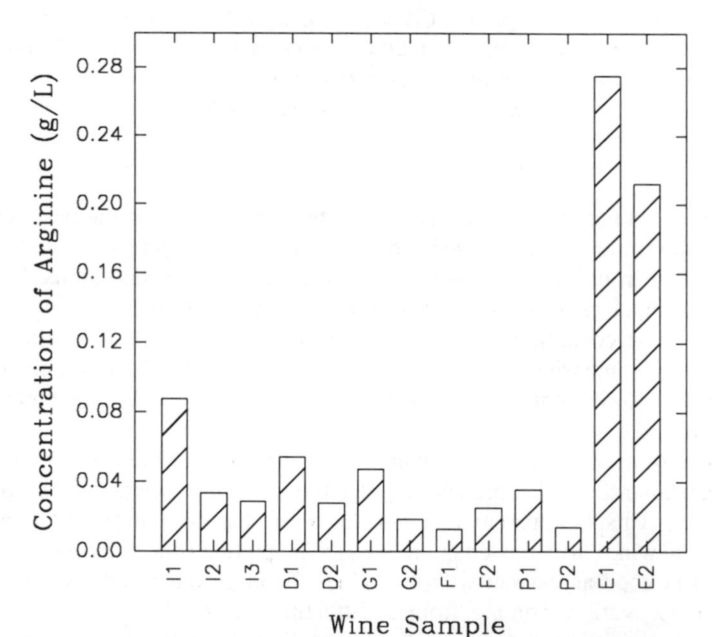

Figure 4 *Concentration of arginine (g/l) of the different wines in this study.*

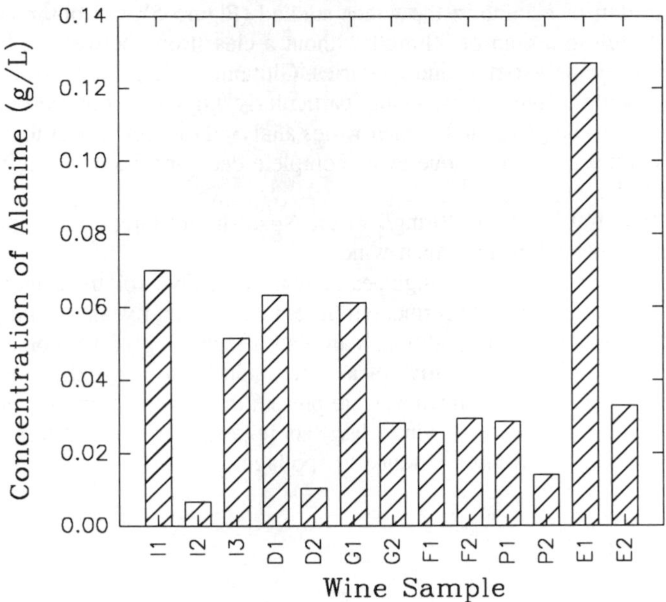

Figure 5 *Concentration of alanine (g/l) of the different wines in this study.*

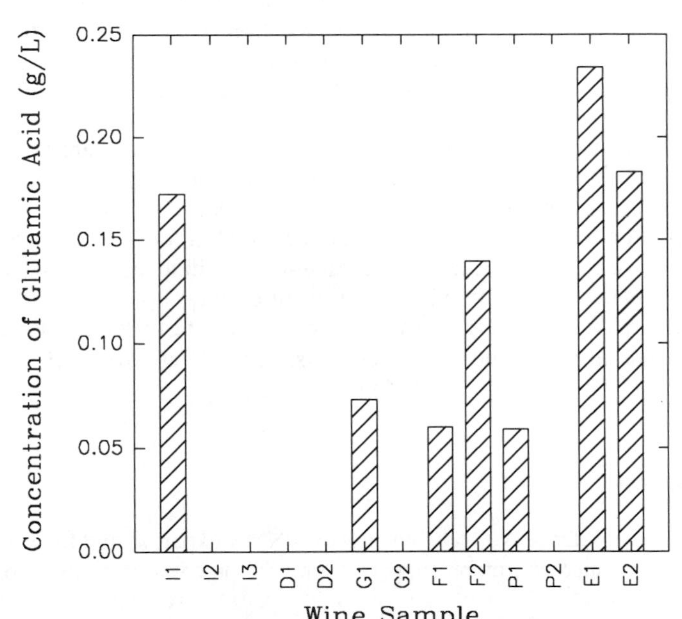

Figure 6 *Concentration of glutamic acid (g/l) of the different wines in this study.*

The concentration of alanine in the wines studied (Figure 5) was observed to range from a low of 10 mg/l to a high of 70 mg/l without a clear trend between red and white or between different grape varieties and countries. Glutamic acid (Figure 6) was detected in some red and some white wines; being particularly higher in concentration in the Spanish red wines (≈200 mg/l). The German wines analysed did not appear to contain this amino acid. This could possibly be due to its complete decomposition during the growth of lactic acid bacteria.

Serine was found at levels of 30 mg/l in the Spanish and Greek red wines. On the contrary it was not detected in the Italian wines.

Some spectra showed a relatively high peak intensity for glycerol, the content of which is affected by many factors such as fermentation temperature and yeast strain; pH, initial sugar concentration and aeration conditions can also affect the glycerol content. A high amount of glycerol can also be indicative of a wine made from botrytized grapes.

Peaks related to organic acids in wines were present in each spectrum, ie. malic, lactic, citric acids, etc. Their presence in wine plays an important role in determining such characteristics as taste, colour and resistance to spoilage.

4 CONCLUSIONS

The use of ^{13}C-NMR as a method for directly determining in a single experiment, and without need for elaborate sample preparation, the amino acid content of wine has been demonstrated in the present work. Through the quantitative analysis of proline and other amino acids a trend may be observed which might aid the detection of fraud. This will, of course, necessitate the comparison of results obtained from wine samples from at least three vintages. The fact that many other compounds present in wine like polyalcohols, sugars, organic acids, etc can also be quantified with one measurement is another advantage of using this technique.

NMR is the only analytical method which is able to detect, compare and quantify different compounds (isotopomers of a chemical species) without any previous separation, provided of course, the isotopic nucleus has a magnetic moment. This compares favourably with HPLC, GLC and other classical method commonly used in food-control laboratories.

^{13}C-NMR spectroscopy in combination with pattern recognition techniques[9] has proven a very powerful technique to classify wines of different origin.

In the near future we will apply a novel method, at present under development, for measuring compounds in wines and other foodstuffs, at levels as low as 10 μg, by the hyphenation of HPLC with proton NMR (HPLC-NMR).

Acknowledgment

The authors wish to thank Dr S. A. Korhammer for his NMR expertise and to Dr A. Perujo for his invaluable assistance and moral support during the preparation of this work.

References

1. H.L.Schmidt, Fresenius Z Anal. Chem., 1986, **324**, 760.
2. G. J. Martin and M. L. Martin, 'The Site-Specific Natural Isotope Fractionation-NMR Method Applied to the Study of Wines', In Modern Methods of Plant Analysis, Edited by H.F. Linskens and J.F. Jackson, Springer-Verlag, 1988, Vol 6.
3. G. Würdig and R. Woller, 'Handbuch der Lebensmitteltechnologie. Chemie des Weines', Ulmer, Stuttgart, 1989.
4. C. S. Ough and M. A. Amerine, 'Methods for Analysis of Musts and Wines', 2nd edition, Wiley and Sons, New York, 1988.
5. C. S. Ough, 'Acids and Amino Acids in Grapes and Wines', In Modern Methods of Plant Analysis, Edited by H.F. Linskens and J.F. Jackson, Springer-Verlag, 1988, Vol 6.
6. W. Ooghe and H. Kastelijn, Ann. Fals. Chim., 1984, **831**, 467.
7. A. Rapp, A. Markowetz and H. Niebergall, Z Lebensm Unters Forsch, 1991, **192**, 1.
8. Z. Huang and C. S. Ough, Am. J. Enol. Vitic.,1991, **42**, 261.
9. J. T. W. E. Vogels, A. C.Tas, F. van den Berg and J. van der Greef, Lab. Inf. Manage. 1993, **21**, 249.

Rapid Cooking Control of Cakes by Low Resolution NMR

A. Davenel[1], P. Marchal[1], and J. P. Guillement[2]

[1]CEMAGREF, DEPARTMENT OF AGRICULTURAL AND FOOD ENGINEERING, 17, AVENUE DE CUCILLÉ, 35044 RENNES CEDEX, FRANCE

[2] DEPARTMENT OF MATHEMATICS, CNRS URA 758, 2, RUE DE LA HOUSSINIÈRE, 44072 NANTES CEDEX 03, FRANCE

1 INTRODUCTION

Food industry places increased emphasis on the development of sensors which can be used for monitoring processes in plant. French bakery companies want to test the capability of different rapid physical techniques to follow starch modifications during cooking of baked products. Starch granules can undergo different physical transformations. Gelatinisation refers loosely both to the loss of order (measured by birefringence and X-ray crystallinity techniques) and the swelling of the granules. However in limited-water systems like cakes, gelatinisation would be considered more as a melting of the crystallites and a very partial swelling occurring at high temperatures (90-110°C).

X-ray diffraction technique could be the best instrumental method if the samples have not to be preequilibrated to the same water activity in order to compare their crystallinity[1]. Microscopic techniques were the first ones largely used for studying starch damages and effects of different sugars and shortenings on cooking of cakes[2-4]. Though they provide relevant observations of the changes undergone by the baked materials, the difficulty to quantify observations and the need of skill restrict their use. Differential scanning calorimetry has been used to determinate the temperatures and the enthalpies of melting as modified by flour components and dough ingredients[5-7]. However the scan of intact cakes generally showed no more significant endothermal transition in the gelatinisation range. To determinate the amount of starch gelatinised during baking of very limited water system like cookies (22.75%, 60% sugar based on flour weight), Abboud[8] had to scan samples with an excess of water.

Damaged starch granules are more easily hydrolysed than intact ones: alpha-amylolysis methods would be very sensitive to discriminate products with partial damaged starches. Despite these methods are time consuming, one of them was chosen to provide a reference method to determinate starch damage in some samples of this study.

Several previous works showed that, in high moisture systems, NMR relaxation times of protons were significantly affected by swelling and gelatinisation mechanisms during heating[9-11]. Leung[12,13] observed two liquid components in the transverse relaxation of flour doughs. However, it would be hazardous to want interpret these variations of relaxation times in terms of water mobility. Proton relaxation is strongly affected by chemical exchange between water protons and exchangeable protons of macromolecules but also by cross relaxation. Oxygen-17 NMR would be the only NMR technique that reports purely on the behaviour of water[14]. Several publications report the water states by this technique in native starch products[15], in starch/sucrose systems[16] and in starch gels[17-20]. However proton relaxation measurements in aqueous systems can potentially give relevant information about the state of biopolymers[21]. More, they are much more available to obtain rapid information in food plants. Using pulsed low resolution proton NMR (20MHz) Schierbaum[22] found a remarkable linear relation between the solid-liquid ratio from NMR and X-ray crystallinity in thermally reversible maltodextrin gels. Teo[23] applied the same method to study the starch retrogradation.

In this work, a similar method associating the [liquid/(liquid+solid)] ratio from free relaxation induction decay together with an analysis of the transverse relaxation behaviour of liquid protons from spin echoes sequence was used to study the effects of the formulation and process parameters on the cooking of cakes.

2 MATERIALS AND METHODS

2.1 Preparation of Cakes.

NMR analyses were realised with small cakes cooked in the laboratory to test some effects of the formulation and with industrial cakes to test some process effects. All cakes were made from commercial soft wheat flour containing 9.9 % protein and 13 % water (ruban violet from Grands Moulins de Paris). The standard batter formula is listed in Table 1. Some formulas were completed by water.

Sucrose and shortening were mixed in a Kitchen KM 230 mixer (Kenwood, Havant, Britain) at speed 4 for 2 min. Eggs were added and mixed at speed 4 for 3 min. Then the mix of dry ingredients (flour, baking powder) was added and mixed for 1.5 min at speed 4. Portions of batter (12-13g) in aluminium baking cups were placed in an ID53 electric oven (Firlabo, Lyon France) at about 220°C.

Table 1 *Combinations of compositional factors expressed as coded and decoded values*

coded variables		Ingredients (% wet basis)					Chemical composition (% w.b.)			
W	S	Sugar	Flour	Eggs	Fat	Powder[a]	Moist.	Starch	Fat	Prot.
Central points (standard formula)										
0	0	22.6	31.6	27	18.0	0.8	27.1	19.6	18.0	6.1
Axial points										
-1	-1	19.3	40.1	20.1	19.5	1.0	23.9	24.8	18.6	6.2
-1	+1	29.4	28.2	22.1	19.4	0.9	23.7	17.4	18.7	5.3
+1	+1	25.9	24.8	24.8	17.1	0.6	31.0	15.3	17.0	5.3
+1	-1	16.9	35.0	24.8	17.0	0.7	32.0	21.7	17.0	6.3
Star points										
0	$+\alpha$	30.2	21.4	27.0	18.6	1.0	27.6	13.3	18.6	5.3
$-\alpha$	0	25.2	35.3	18.4	20.1	1.0	22.4	21.8	18.9	5.6
0	$-\alpha$	15.9	39.4	25.8	18.0	0.9	28.0	24.4	18.0	6.8
$+\alpha$	0	20.8	29.2	24.2	16.6	0.7	33.7	18.0	16.6	5.7
Extra points (cakes without sugar)										
flour +		0.0	58.5	22.4	18.3	0.8	27.3	34.5	17.8	8.0
water +		0.0	31.6	26.2	18.0	0.8	50.5	22.5	18.0	6.6

[a] Baking powder

For each baking of a standard formula and a modified formula, cups were sampled at 0 (batter), 2, 3, 4, 5.5, 7, 9.5, 11 and 13 minutes during heating.

We set up two central composite designs[24] to analyse the effects of two compositional factors, the water content and sucrose/(sucrose + dry flour) ratio, and the effects of two process factors, the cooking time and temperature. Shortenings play an important role in the texture of baked products but would not affect the starch gelatinisation temperature[25]. In this work, the shortening content was kept at a constant percentage of the flour plus sugar weight.

In the study of the two compositional factors, our baking experiment tested four axial points, four star points, and eight central points (standard formula). Water [W] and sucrose/(sucrose+dry flour) [S] levels were coded in a ± 1.41 interval (Table 1). In order to analyse better the effect of sugar on transverse relaxation behaviour, we also tested two extra points. In these batters without sugar, water (water+) or flour (flour+) replaced sugar.

In the study of the two process parameters, the experiment on standard formula was realised by CTUC company with larger cakes (180g). The number of central points, called 'standard process' (220°C, 35mn) was limited to four. Other values were 200-206-234-240°C for cooking temperature and 25-28-42-45 minutes for cooking time.

2.2 NMR Meaurements

The NMR measurements were carried out with a Bruker Minispec pulsed NMR spectrometer operating at 20 MHz. The sample coil was 10mm diameter. NMR tubes were filled to about 1.5 cm height. In this preliminary study, samples were previously stabilised at 40°C for about 10mn before measurements. In the experiment on the compositional effects, the NMR measurements were realised on samples of batters and central parts of cakes. To display the effects of process parameters on transverse relaxation, NMR data were also acquired in duplicate with samples of the non-cooked batter gently dried at 50°C to cover the same moisture range of cooked samples. In this second experiment, batters and central parts of cakes were sampled for measuring NMR transverse relaxation and enzymatic susceptibility to alpha-amylase by a kinetic method[26]. In order to analyse some heterogeneity in cakes, other measurements also were carried out on samples taken at about 1 cm from the edge of each cake and on the outer part of three cakes.

Free relaxation decays were registered at 12µs (FID12) and the liquid-like free relaxation decay each millisecond between 50µs and 100µs after the beginning of the first pulse. The 845 spin echoes amplitude maxima (0.2ms apart) of a spin echoes CPMG pulse sequence were also digitised. In the present work, each signal intensity measured was an average of 49 repetitions and the delay between repetitions was 1s. To eliminate the effects of the variations of bulk density between the samples, all NMR intensities measured were divided by the FID12 value, approximatively proportional to the total number of protons.

2.3 Data Analysis

Amplitude L of the liquid phase was extrapolated by fitting the first part of the liquid-like free relaxation with an exponential curve and used to calculate a liquid/(solid+liquid) ratio (L/FID12).

Principal component analysis (PCA) has been used to transform the numerous collinear variables from infrared spectra or NMR relaxation curves of large sets of samples into a limited number of new synthetic orthogonal centred data[27,28]. They are linear combinations of the original ones and called 'factorial coordinates' (FC). This technique was separately tested on each our two sets of NMR data issuing from the two experiments. Variables were the normalised NMR data, free relaxation values sampled between 50 and 100μs and spin echoes amplitudes. Results showed that the variances of the factorial coordinates obtained decreased very sharply. The representation of NMR data could be limited to their four first FC, indeed to the plot of the two first axes. A partial explanation of the location of each sample in this plot, in terms of proton relaxation behaviour, may be obtained by an examination of the 'component transverse relaxation curves' (eigenvectors). Each relaxation curve sample can be approximated by a linear combination of these vectors weighted by the FC of this sample and added to the average relaxation curve of the set whose all factorial coordinates equal zero value.

A better explanation of these locations was obtained by analysing some typical relaxation curves as a sum of discrete exponential terms. However the theory for the interpretation of proton relaxation in water limited products is much less developed than for diluted ones. It would be hazardous to presume the number of exponentials in those systems. More in complex systems like batters and cakes the true relaxation spectrum $F(T_2)$ is more probably to be continuous, reflecting some heterogeneity in the distribution of water and fat. We also used the same maximum entropy method (MEM) as Tellier[29] applied to investigate the distribution of water during syneresis of curds. It allows no restrictive prior assumptions about the number of exponentials and introduces the least correlations in the reconstructed distribution from noisy data. Because MEM software used here can only perform data sampled with a constant spacing in the time domain, fittings were limited to spin echoes data.

3 RESULTS AND DISCUSSION

3.1 Effects of Compositional Factors on Proton Relaxation

PCA was rapidly performed, within two minutes, on the set of the 128 samples, batters and cakes, issuing from the 16 bakings of the experimental design (extra points non included). The factorial coordinates from the four first components are plotted versus each other in Figure 1. Biplot of the two first components, accounting for the greatest proportion of the total variance (98%), shows very discernible patterns of the two 'orthogonal' evolutions, the cooking of batters following by the dehydration of cakes. More, in this plot, the path of each formula is generally significantly different between each other. The central part of all batters was cooked after five minutes of baking. Some other information would be available from the third and fourth components (figure 1b and 1c), but global variance between formulas was barely larger than this one of the repetitions of the standard formula. In the principal plot, cooking effect is all the more discernable than formulas were higher in water and sugar content. Some formulas (-w/-s, -w/+s, --w/0) have minor variations particularly following the second axis. After cooking, the cakes roughly follow a similar curved line leading to an almost constant value of the first FC.

We previously noted that PCA performed centred data: the imaginary average sample whose FC equals zero should be roughly a mid-cooked cake from standard formula. The first eigenvector (figure 2) is related to a global variation of the relaxation time that discriminates batters in function of water and sugar contents. The first component increased rapidly during the cooking phase, then more slowly during the dehydration phase. It was negligibly correlated to a variation of the amplitude of the liquid phase. The origin of this eigenvector is roughly equal to zero.

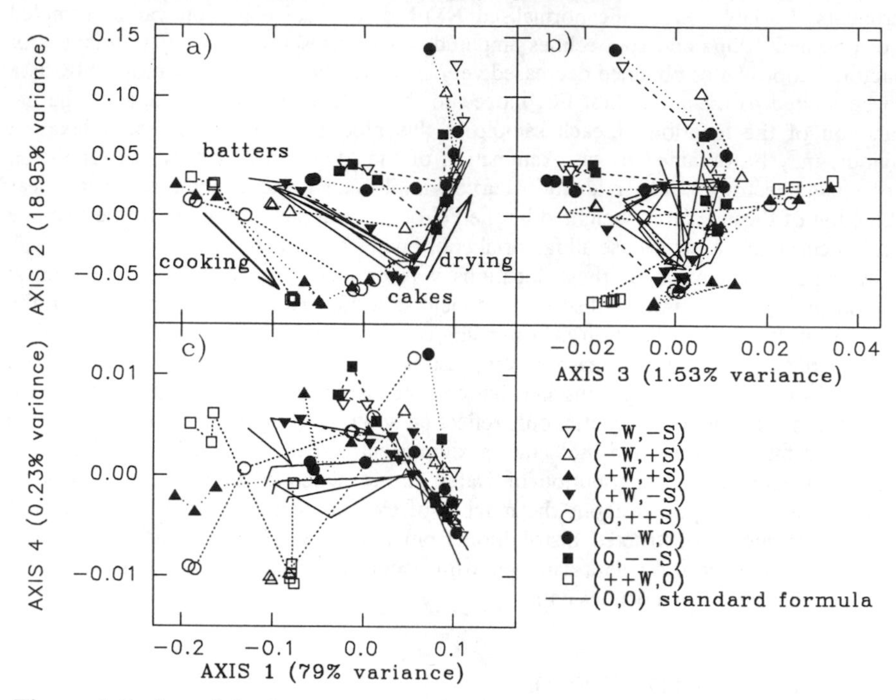

Figure 1 *Biplots of the first components from transverse relaxation data: variations during cooking of batters with different moisture levels (W) and sugar/(sugar+starch) ratios (S)*

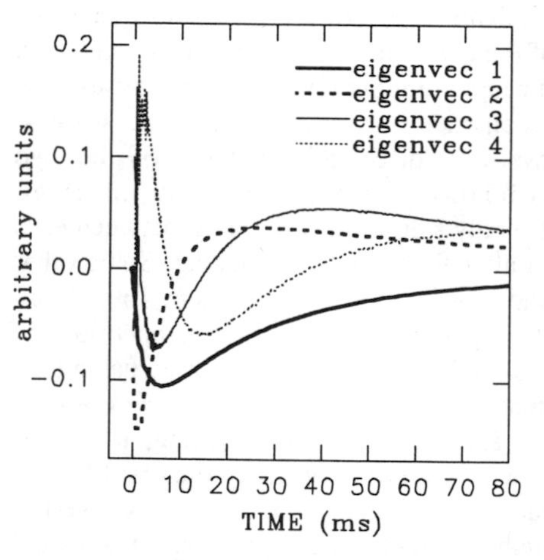

Figure 2 *First eigenvectors from experimentation on batters with different formulas*

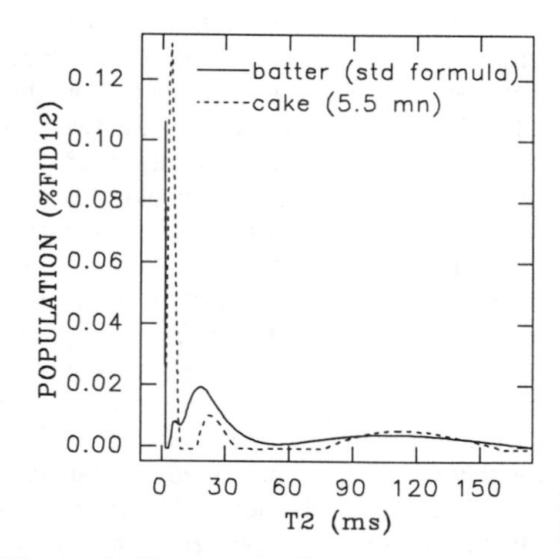

Figure 3 *Continuous T_2 spectra of standard batter before and after cooking (25 % moisture)*

On the other hand, L/FID12 ratio was significantly negatively correlated to the second component (r^2=0.85). The third component significantly improved this correlation (r^2=0.96). It would particularly help to discriminate some batters. After cooking, the L/FID12 ratio was almost completely correlated to the only second component (r^2=0.97). This second component decreased during the cooking phase, particularly for batter higher in water and sugar, then increased during the drying phase.

In spite of moisture was sligthly reduced by about 0.2-2% during the about 5mn cooking phase (Table 2), the L/FID12 ratio markedly increases between 4 and 12 per cent in the same time. This behaviour could be attributed to a more or less important reducing of crystallinity related to the melting of sugar or starch. Batters without sugar, especially the high moisture content batter, showed also significant rise of the L/FID12 ratio. So there is some evidence that a large part of the starch protons becomes liquid-like during cooking phase.

The prediction equations for different variables measured on batters and cakes at 5.5 mn (extra points non included) and the corresponding r^2 to the response surface are presented in Table 3. L/FID12 ratio was strongly positively correlated to the water content W and sugar/(sugar+flour) S ratio. The major part of the sugar in these batters would be dissolved in water. This experimentation did not allow to exclude the possibility of a small part of sugar could be in solid phase, especially for batters (-1/+1, -α/0) whose the sugar/ water ratio was close to the 0.63 saturation value. During the cooking phase, the increase ΔL of the L/FID12 ratio, though measured with an error of repeatability equal to 8%, was very significantly negatively correlated to the S ratio. This means that the evolution of ΔL is related to the flour content of batters (see ΔL/%starch in Table 2).

Table 2 *Variation of L/FID12 ratio and proton populations in batters and cakes with different formulas*

W	S	Water Loss %w.b.	Sugar/ Water[a]	L/FID12 batters	L/FID12 cakes 5.5mn	ΔL (10^3)	ΔL/ %starch (10^3)	Pop[d] < 2 ms batters	Pop[d] 2-22 ms cakes[b]
Central points (standard formula)									
0	0	1.94 (1.156)[c]	0.49	0.839 (0.0019)	0.920 (0.0046)	81 (6.4)	4.13	0.118	0.470
Axial points									
-1	-1	1.83	0.50	0.794	0.870	76	3.06	0.145	0.226
-1	+1	1.81	0.59	0.844	0.892	48	2.76	0.100	0.438
+1	+1	1.44	0.48	0.898	0.951	53	3.46	0.093	0.485
+1	-1	1.44	0.37	0.837	0.934	97	4.47	0.122	0.512
Star points									
0	+α	1.63	0.54	0.886	0.941	55	4.13	0.085	0.518
-α	0	2.12	0.58	0.804	0.868	64	2.94	0.121	0.299
0	-α	1.72	0.40	0.810	0.909	99	4.06	0.149	0.418
+α	0	0.26	0.41	0.869	0.941	72	4.00	0.088	0.576
Extra points (cakes without sugar)									
flour +		0.86	0	0.733	0.767	34	0.98	0.219	0.126
water +		1.65	0	0.872	0.949	77	3.42	0.104	0.712

[a] amount of water before flour addition, [b] cakes at 5.5 mn, [c] standard error of repetition.
[d] population in per cent of the FID12 value

Table 3 *Prediction equations for L/FID2, ΔL and proton populations in batters and cakes (5mn) with different formulas*

Variable	Equations						r^2
L/FID12 Batters	0.84 (***)	+0.0236.W (***)	+0.0273.S (***)	+ 0.0045.S^2 (*)	- 0.001.W^2 (N.S)	+ 0.003.W.S (N.S)	0.996
ΔL	0.078 (***)	+0.0046.W (*)	- 0.017.S (***)	- 0.006.S^2 (N.S)	- 0.007.W^2 (*)	- 0.004.W.S (N.S)	0.930
Pop1 Batters	0.118 (***)	+ 0.096.W (*)	- 0.020.S (**)	+0.0007.S^2 (N.S)	- 0.005.W^2 (N.S)	+ 0.004.W.S (N.S)	0.939
Pop2 Cakes	0.47 (***)	+ 0.090.W (**)	+ 0.041.S (*)	- 0.010.S^2 (N.S)	- 0.026.W^2 (N.S)	- 0.06.W.S (N.S)	0.906

(***) P=0.0001, (**) P=0.005, (*) P=0.05, NS=not significant.

During melting a part of the starch protons would become liquid-like. However water content or more likely sugared water content would be an important restrictive factor to melting of granules in the low moisture cake without sugar or even in cakes (-1/-1, -1/+1, -α/0). ΔL was often higher in cakes with a low sugar/water ratio.

The variation of ΔL during cooking was also related to a significant modification of the transverse relaxation of the liquid-like protons (Figure 3). MEM allows to identify in standard batter three main populations. Despite a small positive effect of the water content, the population POP1 with T_2 less than 2ms was very negatively correlated to the S ratio (Table 3). This population could be assigned to water and starch exchangeable protons in starch granules or flour particles characterised by a highly anisotropic mobility. It accounted for about fifteen per cent of the liquid-like population of standard batter. The second large population (average T_2, 20ms) counts about the half of the liquid-like population. Its amplitude and average T_2 fluctuated positively with the moisture and sugar contents of the batter (figure 4a,4b). This population would correspond to the larger part of sugared water outside the flour particles: its motion would be far less affected by starch or gluten interactions than the previous one. We note that in the batter without sugar and poor in water, the second population no longer existed. The third population with a large T_2, about 100ms, is mainly due to fat protons. Its amplitude decreased to 10 to 25 per cent after baking. In batters, a part of this population could be allotted to water or sugared water not very affected by starch, for example residual droplets in fat emulsion.

MEM allows to identify another small population (about 15 per cent of POP1) with a remarkably almost constant 6 ms T_2 value situated between the two first water populations. Because it existed also in batters without sugar, this population cannot be assigned to saturated sugared water but this only sometimes merged with it in cakes (-W/-S,-W/+S). This population could be due to outside water affected by the surface of large flour particles or exchanging with water in very small or damaged granules.

After cooking, this three (or four)-modal water distribution led to a more homogenous almost unimodal T_2 distribution (figure 4c,4d). It was characterised by a population POP2 (average T_2, 5ms) situated between the two initial first populations and the T_2 value of which grows with the sugar-water content. We note that its 5ms average T_2value was very close to this of the small population at 6ms in batters. A small part of the population (T2, 20ms) subsisted in standard cakes and cakes high in water/sugar content.

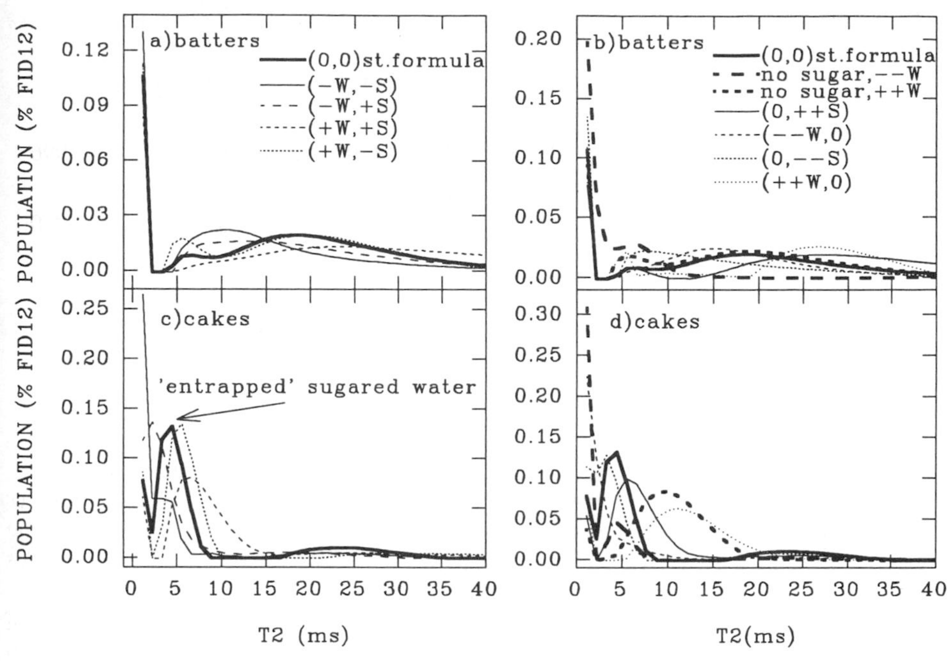

Figure 4 *Variation of continuous T_2 spectra in batters and cakes from different formulas*

This population POP2 would be related to the swelling of the granules and would correspond to entrapped water[11] in gelatinised starch. As would be expected, the amplitude of the population POP2 of cakes with very limited sugar-water content was much smaller than these of other ones (Table 4). In these cakes it is often difficult to discriminate this new population from POP1. This would be due to a limited swelling of starch in these cakes. Analysis of continuous T_2 distribution is very useful to have some comprehensive view of the modifications of relaxation time occurring during cooking, but it is time consuming. This evolution was strongly correlated to the two first PCA components: this technique could provide a rapid method to situate an unknown sample in the transverse relaxation space. Unfortunately these two data analysis methods seem much less efficient to discriminate cakes with different formulations when crumbs are too dry.

3.2 Effects of process factors on proton relaxation

All batters had similar expansion after cooking, but a small depression was observed in the upper part of the cake (220°C, 25 mm). Two samples of this part were also taken for measurements. As would be expected, coloration of cakes fluctuated with temperature and above all, cooking time. The variations of the batters cooked with the different levels of the two process factors are represented in the plot of the two first PCA components that accounted for 99% of the total variance (figure 5). Except for the upper part of the cake (220°C, 25mn), the evolution in the centre and at 1 cm of the edge was very similar to this observed in the previous experience. After 28 mn cooking time, all these samples were 'cooked' and approximatively followed the same 'curve' in the scatterplot of the two first principal components. Samples taken at 1 cm from the edge would undergo similar path but their cooking phase was faster than that of the central parts.

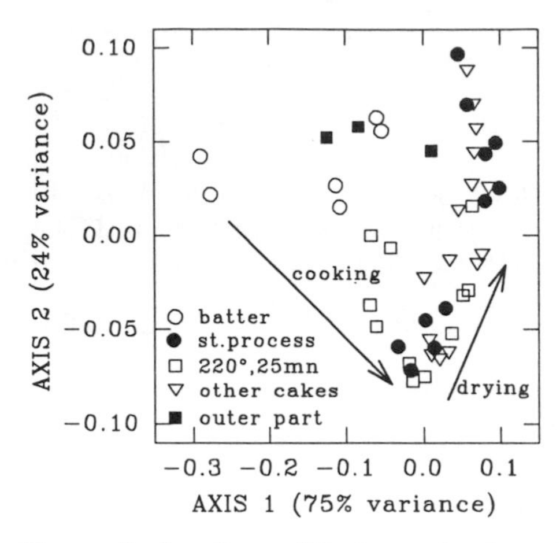

Figure 5 *locations of batter and cakes cooked with different process conditions in the plot of two first PCA components*

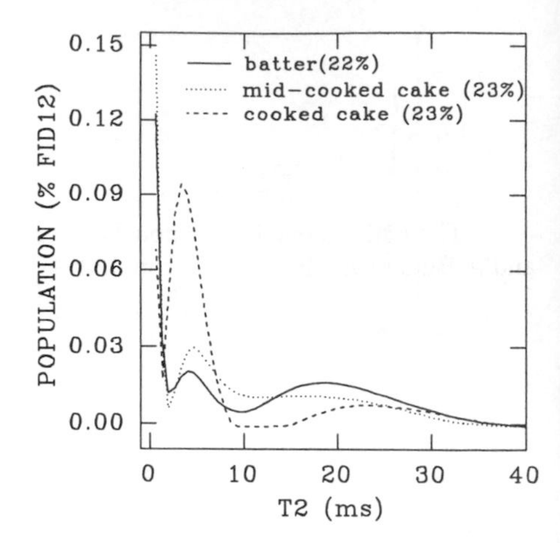

Figure 6 *Continuous T_2 spectrum of upper part of cake (200°C-25mn) compared to batter and cake*

They were more dehydrated (14-22%w.b.) as these, also they are located on the right upper part of the figure. They can still be differentiated from batter samples dried at about 16% moisture, but NMR is less sensitive at this moisture level. Except cakes cooking during 42 mn, the reducing of moisture for central parts was only from 0 to 2%. The average rise of ΔL during the cooking phase was about 7% compared to about 9% for the standard cakes in the first experimentation in which cakes were measured within two hours after cooking. This reducing could be due to some retrogradation of starch in industrial cakes delivered in a frozen state.

The cake cooked during only 25mn was a very heterogeneous product with parts at 1cm from the edge completely cooked and with a large cooking gradient from upper to centre. Mid-cooked samples taken at the upper part, situated in the centre of the figure 5, had undergone an important dehydration (about 22.5% moisture) during cooking phase. MEM fits (figure 6) show that the swelling in the upper part was very limited when we compared to continuous T_2 spectra of batter and cake at about the same moisture level.

Moisture of the samples taken on outer part of cakes was about 17% w.b.. Samples from two brown cakes had a transverse relaxation behaviour very similar to the batter drying to about the same moisture level. The other sample from a less brown cake had an intermediary behaviour. On the basis of X-ray diffraction, starch damage tests or microscopic techniques, Gordon[4] and other authors found that the starch granules from outer part of cake and from the crust of bread showed less evidence of gelatinisation than did granules from inner side.

The starch alpha-amylolysis measurements showed that the easily degradable fraction of starch , after elimination of the contribution of sugar, increased from about 15% in batter to 74-78% in central parts of cakes issuing from the standard process. This value was only equal to 52% for the central part of the cake cooked only 25 mn. Values for others cakes were generally close to this obtained for central points. For cakes (220°C-45mn; 240°C-35mn) and the cake (206°C-42mn), this value was 68% and 82% respectively. But, except for the cake (220°C-25mn), no evident relation was found between these measures of starch damage and the process factors. These limited results showed that modifications observed by NMR were in correct enough agreement with the starch damage determinations based upon enzymatic susceptibility to alpha- amylase.

Acknowledgments

The authors thank K.Benoualid from CTUC company for providing industrial cakes and A.Buleon and P.Colonna from INRA-Nantes for use of their alpha-amylolysis data

References

1. P. Colonna and M. Champ, *Sci. Aliments*, 1990, **10**, 877.
2. G. T. Carlin, *Cereal Chem.*, 1944, **21**, 189.
3. E. J. Strandine, G. T. Carlin, G. A. Werner and R.P. Hopper, *Cereal Chem.*, 1951, **28**, 449.
4. J. Gordon, E. A. Davis and E. M. Timms, *Cereal Chem.*, 1979, **56**, 50.
5. E. E. Hsu, J. Gordon and E. A. Davis, *J. Food Sci.*, 1980, **45**, 1243.
6. R. D. Spies and R.C. Hoseney, *Cereal Chem.*, 1982, **59**, 128.
7. P. O. Lin, Z. Czuchajowska and Y. Pomeranz, *Cereal Chem.*, 1994, **71**,69.
8. A. M. Abboud and R.C. Hoseney, *Cereal Chem.*, 1984, **61**, 34.
9. E. Jaska, *Cereal Chem.*, 1971, **48**, 435
10. J. Lelievre and J. A. Mitchell, *Starch,* 1975, **27**, 113.
11. A. Mora-Gutierrez, *J. Agric. Food Chem.*,1989, **37**, 1459.
12. H. K. Leung, M. P. Steinberg, A. I. Nelson and L. W. Wei, *J. Food Sci.*, 1976, **41**, 297.
13. H. K. Leung, J. A. Magnuson and B. L Bruinsma, *J. Food Sci.*, 1979, **44**, 1408.
14. P. S. Belton, *J. Agric. Food Chem.*, 1990, **2**, 179.
15. S. J Richardson, I. C Baianu and P. Steinberg,, *Starch,* 1987, **39**, 198.
16. P. Chinachoti and T. R. Strengle, *J. Food Sci.*, 1990, **55**, 1732.
17. S. J. Richardson, *J. Food Sci.*, 1988, **53**, 1175.
18. P. Chinachoti, V. A. White, L. Lo and T. R. Strengle, *Cereal Chem.*, 1991, **68**, 238.
19. H. Lim, C. S. Setser, J. V. Paukstelis and D. Sobczynska, *Cereal Chem.*, 1992, **69**, 382.
20. H. Lim, C. S. Setser and J. V. Paukstelis, *Cereal Chem.*, 1992, **69**, 387.
21. B. P. Hills, S. F. Takacs and P. S. Belton, *Molec. Phys.*, 1989, **67**, 919..
22. F Schierbaum, S. Radosta, W. Vorweg, V.P. Yuriev, E. E. Braudo and M. L. German, *Carbohydrate Polym.*, 1992, **18**, 155.
23. C. H. Teo and C. C. Seow, *Starch*, 1992, **44**,288.
24. G. E. P. Box, W. G Hunter and J. S Hunter, 'Response surface methods. Statistics for Experimenters', Wiley and sons, New York, 1978.
25. K. Ghiasi, R. C. Hoseney and E. Varriano-Marston, *Cereal Chem.*, 1983, **60**, 58.
26 M. T. Tollier and A. Guilbot, *Ann. Zootech.*, 1974, **20**,633.
27. D. Bertrand, M. Lila, V. Furtoss, P. Robert and G.Downey, *J. Sci. Food Agric.*, 1987, **41**, 299.
28 A. Davenel, N. Nathier-Dufour, D. Bertrand and P. Marchal, 'Agro-industrie et méthodes statistiques, ASSU, Montpellier, 1992, 152-155.
29. C. Tellier, F. Mariette, J. P. Guillement and P. Marchal, *J. Agric. Food Chem.*, 1993, **41**, 2259.

MRI Studies of Mass Transport during the Drying and Rehydration of Foods

B. P. Hills, V. M. Quantin, F. Babonneau, F. Gaudet, and P. S. Belton

INSTITUTE OF FOOD RESEARCH, NORWICH RESEARCH PARK, COLNEY, NORWICH NR4 7UA, UK

1.INTRODUCTION

M.R.I. is being increasingly used to follow the mass transport of water during the drying, rehydration and processing of foods and recent applications have been been reviewed (1,2). Although conventional 2-dimensional imaging techniques developed for clinical purposes can be applied to food materials there are a number of difficulties that emerge when quantitative studies of moisture transport are required. One such difficulty is the shrinkage/expansion of the food during the processing which destroys the simple linear relationship expected between the proton spin density $M(t=0)$ and the gravimetric water content. Another difficulty is the existence of large internal field gradients generated by susceptibility discontinuities. A third major difficulty is the speed of most food processing operations such as baking which necessitate the development of rapid imaging protocols.

In this paper the problems associated with sample shrinkage and susceptibility discontinuities are first illustrated for a well-characterized model sample consisting of a bed of Sephadex microspheres containing varying amounts of water. A fast multiple echo radial imaging protocol is then presented for overcoming these problems and the technique is used to follow the ingress of water during the rehydration of extruded pasta. The resulting time dependent moisture profiles are then modelled numerically for three types of extruded pasta made with differing percentages of hard and soft wheat. It is shown that the rehydration is an example of non-Fickian diffusion where increasing the percentage of hard wheat shifts the diffusion closer to the Case II diffusion limit.

1.1 Single-echo imaging of foods with varying water content

Figure 1 shows a simple slice-selective single-echo imaging sequence which can be used to obtain an image projection in the x direction. A two dimensional, "spin warp" image is obtained by inserting phase-encoding gradients in the y-direction, but for present purposes these will be neglected. The image intensity, I, observed with this spin-echo sequence can be written as

$$I = M_0 . [1 - \exp(-TR/T_1)] . R(t_a,t_b) . DIFF(t_a,t_b) \qquad [1]$$

where M_0 is the initial water proton magnetization and $R(t_a,t_b)$ and $DIFF(t_a,t_b)$ are the attenuating effects of transverse relaxation and diffusion respectively. The time delays

t_a and t_b are identified in figure 1. The repetition time, TR, is usually chosen to be at least 5 times the longitudinal relaxation time, T_1, so the factor in square brackets can be neglected. For a homogeneous liquid like pure water equation [1] can be made more explicit since

$$R(t_a,t_b) = \exp\{-2(t_a+t_b)/T_2^*\} \qquad [2]$$

and

$$DIFF(t_a,t_b) = \exp\{-\gamma^2 G^2 D\, t_a^2[2(t_a+t_b)-4t_a/3]\} \qquad [3]$$

Equation [3] expresses signal attenuation due to diffusion between the two read-out gradient pulses of amplitude G. Examination of equations [2] and [3] shows that simple variation of t_b at fixed t_a will give an exponential decay but with a time constant that is neither T_2 nor T_2^* since the DIFF term will cause diffusive attenuation from the read-out gradient pulses. This fact is often overlooked when "T_2-weighted" images are generated with conventional spin-warp imaging. In heterogeneous food materials where diffusion is restricted, it is no longer, in general, valid to assume that relaxation and diffusion are simply factorizable. Nevertheless, for simplicity, this complication will be ignored and it will be assumed that equation [1] remains applicable at all water contents, with M_0, $R(t_a,t_b)$ and $DIFF(t_a,t_b)$ dependent on water content W, defined as weight of water per unit mass of dry food sample. We first test the conventional assumption that the initial magnetization, M_0, provides a useful measure of water content by considering a model sample consisting of packed beds of Sephadex G25-300 with varying water content from the (almost) dry powder to the fully saturated suspension. This sample was chosen because Sephadex exhibits many of the features typical of real foods. For example there is volume shrinkage at lower water contents and the system is compartmentalized on a microscopic distance scale of one to several hundred microns. Like many food biopolymer systems the intrinsic water proton relaxation mechanism is dominated by proton exchange for which quantitative theoretical models are available (3). At lower water contents the presence of air gaps in the sample creates large local internal susceptibility gradients which can cause severe dephasing.

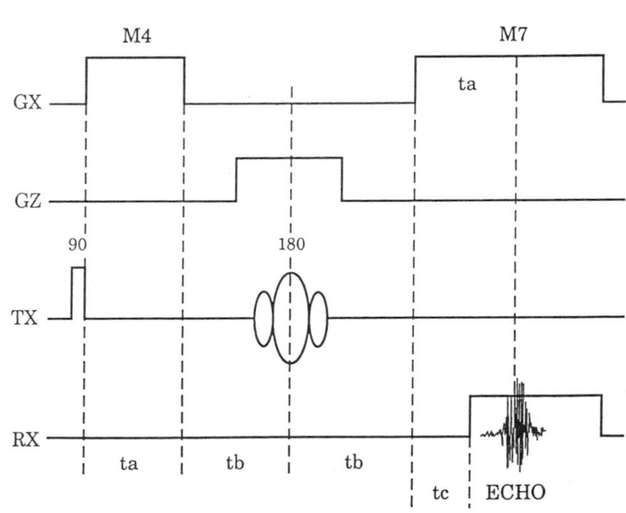

Figure 1. A simple one-dimensional, slice selective single echo imaging sequence.

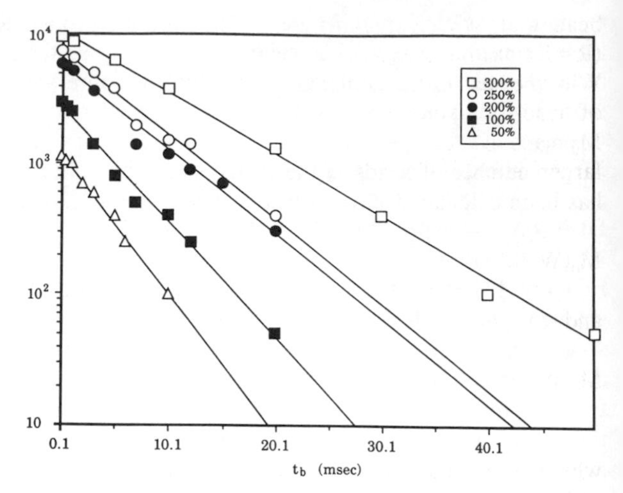

Figure 2. The log-linear dependence of the signal intensity (arbitrary units) obtained with thesingle-echo sequence in figure 1 for packed beds of Sephadex G25-300 beads with the indicated water contents. Intensities are plotted against t_b at fixed t_a (2.1 msec.).

1.2 The initial magnetization, $M_0(W)$, for beds of Sephadex G25-300 with single echo imaging

Figure 2 shows the log-linear dependence of the signal intensity, I, on the increment in t_b at fixed t_a and field gradient, G, for beds of Sephadex G25-300 at several water contents W. The intercepts in figure 2 can be equated with $M_0(W)$ and figure 3 shows a plot of the $M_0(W)$ against water content, W. It is clear that $M_0(W)$ is not a simple linear function of water content but shows some unusual departures from linearity, arising from changes in bead radius and packing density as the water content is changed. Bead shrinkage was analysed in a previous paper (2), where an "Osmotic-capillary model" was developed to account for the dependence of Sephadex bead radius, R, on water content, W, defined as the mass of water per unit mass of dry Sephadex powder. It was shown that

$$(R/Rmin) = [1+ W.n]^{1/3}, \quad 0 \leq W \leq Win \qquad [4]$$

where Rmin is the radius of a dry Sephadex bead, n is the specific density of a dry

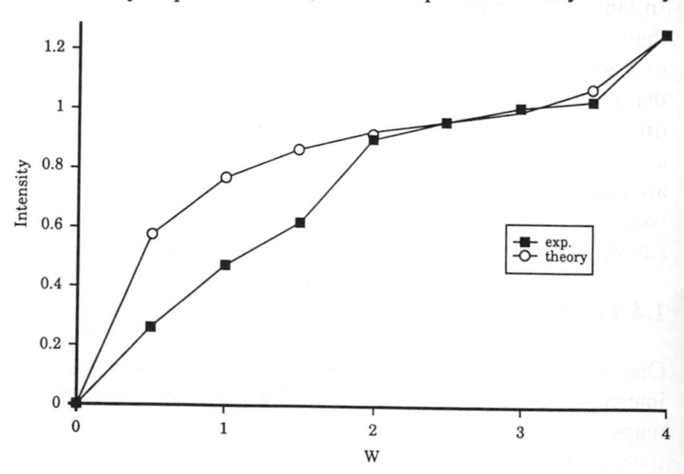

Figure 3. The dependence of M_0 on W derived from figure 2. M_0 has been normalized for $W=W_{in}$. The theoretical curve is calculated with equations [5] and [6].

bead and Win is the critical water content when the beads are fully swollen (R=Rmax) but there is no water outside the beads. For water contents greater than Win, the bead radius remains constant at its maximum value, Rmax. This dependence of bead radius on water content has important implications for the interpretation of M_0 maps in slice-selective imaging because loss of water is partly compensated by the larger number of beads in the image sampling volume. This "compensation effect" has been calculated in reference 4 where it was shown that

$$M_0(W)/M_0(Win) = (W/Win).[\{1+nWin\}/\{1+nW\}] , \quad \text{for } 0 \leq W \leq Win. \qquad [5]$$

and

$$M_0(W)/M_0(Wsat) = W/Win, \quad \text{for } Win \leq W \leq Wsat. \qquad [6]$$

where Wsat is the water content of the saturated, packed suspension. The dependence of $M_0(W)$ on water content for Sephadex G25-300 predicted by equations [5] and [6] for the experimental values of Win and Wsat (4) is shown in figure 3. The theory accounts reasonably well for water contents, $W>2$, but overestimates M_0 at water contents below 2. A possible reason for this is seen in the light microscope which shows that at lower water contents the beads not only shrink but tend to clump together in aggregates. If these aggregates pack together randomly as large "pseudo-spheres" the packing density would be considerably lower than the regular close-packed array assumed when deriving equation [5]. Because of these "compensation" and "aggregate packing" effects the initial magnetization, M_0, is clearly not a reliable measure of water content. This is a significant point, because similar effects can occur when using slice-selective pulse sequences with real food systems such as cellular tissue, pastas and gels, all of which undergo shrinkage and possibly structural reorganization as the water content is reduced.

1.3 The relaxation and diffusive terms for beds of Sephadex G25-300 for single-echo imaging.

In table 1 the experimental slopes derived from figure 2 for water contents W less than Win are compared with those calculated with equations [1] to [3] for variation of t_b at fixed t_a and G using bulk measurements of $T_2(W)$ and D(W). It can be seen that the experimental slopes are very roughly twice the theoretical values and that the difference increases with lower water contents. This discrepancy is undoubtedly a consequence of large internal magnetic field gradients generated across the Sephadex-air interface, which are not included in the theoretical calculation. These gradients would be expected to increase as the beads shrink at lower water contents. Similar susceptibility effects are expected in most heterogeneous food materials.

1.4 The Multiple-echo Imaging Protocol

One method for overcoming the difficulties of sample shrinkage/expansion and internal susceptibility gradients is to measure T_2-weighted images and relate the T_2-maps to gravimetric water profiles using a T_2 versus W calibration curve obtained from bulk, uniformly hydrated samples. The multiple echo pulse sequence shown in

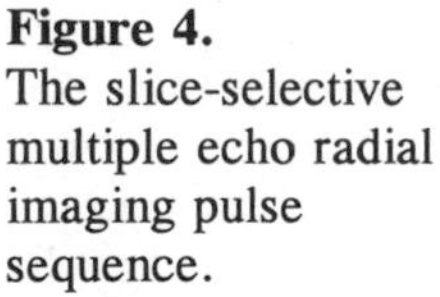

Figure 4.
The slice-selective multiple echo radial imaging pulse sequence.

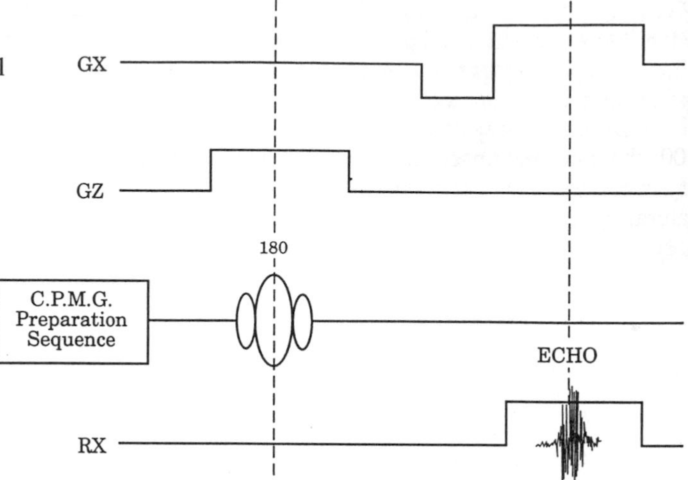

figure 4 is suitable for this since it reduces the dephasing effect of diffusion by placing the initial pulsed read-out gradient of amplitude +GX in the Hahn-echo imaging sequence on the other side of the selective 180 degree pulse. The dephasing effects of inhomogeneous background gradients and internal susceptibility gradients are minimized by keeping t_b short (e.g. 100usec). It is important to note that because of proton exchange and susceptibility discontinuities, T_2 in most foods depends on the CPMG pulse spacing, τ. The image intensity therefore needs to be measured as a function of increasing numbers of echoes in the CPMG preparation sequence, while keeping the pulse spacing fixed at a short value.

1.5 The multiple echo radial imaging protocol

Most food samples can be cut or moulded into a cylindrical geometry, which is the most convenient geometry for MRI purposes since magnet bores, probeheads and NMR sample tubes are usually cylindrically symmetric. In this geometry the read-out gradient in the 1-dimensional multiple echo pulse sequence in figure 4 can be applied either along the axis of the tube or across the tube. In the later case one dimensional Fourier transformation, F{ },of the echo generates an image projection rather than the radial profile which is of interest. In principle, Hankel transformation, H{ }, of the echo yields the radial profile directly, but it is considerably easier to first Fourier transform the echo and then convert the projection into the radial profile with an inverse Abel transform, IA{ } (5). These relationships can be summarized as

$$\text{image}(r) = H\{\text{fid}(k)\} = IA\{\text{proj}(x)\} \qquad [7]$$

where

$$\text{proj}(x) = F\{\text{fid}(k)\} \qquad [8]$$

Here fid(k) is the free induction decay or echo acquired in the field gradient, proj(x) is the image projected along the x axis, image(r) is the desired radial profile and k is the wavevector γGt. This radial protocol increases the speed of image acquisition since no phase encoding gradients are needed.

Figure 5.
Experimental radial profiles of two concentric NMR tubes of 5 and 10 mm. diameters. The inner tube contains a bed of Sephadex G25-300; the outer tube pure water. The top profile corresponds to water-saturated Sephadex; the bottom profile to a W of 100%w/w.

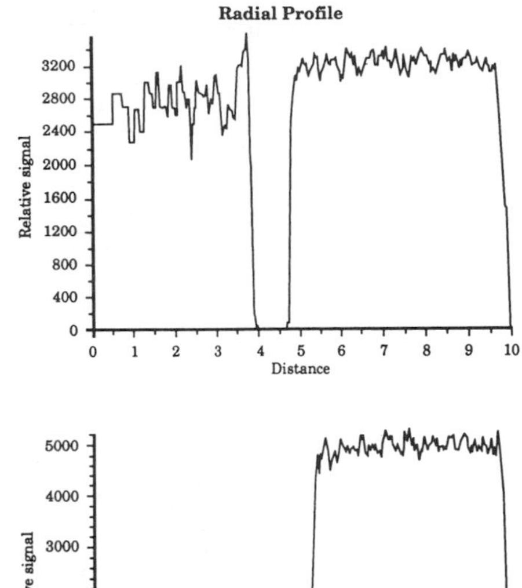

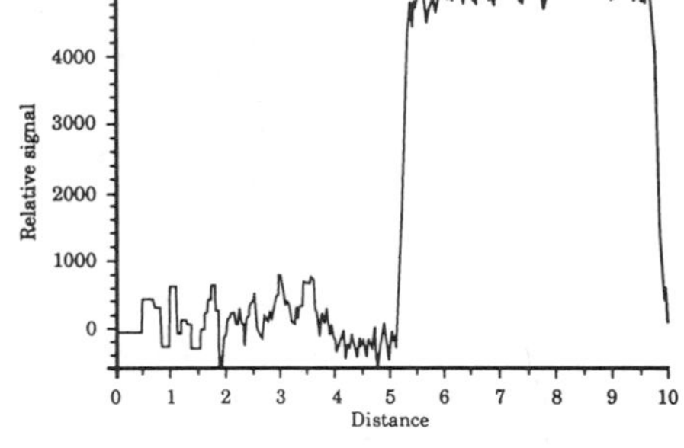

Figure 5 shows the experimental radial profiles derived from two concentric tubes, the outer one containing water as an intensity standard, the inner one Sephadex G25-300 at two water contents. Only the right hand side has been processed in these radial profiles.

2. APPLICATION OF THE MULTIPLE ECHO RADIAL IMAGING PROTOCOL TO THE REHYDRATION OF EXTRUDED PASTA

The multiple echo radial imaging protocol was used to investigate the rehydration kinetics of three types of extruded pasta (spaghetti) differing in the percentage of hard and soft wheat used in their extrusion.

2.1 Sample arrangement during processing

In these experiments a vertical, 2cm wide microimaging probehead, thermostated at 298K, was used throughout, but because pasta expands during rehydration it was important that the sample is able to change its length in the imaging probehead without bending or breaking. This was achieved by attaching the bottom of the pasta to a small cylindrical plastic cap with a diameter slightly smaller than the vertical R.F coil. The sample and cap were then suspended down the centre of the coil. The lower half of the sample (including the cap) was first immersed in water at 83°C for a known time period then removed, cooled in cold water to stop further rehydration, and transferred to the imaging probe head.

2.2 Radial and length expansion rates

The expansion rates of the three pasta samples were measured using a travelling light microscope. The resulting radial expansion provided an independent check on the

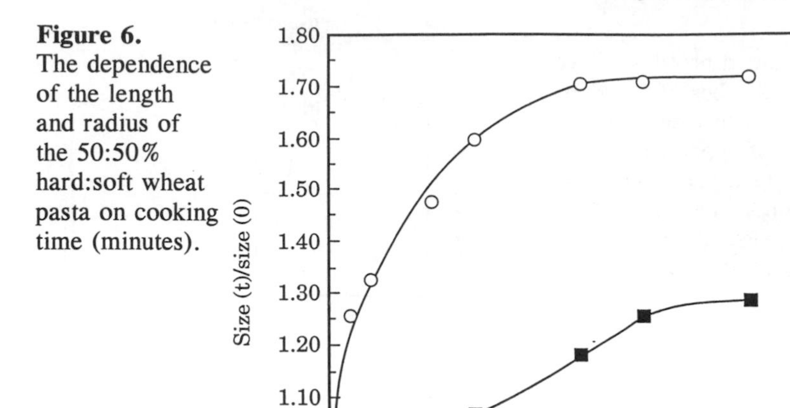

Figure 6. The dependence of the length and radius of the 50:50% hard:soft wheat pasta on cooking time (minutes).

expansion observed with the NMR microimaging. The length expansion could not, of course, be measured in the slice-selective imaging experiment. Figure 6 shows the time dependence of the radius and length of the 50:50% hard:soft wheat sample after immersion in water at 83°C for the specified periods. The length expansion is delayed compared to radial expansion because of the existence of a hard inner glassy core of unhydrated pasta, which is only rehydrated in the final stages of cooking. This difference implies the existence of stress pressures inside the pasta during rehydration and suggests that the kinetics may well depend on the sample aspect ratio. Fortunately no stress-induced cracking or splitting of the sample was found in these samples, though it has been observed when rehydrating pastas of other morphologies. Very similar results were obtained for the pastas prepared from 100% soft and 100% hard wheat. The increase in sample length means that the total integrated signal intensity in the absence of dephasing by relaxation and diffusion in the slice-selective imaging volume is no longer proportional to the water content (by weight). Moreover the changes in signal intensity at any fixed radial distance are caused both by water penetration and by radial expansion which must be taken into account in any theoretical interpretation of the imaging results.

2.3 The bulk rehydration kinetics

Conventional rehydration or drying studies do not have access to moisture profiles and rely exclusively on bulk rehydration/drying curves obtained by weighing the sample as a function of rehydration/drying time. In figure 7 the total weight has been plotted against the square root of the rehydration time. For one-dimensional diffusion into a cylinder, under perfect sink initial and boundary conditions, with a constant penetrant diffusion coefficient and constant cylinder radius the mass increase can be written in the form (6)

$$M(t)/M(t=\infty) = Kt^n \qquad [9]$$

Figure 7.
A comparison of the experimental and theoretical bulk rehydration curves for 50:50% hard:soft wheat pasta. Units of minutes$^{0.5}$ have been used.

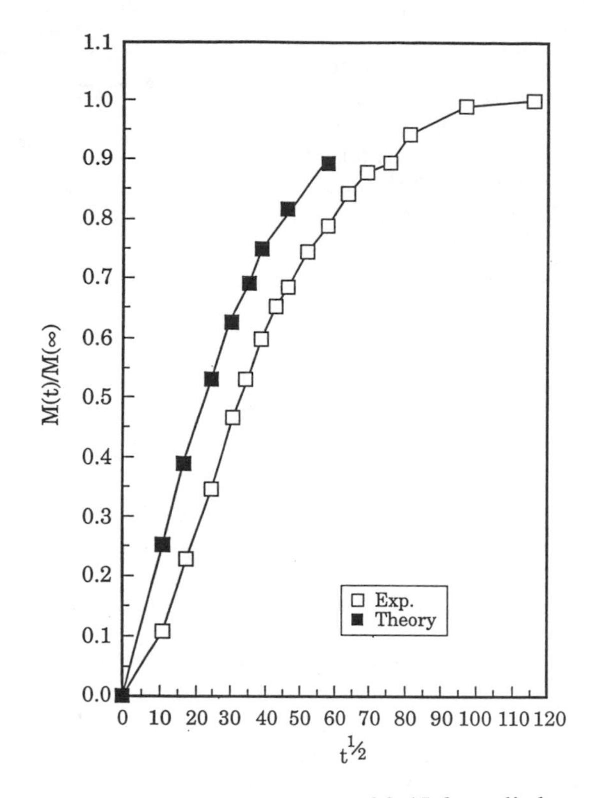

Case I Fickian diffusion is then characterized by an exponent of 0.45 for cylinders, (7), which applies when the starch glass to gel transition rate is fast compared to the water diffusion. In the opposite limit of slow glass to gel transition rates compared to diffusion (Case II diffusion) n=0.89 for cylindrical geometry and there is a transitional solvent "front" at the glass-gel transition that progressively moves into the sample. The experimental values of the exponent n obtained by fitting the initial linear part of the Ln $\{M(t)/M(t=\infty)\}$ versus $t^{0.5}$ plots fall between these values and are characteristic of non-Fickian diffusion (see table 2). Indeed the slight sigmoidal shape of the curve expected for non-Fickian diffusion is apparent in the data of figure 7. Table 2 shows that the hard wheat sample with n=0.71 falls nearest to the pure case II diffusion limit, whilst the 50:50 and soft wheat samples have progressively smaller n values, though even the soft wheat sample (n=0.59) is still markedly non-Fickian. This trend in the exponent n is consistent with a slower glass-to-gel transition rate for the hard wheat sample than for the soft wheat sample. It is worth noting that although the n-values are widely quoted in the literature, they do not have the reliability that is often attached to them. They depend on the morphology of the sample (whether lamellar, cylindrical or spherical) and are calculated assuming constant radii and constant diffusion coefficients. Neither of these assumptions is valid for the pasta samples considered here. Even if these assumptions were valid, an equation of the simple polynomial form Kt^n rarely applies to more than 60% of the fractional solvent uptake after which additional terms are required. To really study the nature of the rehydration kinetics it is necessary to monitor the time course of the moisture profiles.

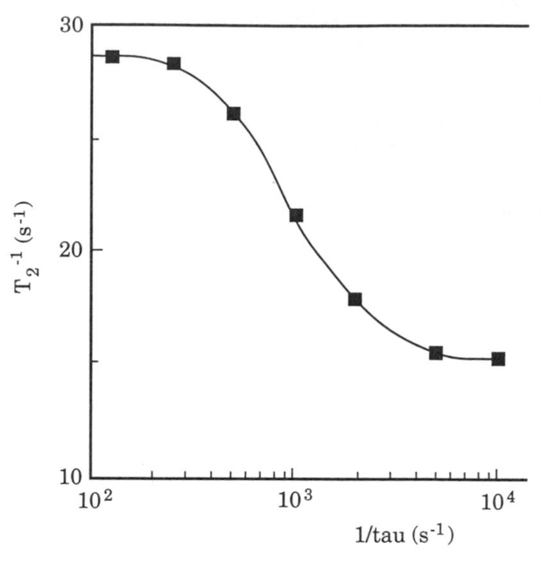

Figure 8.
The experimental water proton transverse relaxation dispersion for 100% soft wheat pasta uniformly rehydrated to 70% water w/w. Here tau is the C.P.M.G. 90-180^0 pulse spacing.

2.4 Non-spatially resolved transverse relaxation

To convert the observed relaxation time radial profiles to moisture profiles it is necessary to measure the dependence of relaxation time on moisture content in bulk, uniformly hydrated samples. Figure 8 shows the transverse relaxation dispersion for the main water peak in the soft wheat pasta rehydrated uniformly to W=0.7 where W is the water content defined as weight of water per unit weight of the hydrated sample. This dispersion was obtained by varying the 90-180^0 pulse spacing in the CPMG pulse sequence, keeping all other parameters constant. The classic sigmoidal frequency dependence is typical of proton exchange between water and mobile carbohydrate components. These mobile components could either be low molecular weight oligosaccharides trapped in the starch matrix, or possibly, plasticized side chains. To avoid ambiguity in the imaging experiments the pulse spacing in the CPMG preparation sequence was therefore fixed at 200μsec. The transverse relaxation rate of the main water peak measured with a 200μsec pulse spacing was found to have a log-linear relationship with solid content, (1-W), such that $\log(1/T_2) = \alpha+\beta(1-W)$. The coefficients α and β are listed in table 2 for the 3 pasta samples. The minimum echo time in the radial imaging sequence is ca. 8 msec, which sets a lower limit of ca. 2-3 msec on the shortest T_2 that can be detected in the imaging experiment, which corresponds to a minimum water content of ca. 10-15% by weight.

2.5 The Radial Moisture Profiles

Figure 9 shows an example of the moisture profile obtained using the calibration data in table 2 and the radial expansion data in figure 6. Note that because of the lower limit of ca. 8 msec in the T_2 measurements that points having "S=0" are actually points with S<0.1 . The radial profiles all show considerable "noise". In part it is caused by the "graininess" of the pasta samples which can be seen in the dry pasta samples and probably arises from wheat fibres and grains that have not been evenly gelatinized by the extrusion process. The other factor is the poor signal to

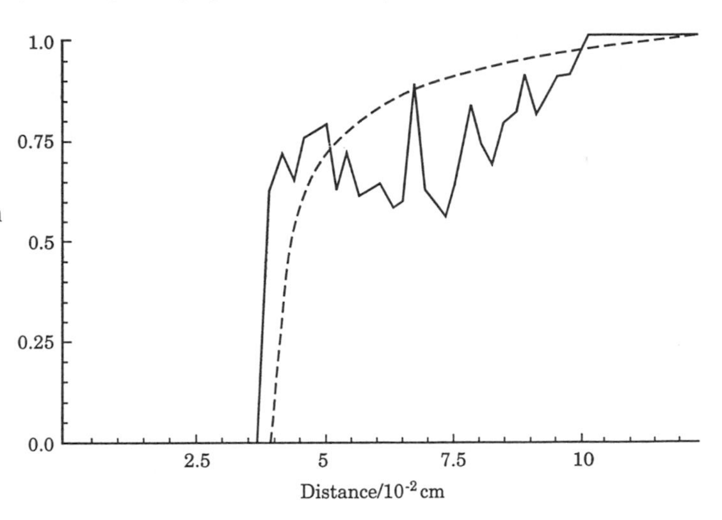

Figure 9. The radial moisture profile for 100% hard wheat pasta after 7 minutes of rehydration. The degree of saturation is plotted against radial distance. Dashed line is the theoretical fit.

noise at low water contents, which is exacerbated by the Inverse Abel Transformation, especially at small radial distances.

3. THEORETICAL MODELLING OF THE MOISTURE PROFILES

The moisture profiles were modelled using numerical, finite-element solutions of the diffusion equation in a cylindrical geometry, taking into account the radial expansion of the sample and dependence of the diffusion coefficient on local water content.

3.1 Modelling radial expansion

A simple model for radial expansion can be introduced by course-graining the calculated moisture profile into groups of 10 mesh points in the simulation program. Explicitly this means that if the radial profile is characterized by a total of 1000 mesh points in the simulation then this profile is course-grained into 100 units of 10 mesh points. If the average degree of saturation S $(=W/W_{sat})$ of a 10-meshpoint group exceeds some critical value S_{crit} then the number of meshpoints in the group is increased by an integral number calculated from the experimental increase in radius $(R_{max}/R_{min})$and all meshpoints in the group are assigned the average degree of saturation. Once a meshpoint group has been expanded it cannot expand again. This ensures that the final, fully hydrated pasta has a radius equal to the measured R_{max} after full rehydration. After an expansion step the new moisture profile is used as input for the diffusional part of the calculation. To ensure consistency the number of expansion steps is increased until there is no significant change in the final moisture profile or radius.

3.2 Modelling non-Fickian Diffusion

The effect of varying moisture content on the local self-diffusion coefficient was incorporated by writing

$$D\{S(r,t)\} = D_0 \exp\{ A.S(r,t) \} \qquad [10]$$

where $S(r,t)$ $(=W(r,t)/W_{sat})$ is the local degree of saturation at time t and radial position r. D_0 is the hypothetical water self-diffusion coefficient in the "dry" pasta (S=0) and A is a constant which can be determined by fitting the experimental profiles. The two parameters D_0 and A are also related to the water self-diffusion coefficient in the fully hydrated pasta, D_{sat}, since $A = Ln(D_{sat}/D_0)$. Figure 9 includes the theoretical profiles used to fit the moisture profile and the radial expansion.

3.3 Theoretical values of the exponent n

As well as fitting the moisture profiles and the radial expansion data the same model was used to calculate the exponent n characterizing the increase in the total mass of water in equation [9]. The theoretical data points in figure 4 were calculated in this way from the integrated areas of the theoretical moisture profiles. Table 3 lists the sets of parameters D_0, A and n(theory) obtained from the model for the three pasta types. It is apparent from these results that as the amount of hard wheat in the pasta increases the rate of water penetration (D_0) decreases and the diffusion becomes more non-Fickian (n increases). Despite these reasonable deductions the theoretical fit to the data can still be considerably improved by refining the model. For example an improved model can be devised by introducing relaxation times associated with the diffusion coefficient and expansion mechanisms (6). This may also explain the apparent delay in the experimental plot in figure 4 compared to the theoretical results. However, incorporating these refinements also increases the number of parameters in the model, so has not been attempted in these preliminary investigations.

4. DISCUSSION

The main objective of the present study has been to demonstrate the practicality of the multiple echo radial imaging approach in a real food processing situation such as the rehydration of extruded pasta. For this reason the study has not tried to be exhaustive and there are many extensions of the radial imaging protocol that have yet to be attempted. For example radial profiles of the apparent water diffusivity could be measured by replacing the CPMG preparation sequence in figure 1 with a diffusion weighted pulsed gradient sequence. In this way it should be possible to directly compare profiles of the water self diffusion coefficient with the moisture content profiles and establish directly the dependence of the water diffusivity on water content. This is preferrable to postulating some dependence such as equation [10] and then deriving parameters by fitting the moisture profiles.

Provided either the relaxation times and/or the water self diffusion coefficient are temperature dependent MRI can, in principle, be used to image temperature distributions (8). This should therefore be possible to measure radial temperature profiles with the above radial microimaging protocol. These, in turn, could be modelled in terms of the thermal diffusivity in a similar way to the water diffusivity in this paper.

5. ACKNOWLEDGMENTS

Samples of the three types of pasta were kindly supplied by Prof. Jean-Claude Autran of the Laboratoire de Technology des Cereales, INRA. F.Babonneau and V.M.Quantin wish to thank the University of Nantes for the award of an ERASMUS studentship to pursue this work. F.Gaudet gratefully acknowledges the support of the Institut National Agronomique, Paris-Grignon for participation in this project as part of his second year study programme.

Table 1. *Theoretical and experimental slopes for single echo imaging of Sephadex G25-300 beds of varying water content, W.*

Water Content W	*Theoretical slope Without susceptibilty Gradients (sec-1)*	*Experimental Slope (sec-1)*
1.0	77.7	203.5
1.5	67.6	172.7
2.0	62.5	149.9
2.5	59.4	150.2
3.0	57.4	109.5

Table 2. *Parameters in the non-Fickian diffusion model*

	$D_0(cm^2s^{-1})(*10^{-9})$	A	n(exp)	n(theory)
Hard	0.3	7.0	0.71	0.59
50:50	3.0	3.7	0.66	0.44
Soft	4.5	5.2	0.59	0.27

Table 3. *Parameters in the T_2 calibration equation*
$Log(1/T_2) = \alpha + \beta(1-W)$

	α	β
Hard	0.28	2.85
50:50	0.43	2.55
Soft	0.47	2.63

References

1. P.S.Belton, I.J.Colquhuon and B.P.Hills., *Ann. Rep. NMR Spec.*, 1993, **26**, 1.
2. M.J.McCarthy, Magnetic Resonance Imaging in Foods, Chapman and Hall, London, 1994
3. B.P.Hills and F.Babonneau, *Mag. Reson. Imag.*, (in press)
4. B.P.Hills and F.Babonneau, *Magn. Reson. Imaging*, vol. **12** no. 6 (1994), in press.
5. R.N.Bracewell, The Fourier Transform and its applications, McGraw-Hill, 1978 .
6. J.Crank, The Mathematics of Diffusion, Oxford Univ. Press, Oxford, 1975.
7. N.A.Peppas and L.Brannon-Peppas, *J.Food Eng.*, 1994, **22**, 189
8. X.Sun, J.B.Litchfield and S.J.Schmidt, *J.Food Sci.*, 1993, **58**, 168.

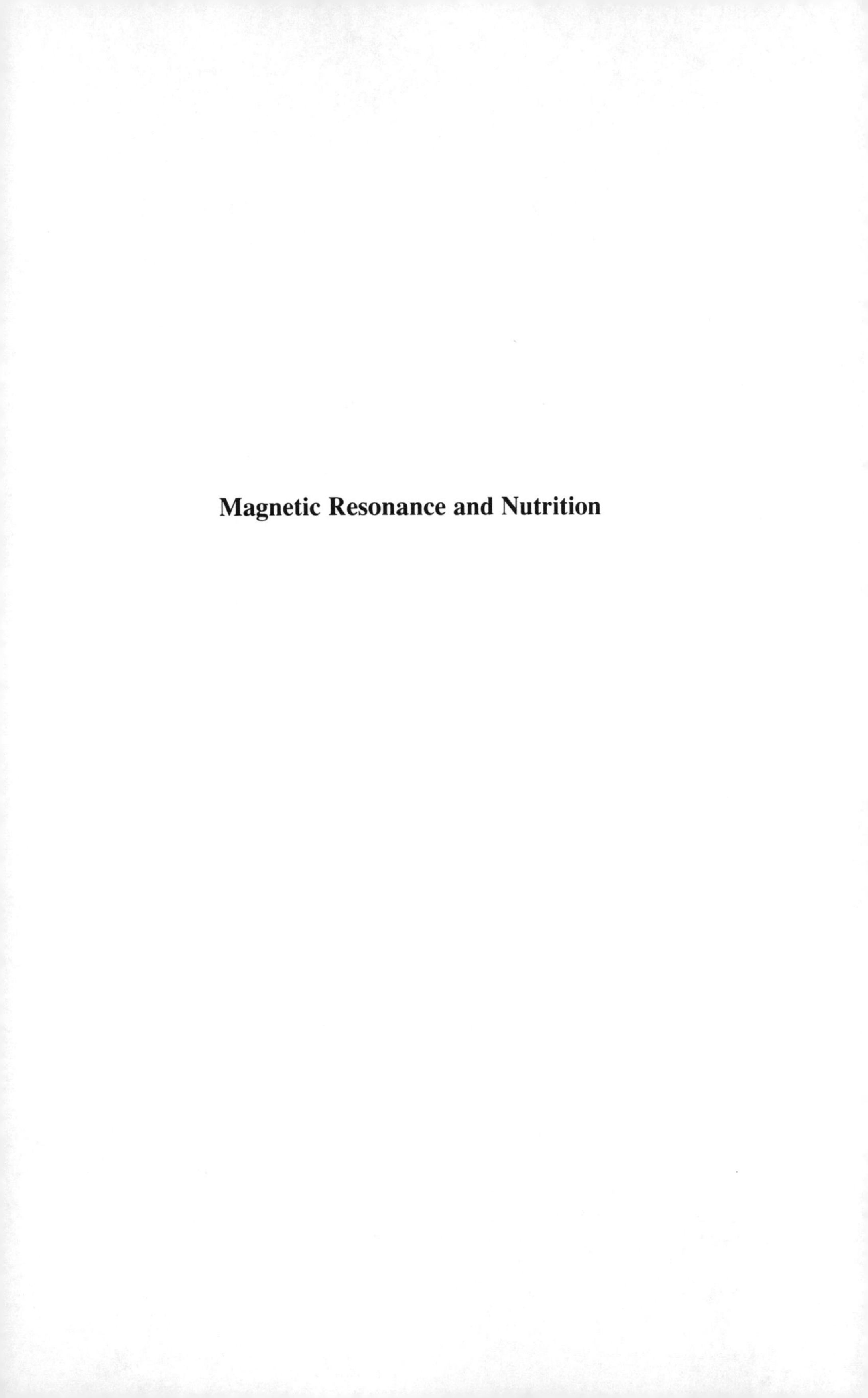

Magnetic Resonance and Nutrition

Important Issues in the Understanding of Human Metabolism in Health and Disease: the Role of NMR

P. J. Aggett

INSTITUTE OF FOOD RESEARCH, NORWICH RESEARCH PARK, COLNEY, NORWICH NR4 7UA, UK

I will outline an overall strategy of nutrition, delineate the principal problems prevalent in nutrition in developed countries and, hopefully, stimulate thoughts and discussion as to areas where nuclear magnetic resonance spectroscopy and imaging could advance the science.

1. THE IMPORTANT NUTRIENTS

The central nutrients in human nutrition are regarded as (i) protein or more specifically amino acids of which eight are essential, and possibly six more, can be regarded as being conditionally essential in that they are essential for optimum health under certain conditions of physiological development and supply of other nutrients in the diet; (ii) carbohydrate, both refined carbohydrate e.g. (glucose) and the broad range of complex carbohydrates which are present in the diet and which are derived from the plant cell wall structures, glycogen etc; (iii) lipids, here again, as with proteins, lipids cannot really be regarded as a single entity. At least two fatty acids are essential these are linoleic (18:2 n-6) and α-linolenic (18:3 n-3).[1,2]

In addition to these three major substrate groups there are a variety of major elements such as calcium, magnesium, sodium, potassium and phosphorus which are essential and micronutrients, which include the vitamins, the trace elements and numerous bioactive compounds, the importance of which are becoming increasingly appreciated. This knowledge has now extended to a situation in which it could be envisaged that in the near future that a number of these compounds such as flavonoids, carotenoids, and glucosinolates may become perceived as being essential or certainly desirable components of the diet.

2. THE 'SCIENCE' OF NUTRITION

It is worth considering just what is involved in nutrition. It is a very applied discipline and it can be briefly defined as "the interaction between an organism and the food components on which it depends". Thus as much as nutrition is an applied science it is also an art: the art of drawing together knowledge and expertise of many seemingly unrelated disciplines. Within the definition given above the entire food chain, from bedrock, geochemistry, hydrology, soil chemistry through botany, agriculture, food processing and delivery, to human socio-economic, political and biological sciences all can be envisaged to be in the legitimate realm of nutritionists of some ilk or other. All of these disciplines can inform nutritional science and policy. Thus, on one hand nutritionists are dependent upon those who characterise and manipulate the environment, such as geochemists, geologists, civil engineers, and those who control or fashion the socio-economic milieu in which the food supply operates. This would include economists, politicians, non-government organisations, international agencies, those who process and preserve foodstuffs within the food industry; those who try and impart aspects of nutritional education to the consumer through the media and through education; consumer scientists, agriculturalists, botanists, zoologists, geneticists chemists, biochemists, biotechnologists, and those such as the clinical and paraclinical professions, including dieticians who monitor the impact which food has on consumer populations. Finally, and not least, there is, of course, the need to understand the attitude of the consumers themselves. This wide array of expertise has an appreciable influence on many aspects of nutrition. The contribution of NMR in areas of biotechnology and food science would be well appreciated by this audience and it is my intention to focus more on the metabolic interaction of food and the consumer.

3. IMPRESSIONS OF NUTRITION

There are at least two popular sayings concerning nutrition. Both of which contain only an element of truth. The first is that "we eat food, not nutrients". The riposte to this is that although we eat food we absorb nutrients and food constituents, not the foods themselves. It is the function of the gastrointestinal tract to digest ingested food and to ensure the absorption of needed food components. As such the gut should act as a selective barrier, but it is sometimes imperfect in this respect, resulting in excessive exposure to undesirable and potentially toxic food constituents, and sometimes being the victim, itself, of adverse reactions to foods. The gastrointestinal tract also plays an important role in immunotolerance of foodstuffs. One of the most amazing things about our diet is not so much that we occasionally experience adverse immunologically mediated reactions to food components, but that, by and large, we do not do so.

Relatively little is known about the fate of food per se within the gastrointestinal lumen. This is because the gut is not easily accessible to techniques for characterizing intestinal flow, digesta mixing and digestion.

Of particular importance is the actual nature of ingested food and digesta through the intestinal tract. The presence of folds in the intestinal lining create a barrier to flow which, as well as increasing surface area, might induce a turbulence which would be contrary to the common model of assuming that the intestinal contents follow a characteristic laminar flow. The gut mucosa's surface area is increased also by finger-like projections (villi) on the mucosa which increase the surface area 12-fold, and the surface area is yet further increased by the presence on the absorptive cells, the enterocytes, of numerous small fingerlike projections (microvilli), overall these devices increase the surface area available for nutrient absorption by about 1600-fold. A number of specific mucosal carriers have been identified in the polar and baso-lateral surfaces of the enterocytes. These effect the selective uptake and transfer of nutrients to the body (i.e. absorption); but relatively little is known about the actual processes involved. This applies, in particular, to lipids and to liposoluble compounds. In fact the entire process of the formation and stabilisation of intestinal luminal emulsions and the subsequent formation of micelles involving lipolytic products and bilesalt surfacants is understood poorly. Clearly the mechanisms involved and the subsequent mucosal uptake of lipophilic compounds lend themselves to investigation by biophysical techniques particularly NMR spectrocopy.

The characteristics of the flow of food digesta through the intestinal luman is also poorly understood. The possibility that mucosal folds induce a turbulent rather than a laminar flow (see above) occasions interest in the potential of NMR techniques in exploring the existence or nature of an unstirred water layer at the mucosal surface which might present a barrier to the diffusion and uptake of nutrients by the intestine.

The second popular but inaccurate saying is that "we are what we eat". Numerous nutrient gene integrations occur to achieve specific mechanisms for transformation of absorbed dietary constituents into substrates for energy and production of tissue components. These are the process of metabolism, but the precise nature of the nutrient gene interactions involved has not been characterised and so little is known about the mode in which carrier and control mechanisms are up-regulated and down-regulated to ensure the optimum supply, utilisation and degradation of absorbed dietary constituents. It is this fascinating aspect of metabolism which underlies the sustenance of the normal healthy interaction with diet, or leads, with inappropriate dietary intake, to chronic disease states. The involvement of numerous nutrient gene interactions clearly raises the possibility, with such polygenetic processes of idiosyncratic and inherited defects in metabolism arising from genetic polymorphism. Some of these are evident in simple autosomal inborn areas of metabolism and, more recently, the polygenetic nature of disturbances in the metabolism of lipids and in the synthesis and function of their carrier molecules, apolipoproteins, have become evident. Here again the nature of the systemic carrier molecules and their interactions with specific receptors at cell surfaces and intracellular organelles is a rich area for research using biophysical expertise.

4. CONCERNS IN NUTRITION

The major issues in nutrition can be summarised as those basic questions relating to the quantitative and qualitative requirements for specific nutrients. These are very poorly characterised and although many committees have sat to determine ideal nutrient intakes and have published recommendations, such as Dietary Reference Values, and Recommended Dietary Intakes, or Recommended Dietary Allowance, the evidence on which these are based is quite sparse.[2] Such recommendations are essentially inductive and are frequently extrapolated from studies done on limited numbers of individuals and, frequently, just on male subjects. The better determination and characterisation of ideal nutrient intakes to maintain optimum health would depend, in turn, on a better characterisation and understanding of human metabolism. The basic questions in this area are essentially:

What do individual nutrients and non-nutrient food constituents do to the gut or to the body or both?

How much of these individual nutrients do we need or can we tolerate?

How does the body handle these various nutrients and foods constituents?

How are such processes affected by age and other nutrients and food structures?

How can we detect inappropriate intakes?

The only way to try and resolve these problems is by gaining a better understanding of nutrient metabolism and deriving from this means of detecting early deficiency and excess which would allow us to establish accurate requirements and enable us better to screen individuals and populations to achieve a better appreciation of the function and interaction of dietary nutrients.

These issues can be seen within the broad spectrum of what is commonly known in nutrition as "status". The spectrum of status can be illustrated as follows:

DEATH

Toxicity	**homeostasis overwhelmed metabolic defects**
	compensatory catabolism,
	excretion,
Excess	**sequestration,**
Adequacy	**acquisition, retention, disposal effective**
Deprivation	
	increased absorption and retention
	homeostasis inadequate
Deficiency	**function defects, metabolism of other nutrients**
	disrupted

DEATH

From this it can be appreciated that gross deficiencies and excesses can be determined readily. However it is at the level of marginal toxicity or deprivation both in populations and individuals that specific and sensitive diagnostic procedures and tests are needed to inform food policy and other strategic interventions. By and large such tests do not exist and they will probably only emerge as the homeostatic mechanisms influencing nutrient metabolism are elucidated. That is the enormity of the task facing anyone or any agency involved in Human Nutrition.

The second set of issues in nutrition relate to the interactions between nutrition and disease. The major problems in this context are obesity; cardiovascular disease involving arteriosclerosis which leads to cerebral, renal, intestinal, skeletal, and cardiac muscle ischaemia, with consequent impaired organ function and infarction, (the latter, as well as involving the myocardium classic heart attack, can also affect gut and kidneys); cancer (in particular involving the lung, large intestine, stomach, breast and prostate); osteoporosis; ageing and hypertension. There is an increasing appreciation that perhaps a better type of diet might improve the quality of life.

Much of the evidence of a relationship between diet and the above clinical phenomena is derived from epidemiological studies. Such studies raise questions and hypotheses rather than demonstrating definitive casual relationships. There is, as yet, very little definitive evidence linking or showing clearly the nature of the interaction between diet and the processes which lead to these diseases. Nonetheless, the relationships mentioned have been consistent and in many cases quite strong, to the extent that one can list a number of important questions which should legitimately be explored from the aspect of having a nutritional basis. The questions raised are as follows:

Why are meat-based diets associated with bigger risks of cancer?

Why do plant-based diets appear to be beneficial?

What are the inherent properties of the Mediterranean diet which protect against early death and possibly from the adverse effects smoking and alcohol?

Why is a low alcohol intake protective?

What are the apparent beneficial constituents of nuts, olive oil etc.?

There are a number of candidate nutrients responsible for these various benefits. These include monounsaturated fatty acids, oleic acid, polyunsaturated fatty acids. (but the possibility has been raised that polyenoic fatty acids are more prone to oxidative damage), antioxidants (Vitamin A, E, C, carotenoids, flavonoids, sulphydryl compounds), complex carbohydrates ("fibre"), and plant metabolites which induce antioxidant enzymes, and which inhibit mutagens and carcinogens.

It must again be emphasised that epidemiology is a blunt instrument, in that it creates hypotheses rather than proves them. For example, many studies have shown potentially beneficial effects of plant-based diets in reducing the incidence of cancer. It has also been shown that individuals who eat a lot of plant-based diets have higher circulating levels of betacarotene in their bloodstream. This generated the hypothesis that betacarotene was potentially a beneficial component of the diet, yet ironically in a recent intervention study no benefit, indeed a possible disadvantage, was observed of betacarotene supplements on the incidence of cancer of the lung[3]. Clearly, betacarotene may be acting as a surrogate marker for other more effective anti-cancer compounds in plants. Alternatively, one must accept the possibility that some of the original epidemiological studies were themselves flawed. This type of dilemma illustrates the difficulties which arise from using study outcomes which are essentially remote from the actual site and action of the candidate nutrients of interest. This again is another issue which can only really be resolved by better appreciation of the tissue localisation, metabolism and function of nutrients of interest. One must also bear in mind that "no nutrient is an island". Any study involving a particular nutrient has to take into consideration the knock-on effects which an altered supply of a nutrient might have on the utilisation of, or the requirements for, other essential dietary components.

5. THE PLACE OF NMR

Given the broad remit which has been claimed for nutrition it is not surprising that may of the involved disciplines have already benefitted from information derived from NMR based techniques.

NMR imaging has greatly facilitated the assessment of body content of fat and lean tissue[4,5] and of fetal composition in utero.[6] Fast data acquisition in real time with appropriate resolution will enhance studies of mobile organs and gut motility and are being applied to the study of the atherosclerotic narrowing of coronary arteries but as yet probably do not have the resolution to measure directly changes in vascular calibre as an outcome variable in the study of nutritional interventions.[7] However, in cardiac and skeletal muscle, and in brain indirect markers of vascular supply, such as perfusion, can be measured with sufficient precision to enable longitudinal studies with replicate observations.[8-10]

NMR spectroscopy in vivo and ex vivo is making a substantial contribution to nutrition and medicine. High resolution NMR has been applied to the entire range of body fluids, faeces and urine.[9,10] It provides a valuable means of detecting both expected, and more importantly, unexpected products of metabolism in health and disease. Undoubtedly even higher resolutions coupled with innovative data analysis (J Nicholson: this volume) will advance this area appreciably.

The investigation of lipid metabolism is being aided by the use of NMR to determine the degree of unsaturated[11] and linoleic acid[12] in body adipose tissue; however as yet these have only been studied with marked dietary changes. It is not yet clear if the current sensitivity will detect *in vivo* subtle changes following small dietary modifications. However sufficient sensitivity exists to monitor the changes in the hepatic and muscle metabolism of glycogen in vivo and ex vivo.[13,14]

6. SUMMARY AND CONCLUSION

I have reviewed the major nutritional issues current western counties. NMR spectroscopy and imaging techniques have stated, and undoubtedly will continue, to play a major role in metabolic and nutritional research.

References

1. J.S. Garrow and W.P.T. James, *Human Nutrition and Dietetics*, Churchill Livingston, London, 9th Edition, 1993.
2. Committee on Medical Apsects of Food Policy Dietary Reference Values for Food Energy and Nutrition for the United Kingdom. Department of Health Report on Health and Social Subjects No 41. H.M.S.O., London, 1991.
3. The Alpha Tocopherol, Beta Carotene Cancer Prevention Study Group. *New Eng. J. Med.*, 1994, **330**, 1029.
4. G.McNeill, P.A. Fowler,R.J. Maughan, B.A. McGaw, M.F. Fuller, D.Gvozdanovic and S. Gvozdanovic, *Br. J. Nutr.*, 1991, **65**, 95.
5. K. Fox, D. Peters, N. Armstrong, P. Sharpe and M. Bell, *Int. J. Obes, Relat. Metab. Disord.*, 1993, **17**, 11.
6. L.Jovanovic-Peterson, J. Crues, E. Durak and C.M. Peterson, *Am. J. Pertinatol*, 1993, **10**, 432.
7. P.A. Bottomley, *Radiology*, 1994, **191**, 593.
8. J.S. Wyatt, *J. Royal College Physicians London*, 1994, Vol.28, 126.
9. J.D. Bell, J.C. Brown and P.J. Sadler, *NMR in Biomedicine*, 1986, Vol. 2, 246.
10. D.J. Bell and P.J. Sadler, *Chemistry in Britain*, 1993, 597.
11. N. Beckmann, J-J. Brocard, U. Keller and J. Seelig, *Mag. Res. in Medicine*, **27**, 97.
12. C.T.W. Moonen, R.J. Dimand and K.L. Cox, *Mag. Res. in Medicine*, 1988, **6**, 140.
13. P.G. Morris, D.J.O. McIntyre, R. Coxon, H.S. Bacheland, K.T. Moriarty, P.C. Greenhoff and I.A. Macdonald, *Proc. Nutr. Soc.*, 1994, **53**, 335.
14. G.I. Shulman, D.L. Rothman, T. Jue, P. Stein, R.A. DeFronzo and R.G. Shulman, *New England J. Medicine.*, 1990, Vol. 322, 223.

^{1}H NMR Spectroscopy of Biological Fluids and the Investigation of Perturbed Metabolic Processes

J. K. Nicholson[1], P. J. D. Foxall[1], E. Holmes[1], G. H. Neild[2], and J. C. Lindon[3]

[1] DEPARTMENT OF CHEMISTRY, BIRKBECK COLLEGE, UNIVERSITY OF LONDON, GORDON HOUSE, 29 GORDON SQUARE, LONDON, WC1H 0PP, UK

[2] THE INSTITUTE OF UROLOGY AND NEPHROLOGY, ST PETER'S HOSPITAL, UNIVERSITY COLLEGE AND MIDDLESEX HOSPITAL MEDICAL SCHOOLS, MORTIMER STREET, LONDON W1N 8AA, UK

[3] DEPARTMENT OF PHYSICAL SCIENCES, WELLCOME RESEARCH LABORATORIES, LANGLEY COURT, BECKENHAM, KENT BR3 3BS, UK

1. INTRODUCTION

1.1 Biological fluid composition in relation to metabolic processes

Biological fluids perform a variety of physiological and biomechanical functions in the body and can be subdivided into various categories according to their provenance and biochemical composition (Table 1). The relationships between the activities of the many different biochemical pathways operating in the fluid secreting tissues and the body fluids themselves are diverse and complex. Biological fluid compositions are also modulated according to the current functional integrity of the tissues and hence variations in fluid composition gives information on the processes of cellular and organ dysfunction. However, the biochemical profiles of many biofluids, especially urine and plasma, also vary according to the whole body homeostatic demands. Thus the function of organs not directly concerned with the production of a particular fluid may influence the observed metabolite composition.

Table 1. Major Types of Biological Fluids and their Physiological Functions in Mammals.

Biofluid	**Secretory organs**	**Functions**
Amniotic	placenta/foetus	mechanical protection of the foetus
Bile	liver	excretion, digestion
Blood plasma	multiple	solute/metabolite transport, homeostasis, mechanical
Cerebrospinal	CNS ultrafiltrate	solute/metabolite transport, homeostasis, mechanical
Gastric	stomach	digestion
Pancreatic	pancreas	digestion
Saliva	salivary glands (various)	excretion, digestion
Seminal	testis/prostate/seminal vesicles	biochemical and mechanical support for spermatozoa
Sweat	sweat glands	excretion, homeostasis
Synovial	joints	mechanical
Urine	kidney	excretion, homeostasis

Dietary and diurnal variations also influence the metabolic profile of biological fluids as well as many pathological processes. Abnormal biofluid metabolite profiles are strongly reflected in their high field ^{1}H NMR spectra, and hence NMR spectroscopy of biofluids offers a potentially powerful new approach to clinical diagnosis and for investigating disease mechanisms in relation to pathophysiological processes[1-9]. NMR spectra of biofluids are always highly complex and this has led to the application of automatic data reduction and pattern recognition methods to aid spectral interpretation and sample classification according to the physiological or pathological state of the donor[10-16].

1.2 High Resolution ^{1}H NMR Spectroscopy of Biological Fluids

1.2.1 Background: High resolution ^{1}H NMR measurements on the low M_r metabolites in biological fluids have been performed since the late 1970s, and these have resulted in a number of interesting and important applications in clinical and metabolic biochemistry[1-9]. NMR spectroscopic studies on biological fluids can involve both purely analytical approaches and investigations into dynamic molecular interactions of metabolites. Analytical studies are concerned with the collection, assignment, and quantitation of NMR spectra from biological fluids and their biochemical interpretation. Dynamic NMR studies involve the investigation of the interactions of the components in the whole biofluid matrix, including enzymatic reactions, metal complexation, micellar compartmentation and chemical exchange phenomena, and studies on the binding of small molecules by macromolecules[1]. The scientific methodology and approaches that have been developed for the study of biological fluids in recent years also have a wide range of applications in other situations involving the analysis of organic molecules in complex biomixtures including cell and tissue extracts, and studies on foodstuffs, fruit juices and related products.

Biological fluids are effectively isotropic in terms of magnetic susceptibility and are much more magnetically homogeneous than intact organ, tissue or isolated cell preparations. In theory, this should render the collection and interpretation of high resolution ^{1}H NMR spectra of biofluids more simple than the procedures involved in*in vivo* NMR spectroscopy or for NMR spectroscopy of isolated cells. The information content of biofluid ^{1}H NMR spectra is very high in terms of the number of detectable metabolites and with respect to the range of intermolecular interactions that can be observed. In principle, it should also be much easier to conclusively assign and quantify ^{1}H NMR signals from biofluids than from cells or tissues, particularly as it is possible to spike fluids with candidate compounds for assignment purposes. It is, however, paradoxical that the complete assignment of the ^{1}H NMR spectrum of most biofluids is not possible due to the biochemical complexity of the matrix.

NMR signal assignment problems vary considerably between biofluid types. For instance, seminal fluid in normal individuals is highly regulated with respect to metabolite composition and concentrations and the majority of the signals have been assigned in 600 and 750 MHz ^{1}H NMR spectra[6,7]. Urine is much more variable than other biofluids even in normal individuals because its composition is specifically and interactively altered by the kidneys in order maintain whole body homeostasis. There is also enormous variation in the concentration range of ^{1}H NMR-detectable metabolites in biofluids such that some metabolites may be present at 10-100 mM and others at 1 μM or less (i.e. near the current limit of detection of a high field NMR spectrometer). Clearly, metabolites present in concentrations close to the limits of detection pose severe signal assignment problems. Thousands of resonance lines may be apparent in a single pulse high field spectrum of a biofluid (Figure 1) and because metabolite concentrations vary significantly according to biological rhythms and diet, the total assignment of the ^{1}H NMR spectrum of some biofluids, even from "normal" individuals, is not practical and, indeed, may be impossible. This does not, however, reduce the diagnostic potential of the technique, merely drawing our attention to problems of biological variation and the possible limitations in the uses of ^{1}H NMR in clinical diagnosis and other biochemical investigations.

We have previously argued[1] that NMR studies of body fluids should ideally be performed at the highest frequency available to obtain maximal dispersion and sensitivity and most of the work previously published[e.g. 1-9] has been performed at 400 MHz or more. Although lower frequency measurements, e.g. 200 MHz, can be useful for the detection of the most abundant metabolites, and in certain circumstances, give quantitatively accurate results, e.g. in the quantitation of certain drug metabolite signals, comparable higher frequency ^{1}H NMR measurements ($\geq$400 MHz) are usually more accurate[17]. Even the dispersion gain on going from 600 MHz to 750 MHz is significant in ^1H NMR spectroscopy of biofluids and allows more signals to be assigned, considerably easing the analysis of complex biofluid spectra[6-8].

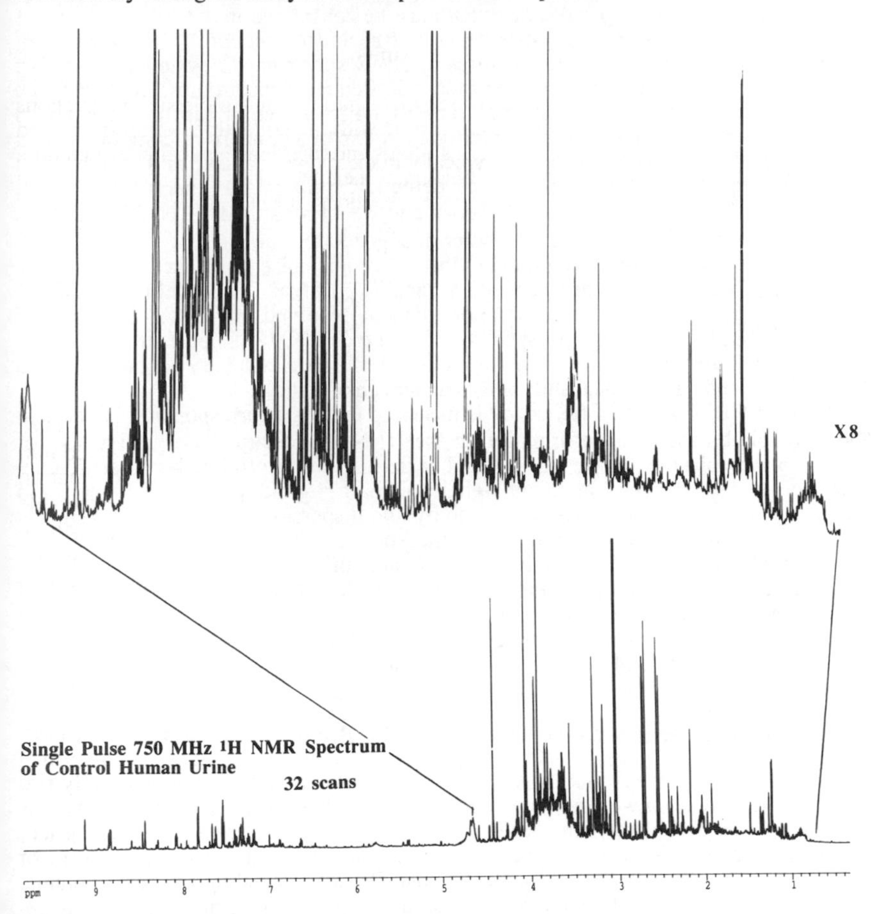

Figure 1: 750 MHz ^1H NMR spectrum of control human urine with x8 vertical expansion of aliphatic region to show complexity and signal overlap. Spectrum obtained with presaturation of the water signal.

1.2.2. Biochemical Information Content of ^1H NMR Spectra of Biofluids. There are two types of accessible biochemical information available in an NMR experiment on a biological fluid termed *Latent* and *Patent* Information. The latter is defined as being that which can be measured quantitatively in a single pulse experiment (irrespective of probe nucleus). Latent information in an NMR spectrum measured at a particular field as we define it, is *not* available in the single pulse spectrum and the biochemical data contained therein can only be obtained by careful selection of an appropriate multipulse experiment to achieve either spectral editing or frequency dispersion in a second or higher dimensions. Latent information can also be transformed into patent data by increasing the measurement field strength. Biochemical information can be latent *via* multiple peak overlap or *via* dynamic molecular interactions of metabolites. In the former case, data may be made patent either by increasing the observation frequency or using appropriate multi-dimensional or multi-pulse methods: in the latter case sample (e.g. pH) or spectrometer conditions (e.g. temperature) must be changed.

There are two basic types of noise that must be considered in the evaluation of ^{1}H NMR spectra of biofluids, i.e. *Instrument Noise* and *Chemical Noise*. Instrument noise and sensitivity are general features of NMR spectrometer systems and are of course influenced by the performance of the radiofrequency electronics, amplifiers and probes. Chemical noise is a special type of noise affecting biofluid spectra in particular, and this results from the extensive overlap of signals from compounds that are individually low in concentration (typically in the range 1-10 μM) and as individual moieties often close or even below the detection limit of the spectrometer, but, collectively, giving rise to apparently broad and weakly-featured ^{1}H NMR responses due to superimposition of their frequencies. It is generally true that for ^{1}H NMR work on biofluids it is the chemical noise rather than electronic noise that is the factor limiting spectral information content and accurate metabolite quantitation[1]. By definition, no one component of a chemical noise envelope can be resolved nor useful information extracted from this envelope, irrespective of the type of NMR experiment performed. The problems associated with chemical noise interference also vary according to the biofluid type and the chemical shift ranges under consideration[1].

As the NMR observation frequency is increased there is a corresponding increase in the available biochemical information in the ^{1}H NMR spectrum. For a given set of patent endogenous metabolite signals, higher field strengths offer better measurement precision. For substances with latent signals, remeasuring the spectrum at higher field strength with associated increased sensitivity and dispersion may lead to the signals becoming patent. In general higher field measurement enables a larger selection of patent biochemical markers to be detected and this consequently improves the biochemical understanding of the system. It is clear that at the present level of technology in NMR spectroscopy, it is not yet possible to detect many important biochemical substances, e.g. hormones, in body fluids because of problems with sensitivity, dispersion and dynamic range, and this area of research will continue to be technology-limited.

1.3 Dynamic Interactions of Metabolites in Biological Fluids

Although NMR spectroscopy of biofluids is now a well-established analytical technique for probing a wide range of biochemical problems, there are still many poorly understood chemical phenomena occurring in biofluids, particularly the subtle physico-chemical interactions occurring between small molecules and macromolecules or between organised multiphasic compartments. The understanding of these dynamic processes is of considerable importance if the full diagnostic potential of biofluid NMR spectroscopy is to be realised as such interactions may carry important physiological messages. The term dynamic is used in its widest sense here to include all types of chemical *activity* that occur in biological fluid matrices. There are basically four types of dynamic processes occurring in biofluids that can be studied by NMR spectroscopy, i.e. enzymatic reactions, macromolecular binding and compartmentation of metabolites, metal complexation reactions of metabolites and chemical exchange reactions of metabolites. Some of these processes can act on the same molecules simultaneously, complicating interpretation of their NMR spectra.

1.3.1 Enzymatic Processes and Reactions. Many biological fluids contain significant amounts of active enzymes. This may be either because the enzymes fulfil a biological function in the fluid, e.g. the esterases and peptidases present in prostatic fluid[7]. Alternatively, enzymes may have leaked into the fluid due to disease or toxin-induced organ damage, e.g. plasma alanine transaminase levels raised in liver and kidney disease, and *N*-acetylglucosaminidase levels raised in urine in kidney disease[18]. Clearly when provided with the appropriate substrates these enzymes will manufacture new products which may also be NMR-detectable. Collection of sequential NMR data may then allow the time-course of this enzymatic conversion to be followed. This may yield important kinetic data on the activity of the enzyme in a "real" biological medium and may also provide indirect NMR evidence of organ damage.

1.3.2 Macromolecular Binding and Compartmentation of Metabolites. Certain biofluids, particularly blood plasma, seminal plasma and bile contain high levels of protein and/or lipids which give these fluids multiphasic properties[1]. Small molecules can then either be free in solution or bound to a macromolecular structure. In the extreme case of micellar aggregations, e.g. in bile, this can be regarded as compartmentation as the small molecule can be either in the micelle wall or within the micelle itself. Many small molecules are more freely soluble in biological fluids than they are in water alone, e.g. cholesterol and its esters in blood plasma. In all cases, binding or compartmentation of small molecules results in changes in the rotational correlation times and hence the magnetic relaxation properties of their nuclei with respect to free solution conditions, and in some cases changes to the chemical shifts and coupling constants as well. Observation of such species may be further complicated by chemical exchange reactions. Visualisation of the signals from protein- or lipid-bound molecules frequently requires some physical perturbation of the sample to make the NMR lines sharp enough to be detected, e.g. alteration of pH to observe aromatic amino acids in plasma bound to albumin[19], or methanol extraction to observe cholesterol in blood plasma[4]. Binding of aromatic amino acids to blood plasma components appears to occur by both non-polar and electrostatic interactions, but only electrostatic interactions are important for the protein binding of lactate, citrate and acetate[19,20].

1.3.3 Metal Complexation Reactions. All biological fluids contain a variety of potential metal chelating agents, sometimes at very high concentrations. The most ubiquitous metal chelators in biofluids are the free amino acids (especially glutamate, histidine and aspartate) and organic acids such as citrate and succinate. Ca^{2+}, Mg^{2+} and Zn^{2+} are the main endogenous metal ions involved in complexation reactions with organic biofluid components in untreated biological fluids and many of these reactions can be studied by NMR[1,4]. Paramagnetic ions such as Gd^{3+} and Mn^{2+} can also be used to effect chemical editing of NMR spectra (and spin-echo based solvent suppression) by selectively binding to endogenous metal chelating agents such as citrate and hence broadening their signals[21]. Ethylenediaminetetraacetic acid (EDTA) is a very effective chelating agent for many di- and trivalent metal ions and can be added to biological fluids to remove metal from the endogenous chelating agents with consequent changes in the NMR signal pattern and the appearance of signals from metal EDTA complexes[4]. EDTA addition also results in a general sharpening of NMR signals in biofluids because of the complexation of trace levels of endogenous paramagnetic ions such as Fe^{3+} and Cu^{2+}.

1.3.4. Chemical Exchange Phenomena. Biofluids contain many endogenous species that participate in a variety of chemical exchange processes covering a variety of exchange time-scales. These processes may be connected with macromolecular binding or with metal complexation reactions or, more simply, involve exchange of protons with each other and/or solvent water. Some molecules may be involved with all three types of chemical exchange phenomena, e.g citrate, and hence signal positions vary considerably according to solution conditions. Depending on the exchange rate, spectral lines may be broadened or shifted from those of the molecules in their free solution condition. Citrate signals are generally broadened in the presence of metal ions (especially Ca^{2+}, Mg^{2+} and Zn^{2+}), and this can be reversed by the addition of EDTA. Amino acids such as glycine and glutamate are in fast exchange with their Ca^{2+} and Mg^{2+} complexes in both urine and CSF and give sharp NMR signals at the their appropriate averaged chemical shifts. High levels of bicarbonate can occur in biological fluids and this can result in the formation of stable amino acid carbamate complexes that are in slow exchange with the free counterparts[14].

2.1 ^{1}H NMR Spectroscopic Studies on the Composition of Urine

2.1.1 Introduction: The kidneys have a key role in the maintenance of body homeostasis, and effectively regulate whole body physiology by the elimination a wide variety of unwanted organic and inorganic compounds in the urine, i.e. the constancy of the internal environment is maintained at the expense of varying urinary composition. This variation in urinary composition is, therefore, a reflection of a large number of biochemical processes taking place both in the kidney itself and elsewhere in the body, and as such ^{1}H NMR urinalysis can offer useful or novel diagnostic information. The composition and physical chemistry of urine is complex and highly variable. A wide range of organic acids and bases, simple sugars and polysaccharides, heterocycles, polyols, low M_r proteins and polypeptides are present together with a wide variety of inorganic ions. The nephron is functionally and morphologically subclassified into several distinct units which perform specific physiological and biochemical tasks and modify the urinary composition in a characteristic way. This has considerable importance when abnormal renal function is considered, e.g. after exposure to a toxin, as malfunction of different nephronal segments leads to characteristic perturbation of the ^{1}H NMR spectroscopic metabolite profiles in urine[22].

2.1.2 One and Two dimensional ^{1}H NMR Spectroscopy of Urine. The region of the ^{1}H NMR spectrum from δ 0.7 to δ 4.7 in urine has a particularly rich profile of resonances from low MW compounds[23] and this region from a typical 750 MHz ^{1}H NMR spectrum of control human urine is shown in Figure 1. Although there is still extensive peak overlap in many regions of the spectrum, the improved dispersion of 750 MHz ^{1}H NMR spectra (even over 600 MHz) is useful in that it will provide a less ambiguous set of NMR descriptors of metabolic perturbation. However, where faced with such complexity it is usually necessary to employ 2 dimensional NMR methods to reduce signal overlap further and to gain more information on coupling constants and spin-spin connectivities in order to facilitate assignment. Detailed analysis of such spectra including the application of J-resolved (JRES) and homonuclear correlation spectroscopy experiments (e.g. COSY) promises to increase the number of assigned compounds significantly. High field JRES spectroscopy gives very good signal resolution, and this appears to be one of the most time-efficient methods for acquiring spectra and minimising spectral overlap for low molecular weight compound mixtures at high field[6,8,24]. The JRES experiment has been largely abandoned for macromolecular studies because of the insuperable problems of short T_2 relaxation times and those problems associated with the complexity of JRES patterns for second-order or strongly-coupled signals which lead to extra peaks in the 2-dimensional spectrum. In urine, most of the low M_r compounds do not have significant constraints placed on their molecular motions (and hence have relatively long T_2 relaxation times), and, at high frequency, the vast majority of the endogenous metabolites have first order spin systems with simple coupling patterns. When 750 MHz JRES spectra of urine are measured there is high sensitivity and excellent dispersion, with a reasonable total spectral acquisition time of <30 mins, together with good signal digitisation because of the relatively narrow second frequency axis in which the J coupling information is displayed[24]. Furthermore JRES projections give one line per resonance, simplify spectra, reduce overlap and remove degeneracy of information. JRES is often more effective than 2 dimensional correlation techniques such as COSY because they are less susceptible to cross T_1 noise and overlap that is characteristic of spectra containing mixtures of compounds in very different concentration ranges.

2.1.3 Effects of Diet on Urinary Metabolite Profiles. Dietary composition also affects the urinary metabolite profiles, and it is important to distinguish these from disease-related processes in clinical or toxicological studies. For example persons consuming large quantities of meat/poultry soon before a urine collection may have NMR detectable levels of carnosine and anserine in their urine; consumption of cherries is associated with elevated urinary fructose; consumption of shellfish and fish are associated repectively with high levels of betaine and trimethylamine in the urine[1]. In experimental animals the effects of diet on urinary metabolites can be quite marked. For instance rats fed on diets depleted in sulphur amino acids excrete low levels of taurine in their urine because of the depletion of intermediates in the biosynthetic pathways leading to taurine and glutathione production. Rats fed on diets containing high levels of the protein casein excrete large amounts of trimethylamine-N-oxide, betaine and dimethylamine due to the catabolism of excess dietary choline. The age of a laboratory animal (and probably humans, although little work has been done in this area) can also influence the excretion profiles of urinary metabolites, for instance increasing age of a rat is associated with increased urinary taurine but a decrease in urinary citrate[25]. These variations due to age or diet must be accounted for in biochemical or toxicological studies, in which it is very important to match controls as closely as possible to the experimental subjects with respect to age and weight.

¹H NMR spectroscopy of urine can also assist in the study of subtle interactions of dietary factors with the metabolism and toxicity of xenobiotic compounds (Figure 2).

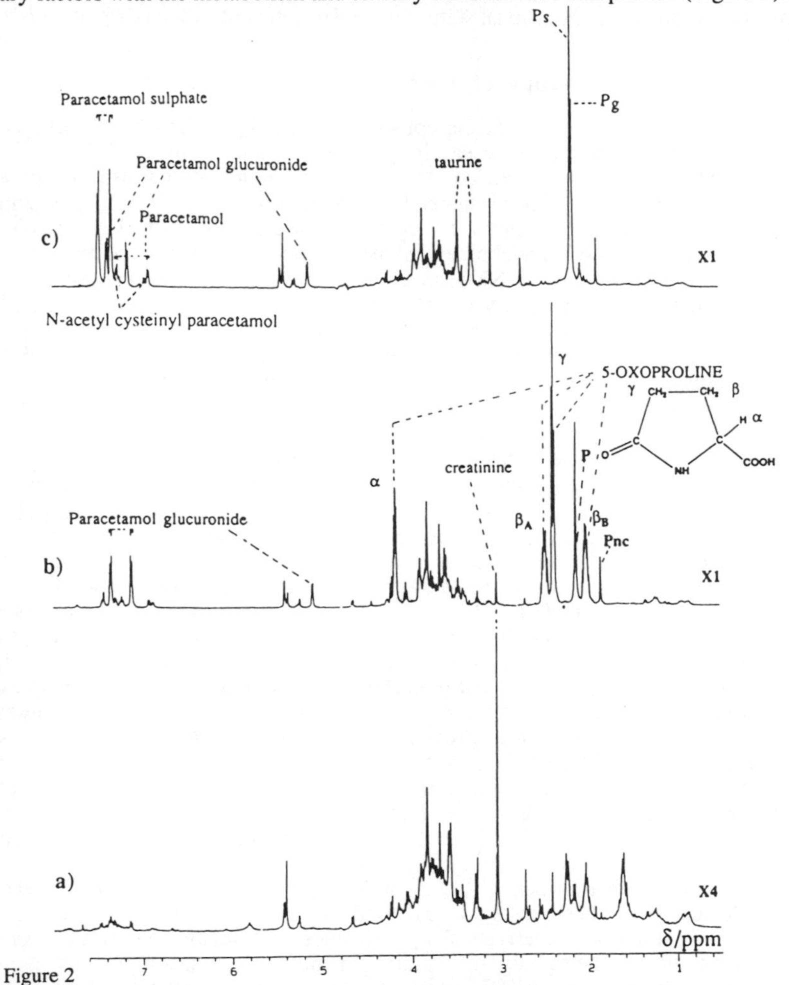

Figure 2

Single pulse 500 MHz ¹H NMR spectra of 6 hr pooled urine collection from (a) control rats (×4 vertical expansion) (b) rats fed on 1% paracetamol diet for 10 weeks (×1 vertical expansion) and (c) rats fed on 1% paracetamol + 1% methionine diet for 10 weeks (×1 vertical expansion). Key: P, Pg and Ps, *N*-acetyl signals of paracetamol, paracetamol glucuronide and paracetamol sulphate, respectively (the resolved aromatic signals of these and the parent drug are labelled separately), Pnc is the resolved cysteinyl side chain *N*-acetyl group.

Rats fed on diets containing 1% paracetamol become anorexic, but supplementation of the diet with methionine protects against the weight loss[26]. Insight into the biochemical basis of this phenomenon was given by ^{1}H NMR urinalysis studies (Figure 2) which showed that rats dosed chronically with paracetamol in their drinking water excreted large amounts of 5-oxoproline (prevented by dietary methionine supplementation) The effect of paracetamol was to cause chronic depletion of sulphur amino acids because of the heavy requirement for these imposed by the metabolism of paracetamol to sulphate, glutathione and cysteine conjugates[26]. In the normal functioning of the glutathione cycle glutamylcysteine is produced as a precursor of glutathione. The depletion of cysteine in the liver and kidneys results in the excessive production of other glutamyl-dipeptides which are cleaved to give 5-oxoproline which is released into the urine in large quantities[26]. Studies such as these show the value of using NMR spectroscopy to explore poorly understood metabolic events related to pathological processes and also show the importance of nutritional status on the development and manifestation of toxic lesions.

3.1 ^{1}H NMR Spectroscopy of Blood Plasma

3.1.1 Introduction: Blood plasma consists of several physico-chemically distinct domains including the bulk, essentially isotropic and disordered free-solution environment, in which are suspended highly-ordered colloidal structures and aggregates of macromolecules including proteins and lipoproteins. ^{1}H NMR measurements on blood serum and plasma can provide a plethora of useful biochemical information on both low molecular weight metabolites and macromolecular structure and organisation[1-5] Characteristic changes in ^{1}H NMR spectral patterns from plasma can also be related to abnormal tissue intermediary metabolism, plasma enzymatic processes and the presence of a variety of pathological conditions in the donors[1].

3.1.2 Comparative Biochemistry of Blood Plasma and Assignment of ^{1}H NMR Spectra. Single pulse spectra of human blood plasma are very complex and resonances of low MW metabolites, proteins, lipids and lipoproteins are heavily overlapped even at 750 MHz ^{1}H observation frequency (Figure 3a). Most blood plasma samples are quite viscous and this gives rise to relatively short T_1 relaxation times for small molecules compared to simple aqueous solutions allowing relatively short pulse repetition cycles without signal saturation. The complex spectral profile given in the single pulse ^{1}H NMR spectrum of blood plasma can be simplified by use of spin-echo experiments with an appropriate T_2 relaxation delays to allow signals from broad macromolecular components and compounds bound to proteins to be attenuated. The effect of applying the Carr Purcell Meiboom Gill (CPMG[27]) spin-echo pulse sequence to blood plasma at 750 MHz is shown in Figure 3b. A substantial reduction in the contribution from the albumin, lipoprotein and lipid signals is achieved with the use of only a short total spin-spin relaxation delay ($2n\tau$ = 88ms). Spin echo techniques have been used to help assign spectra of blood plasma and have been successfully applied in clinical studies on perturbed low MW metabolite profiles associated with various organic disease processes. By the early 1980 s many metabolites had been detected in normal blood plasma using 400 MHz spectrometers; these included alanine, isoleucine, valine, lactate, acetate, creatinine, creatine, glutamine, 3-D-hydroxybutyrate, choline, glucose, acetoactetate and acetone[4,5]. Assignments were, in general, based on the observation of only one or two resonances for each metabolite. In addition, peaks from certain macromolecules such as α_1-acid glycoprotein *N*-acetyl neuraminic acid and related sialic acid fragments have been assigned and used diagnostically, in particular *N*-acetyl groups which give rise to relatively sharp resonances presumably due to less restricted molecular motion[28]. The signals from some lipid and lipoprotein components, e.g. very low density lipoprotein (VLDL), low density lipoprotein (LDL), high density lipoprotein (HDL) and chylomicrons, have also been partially characterised[29] but assignment has been limited by the use of lower field strength spectrometers, their extensive chemical shift overlap and the broadness of their signal due to the short proton T_2 relaxation times of these large supramolecular species.

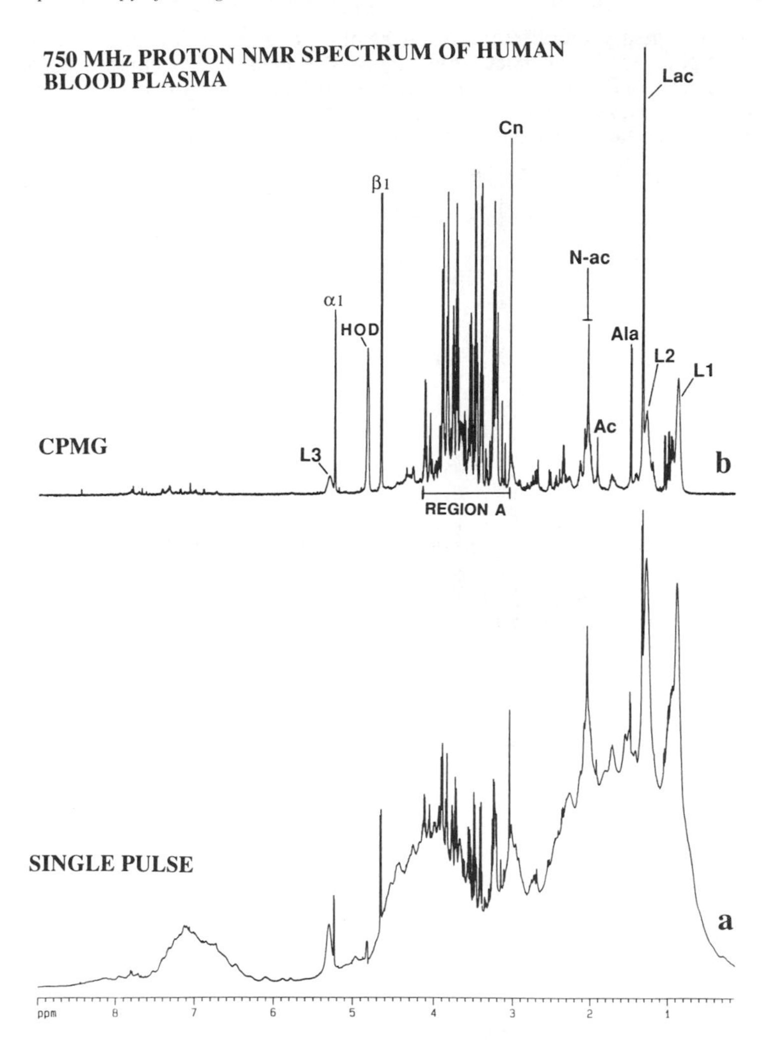

Figure 3: 750 MHz ^{1}H NMR spectra of control human plasma a) single pulse experiment (64 scans) showing heavy overlap of broad signals from macromolecules with sharp signals from low molecular weight metabolites superimposed, and b) CPMG spin-echo spectrum of same sample ($2n\tau$= 88ms) showing attenuation of broad lines. Both spectra were obtained with presaturation of the water signal. Abbreviations: L1, L2 and L3, methyl methylene and olefinic methine signals from low density and very low density lipoproteins; Lac, lactate; Ala, alanine, N-ac, N-acetyl signals from α_1 acid glycoprotein, Cn, creatinine; α_1 and β_1 glucose anomeric proton signals; HOD, residual water signal, Region A contains many signals from glucose, polyols and amino acids.

The application of the JRES spectroscopy results in a dramatic simplification of the blood plasma spectrum, and does not suffer from phase modulation of multiplet peaks in the FT spectrum because the experiment is always performed using a magnitude calculation. This considerably improves the resolution of the complex overlapped resonances in the chemical shift range from δ 3 - 4 in blood plasma[8]. Furthermore, the protein resonances are attenuated as effectively as was seen in the application of the simple spin-echo experiment. The skyline projection through the JRES contour map results in a greatly simplified spectral profile of the effectively ^{1}H-decoupled ^{1}H spectrum of the motionally unconstrained low MW metabolites in plasma. Such skyline projections enable the facile comparison of spectra from subjects with metabolic diseases, and an example showing a projection from a normal human subject and a patient with chronic renal failure (uraemia) is shown in Figure 4.

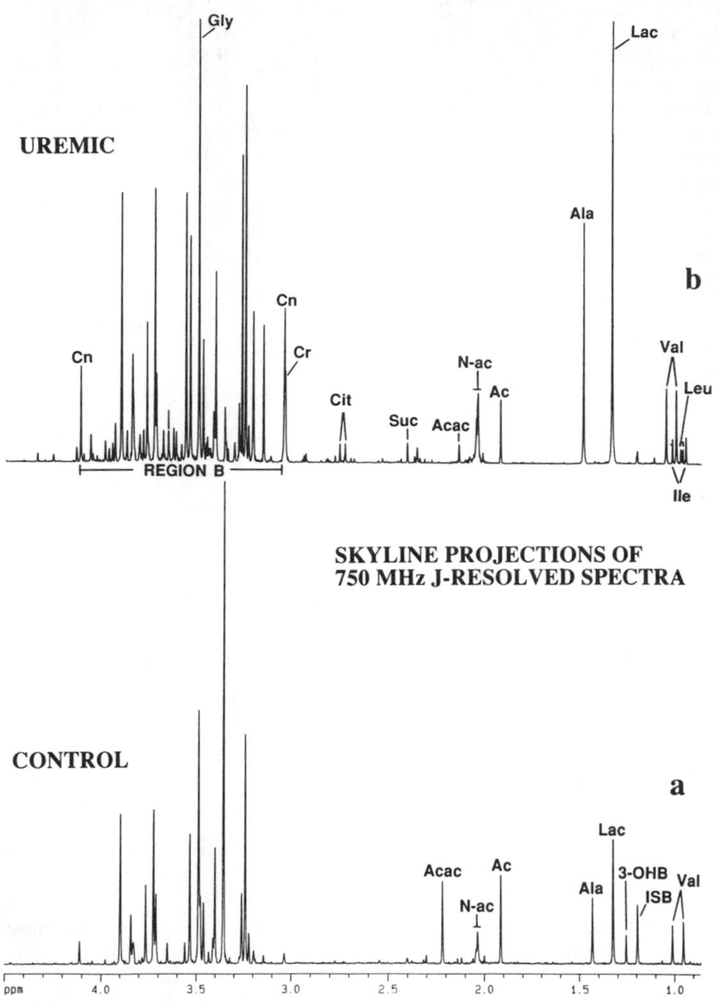

Figure 4: Skyline projections of 750 MHz ^{1}H J-resolved NMR spectra of human plasma. a) from normal subject b) from a patient with chronic renal failure (uremia). Both spectra were obtained with presaturation of the water signal. Abbreviations: Ac, acetate, Acac, acetoacetate; Ala, alanine; Cit, citrate; Cn, creatinine; Cr, creatine; Gly, glycine; Ileu, ileucine; ISB, isobutyrate; 3-OHB, 3-D-hydroxybutyrate; Lac, lactate; Leu, leucine; N-ac, N-acetyl signals from α_1 acid glycoprotein, Suc, succinate; Val, valine; Region B contains many signals from glucose, polyols and amino acids.

Various types of correlation spectra such as 2 dimensional ^{1}H-^{1}H COSY45 and ^{1}H-^{13}C heteronuclear multiple quantum coherence transfer (HMQC) provide an additional assignment aids. 750 MHz COSY 45 spectra enable the confirmation of the presence of a number of amino acid resonances including all those of valine, alanine, leucine, isoleucine, threonine and glutamine and the majority of the resonances of others especially those in particularly diagnostic regions of the spectrum (Figure 5). These include the γ–δ methylene connectivity in lysine, the β–γ and γ–δ of arginine, the α–β and γ–δ of proline, the α–β of citrulline, tyrosine, histidine, asparagine and glutamate. A number of other small molecules can be readily observed in the COSY spectrum and these include the acids 3-D-hydroxybutyrate, citrate, taurine and lactate as well as polyols such as *myo*-inositol and all other resonances of α- and β-glucose. In addition, the resolving power of the COSY experiment is demonstrated by the connectivities observed for many of the lipidic resonances which are not observable in the JRES experiment because of their short T_2 relaxation times. In addition, high frequency ^{1}H–^{13}C heteronuclear multiple quantum coherence transfer (HMQC) spectroscopy of blood plasma also provides useful assignment data for the lipoprotein, cholesterol and other lipid resonances and those from glycoproteins and albumin domains where there is relatively free molecular motion. When used together JRES, COSY and HMQC spectroscopy can allow the near complete assignment of all the major resonances in human blood plasma and this provides a strong basis for the use of NMR spectroscopy of blood plama to provide clinical/diagnostic information.

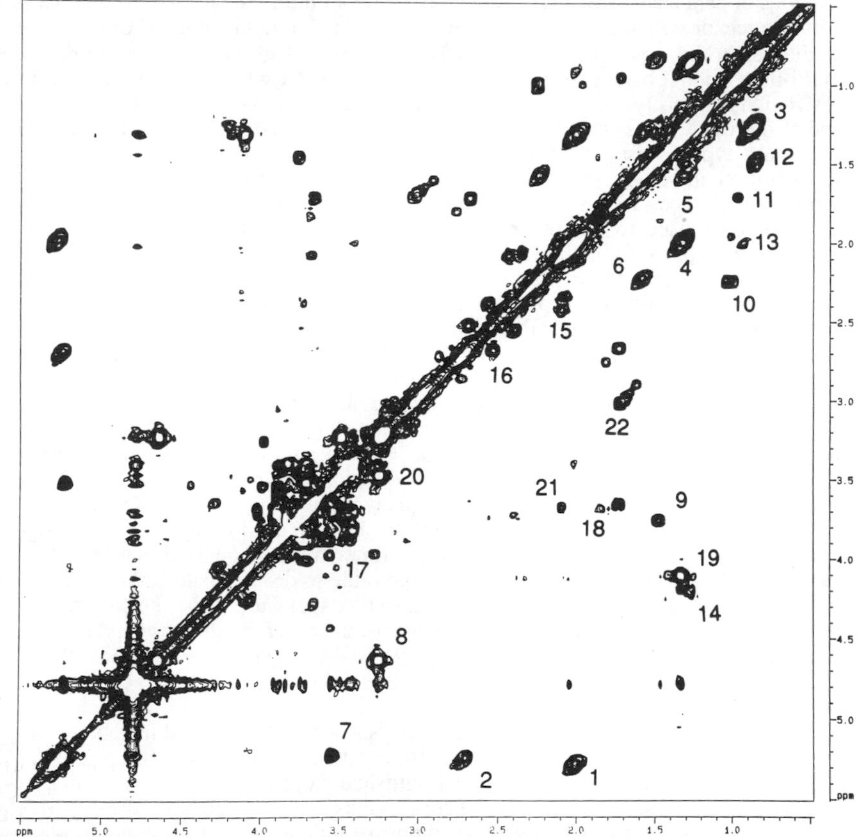

Figure 5. 750 MHz ^{1}H COSY 45 spectrum of control human plasma (symmetrised and linear predicted in F1 and F2), showing region δ 0.45 to δ 5.50 with the one dimensional spectrum of the same sample shown above. Key: FA, fatty acids of lipids; 1, $\underline{CH_2}$.CH= FA; 2, $\underline{CH}.\underline{CH_2}$.CH= FA; 3, $CH_3.CH_2$ FA; 4, lysine; 5, $\underline{CH_2}.\underline{CH_2}.CH_2$.CO FA; 6, $\underline{CH_2}.\underline{CH_2}$.CO FA; 7, α-glucose; 8, β-glucose; 9, alanine; 10, valine; 11, leucine; 12 $\underline{CH_3}.\underline{CH_2}.CH_2.CH_2$ FA; 13, isoleucine; 14, threonine; 15, glutamate; 16, citrate, 17, unknown; 18, arginine; 19, lactate; 21, citrulline; 22, lysine residues in albumin.

4.1 Pattern Recognition Analysis of NMR Spectra and the Investigation of Metabolic Status

In principle ^{1}H NMR methods are capable of detecting and quantifying many hundreds of metabolites in biofluids. The NMR-generated metabolite profile is representative of the concerted action of many biosynthetic and intermediary metabolic pathways and as such reflects the physiological or pathological status of the donor. The metabolite concentrations and/or patterns in an NMR spectrum can be regarded as decribing an *n*-dimensional hyperspace where each of the coordinates is an NMR measurable parameter (such as a metabolite concentration). In the case of a disease process, toxic insult or even a mild physiological perturbation it is expected that the position of the individual or object within the metabolic hyperspace will change. In order to use as much of the NMR spectrum of the biofluid to describe the metabolic status of the donor it is necessary to use a range of pattern recognition (PR) methods to enable data compression and sample classification.

PR methods give a visual representation of the multidimensional parameter space and are well established in the statistical literature. PR is a general term applied to methods of data analysis which can cope not only with fully quantitative data, but also with discrete or scored data, for example, that obtained from histological or behavioural studies or in this case quantitative ^{1}H NMR studies. The methods can work in the multi-dimensional parameter space (in the case of NMR where each dimension is an NMR signal intensity, or a measured concentration or excretion rate of one metabolite) and also provide dimension reduction techniques for display purposes. Methods are of two main types: unsupervised mapping or dimension reduction to enable single visualisation and clustering of samples according to the similarity of their metabolite profiles; and supervised statistical methods in which the known sample class of a training set is used to maximise the separation of such clusters. Examples of the unsupervised approach include non-linear mapping (NLM) and principal components analysis (PCA), and an example of the latter is discriminant function analysis. Both approaches can be used in the processing and interogation of NMR generated metabolic data. In simple terms NLM or PCA maps represent compressions of the original data coordinates in n-dimensional space into 2-dimensional or 3-dimensional approximations to the true multidimensional interpoint distances. Two points which are close on the map should be more similar in terms of input variables than two distant points.

PR analysis of biofluid NMR spectra has been shown to be of considerable value in both clinical and toxicological studies. For instance, ^{1}H NMR spectra of the urine of rats exposed to different types of experimental toxins have been classified according to site, severity and mechanism of toxicological damage by use of NLM or principal components analysis[10-15]. These techniques can also be used to monitor toxicological evens in groups of animals occurring in real time (from sequentially-collected urine samples following toxic insult[13]. Thus it is possible to discriminate between different types of pathological event occuring at different stages during the development of a lesion and to allow a biochemical classifcation of the onset, progression and regression (recovery or regeneration) of the pathology. NMR-PR anlysis of cerebrospinal fluid has allowed the discrimination of Alzheimer's disease from non-Alzheimer's subjects based on the low M_r metabolite profiles[14].

An illustration of the use of PR in the analysis of NMR-generated metabolic data is given in Figure 6 in which urine samples from patients with several different types of inborn errors of metabolism can be distinguished from random urine samples from healthy control subjects using principal components analysis. We have also shown that it is even possible to discriminate between normal individuals experiencing mild (non-pathological) physiological variations due to excercise, fasting, water deprivation etc[16]. Once PR classifications have been established it is then possible to statistically interogate the input NMR data set to find which metabolites or descriptors are most closely associated with the discrimination of the object classes, in this way it is possible to obtain new metablic biomarkers of disease and toxic processes[12,13,15].

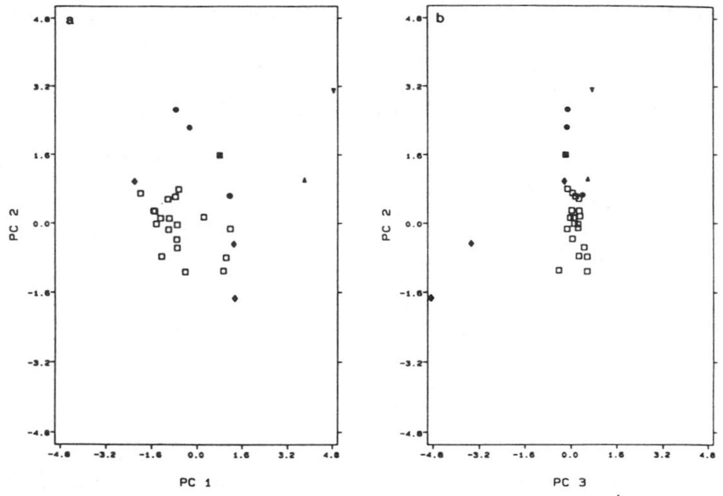

Figure 6. Illustration of the use of PR to effect disease classification based on ^{1}H NMR input data. Principal components maps are shown with metabolite peak intensity data scaled to creatinine and then autoscaled (so that each input variable is equally weighted). The open squares correspond to control early morning urines from healthy individuals. In this PR application NMR data were automatically reduced to a histogram format (500 bars) for each sample and the 500 dimensional data set analysed by PCA. Key to diseases: shaded circles, cysteinuria; shaded diamonds, oxalic aciduria; shaded triangles, porphyria; inverted shaded triangles, Fanconi syndrome; shaded squares, 5-oxoprolinuria. Figure modified from reference 30.

The future use of automatic methods of spectral data reduction (such as the conversion of spectra into high resolution histograms for direct input into PR routines) in the NMR spectrometer host computer or workstation will also help expedite the use of NMR and PR in practical clinical laboratories[30]. The most important single feature of the NMR-PR approach to the study of biofluids is that both NMR and PR are exploratory techniques, which are not dependent on the establishment of a specific scientific hypothesis in order to give useful scientific results. Moreover, NMR-PR results in the generation of novel metabolic hypotheses which can be tested using either NMR or other biochemical methods. Such properties make the NMR-PR techniques ideal for investigating poorly understood metabolic events caused either by disease or drug toxicity and for investigating the response of the whole organism (and its multiple biochemical pathways) to drug therapy or other types of medical intervention or management.

REFERENCES

1. J.K. Nicholson and I.D. Wilson, *Prog. NMR Spectrosc.,* 1989, **21** 444.
2. J.L. Bock, *Clin.Chem.*, 1982 **28**, 1873.
3. M. Traube, J.L. Bock, and J.L. Boyer, *Ann. Intern. Med.,* 1983, **98** 171.
4. J.K.Nicholson, M.J. Buckingham, and P.J.Sadler, *Biochem.J.,* 1983, **211** 605.
5. J.K.Nicholson, M. O'Flynn, P.J. Sadler, A. Macleod, S.M. Juul and P.H. Sonksen. *Biochem. J.,* 1984, **217** 365.
6. M. Lynch, J. Masters, J. Pryor, P.J.D. Foxall, J.C. Lindon, and J.K. Nicholson, *J. Pharm. Biomed.Anal.,* 1994, **12** (**1**) 5.
7. M. Spraul, J.K. Nicholson, M.J. Lynch, and J.C. Lindon, *J. Pharm. Biomed. Anal.* 1994, **12** (**5**) 613.
8. P.J.D.Foxall, M.Spraul, R.D.Farrant, J.C. Lindon, G.H. Neild, and J.K. Nicholson,*J. Pharm. Biomed.Anal.,* 1993, **11** (**4/5**) 267.
9. P.J.D.Foxall, G. Mellotte, M.Bending, J.C.Lindon, and J.K.Nicholson, *Kidney Int.,* 1993, **43** 234.

10. K.P.R.Gartland, S.M.Sanins, J.C.Lindon, C. Beddell, B. Sweatman, and J.K.Nicholson, *NMR Biomed,* 1990, **3** 166.
11. K.P.R.Gartland, C.R.Beddell, J.C.Lindon, and J.K.Nicholson, *J. Pharm. Biomed.Anal.,* 1990, **8** 963.
12. K.P.R.Gartland, J.C.Lindon, C. Beddell, and J.K.Nicholson, *Mol. Pharmacol.,* 1991, **39** 629.
13. E.Holmes, F.W.Bonner, B.C.Sweatman, J.C.Lindon, C.R. Beddell, E.Rahr, and J.K. Nicholson, *Mol. Pharmacol.,* 1992, **42** 922.
14. F.Y.K.Ghauri, J.K.Nicholson, B.C.Sweatman, J.Wood, C.R.Beddell, J.C.Lindon and N.Cairns, *NMR Biomed,* 1993, **6** 163.
15. Anthony, M.L. Lindon, J.C. Beddell, C.R. and Nicholson, J.K. *Mol. Pharmacol.,*1994, **46** 199.
16. E.Holmes, P.J.D. Foxall, G.H. Neild, C. R. Beddell, B.C.Sweatman, E. Rahr, J.C. Lindon, M.Spraul, and J.K.Nicholson, *Anal. Biochem.,* 1994, **220** 284.
17. I.D.Wilson, J. Fromson, I.M.Ismail, and J.K.Nicholson, *J. Pharm. Biomed.Anal.,* 1986, **59** (**2**) 157.
18. M.L.Anthony, C.R.Beddell, J.C. Lindon and J.K. Nicholson, *J. Pharm. Biomed.Anal.,* 1993, **11**(**10**) 897.
19. J.K. Nicholson, and K.P.R. Gartland, *NMR Biomed.,* 1989, **2** 63.
20. J.D. Bell, J.C. Brown, G. Kubal, and P.J. Sadler, *FEBS Lett.,* 1988, **235** 81.
21. S.C. Connor, J.K. Nicholson, and J.E. Everett, *Anal. Chem.,* 1987, **59** 2885.
22. K.P.R.Gartland, F.Bonner, and J.K.Nicholson, *Mol. Pharmacol.* 1989, **35** 242.
23. J.R, Bales, D.P.Higham, I. Howe, J.K.Nicholson, and P.J.Sadler, *Clin.Chem.,* 1984, **30** 426.
24. P.J.D.Foxall, J.Parkinson, I.H. Sadler, J.C. Lindon and J.K.Nicholson, *J. Pharm. Biomed.Anal.,* 1993, **11** 21.
25. J.D. Bell, P.J. Sadler, V.C. Morris and O.A. Levander, *Mag. Res. Med.,* 1991, **17,** 414.
26. F.Y.K. Ghauri, A.E.M. McLean, D. Bealses, I.D.Wilson, and J.K. Nicholson, *Biochem. Pharmacol.,* 1993, **46**, 953.
27. S. Meiboom and D.Gill, *Rev. Sci. Instrum.,* 1958, **29**, 688.
28. J.D.Bell, J.C. Brown, J.K. Nicholson, and P.J.Sadler,*FEBS Lett.,* 1987, **215,** 311.
29. Bell, J.D. Brown, J.C. Kubal, G. and Sadler, P.J. *FEBS Lett..,* 1988, **235,** 81.
30. E.Holmes, P.J.D. Foxall, J.K. Nicholson, G.H. Neild, S.M. Brown, C.R. Beddell, B.C. Sweatman, E.Rahr, J.C. Lindon, M. Spraul, and P.Neidig, *Anal. Biochem.,* 1994, **220** 284.

Proton NMR Studies of Human Brain Metabolism

J. Frahm[1], H. Bruhn[2], and F. Hanefeld[3]

[1] BIOMEDIZINISCHE NMR FORSCHUNGS GMBH, MAX-PLANCK-INSTITUT FÜR BIOPHYSIKALISCHE CHEMIE, POSTFACH 2841, D-37018 GÖTTINGEN, GERMANY

[2] ZENTRUM RADIOLOGIE UND [3] ZENTRUM KINDERHEILKUNDE, SCHWERPUNKT NEUROPÄDIATRIE, GEORG-AUGUST-UNIVERSITÄT, D-37075 GÖTTINGEN, GERMANY

1 INTRODUCTION

The central nervous system not only controls the functioning of vital body functions, cognitive abilities, and mental states, but also represents the source of our consciousness and self-awareness. Over centuries we have mainly learned about its structure from *post mortem* anatomy and about function by relating specific functional deficits to focal brain lesions and surgical interventions. Only for about a decade, X-ray computed tomography and positron emission tomography have contributed to an improved assessment of structural and functional aspects in the living subject. Most recently, however, advances in nuclear magnetic resonance (NMR) technology have opened new and truly noninvasive windows into the functioning human brain. While magnetic resonance imaging (MRI) provides access to high-resolution anatomy (and function), localized magnetic resonance spectroscopy (MRS) offers unique insights into brain metabolism.

This contribution deals with a brief introduction into the use of localized proton MRS to study metabolic characteristics of the intact human brain under various conditions. In fact, although cerebral metabolic profiles of healthy human subjects exhibit only little interindividual variability – provided age and regional differences are taken into account – closer inspection unravels many aspects of metabolism that are directly related to nutrition. To underline this at first glance surprising observation this work summarizes selected examples of both physiologic and pathologic alterations of brain metabolite concentrations from several independent studies.

2 LOCALIZED PROTON NMR SPECTROSCOPY

2.1 Single Voxel Spectroscopy or Chemical Shift Imaging?

Meaningful insights into human brain metabolism in most cases require spatial discrimination of *in vivo* NMR spectra. In principle, there are two different concepts either derived from an imaging or spectroscopy framework: (i) Chemical shift imaging techniques encode spatial information into the phase of the NMR signal along one, two, or even three dimensions, but acquire the frequency or chemical shift information in the absence of a magnetic field gradient. Two-, three-, or four-dimensional image reconstructions are required to obtain metabolic maps, i.e. spatial distributions of individual metabolites as part of the entire data set. (ii) Alternatively, single-voxel localization techniques attempt to acquire the spectroscopic information directly from a restricted volume-of-interest (VOI).

While chemical shift images emphasize the spatial distribution of metabolites, single voxel spectra in general reveal a much more detailed and quantifiable metabolic picture of selected foci (in particular, if proton MRS is concerned). Prominent single voxel advantages are due to the excellent spectral resolution (homogeneity) and water suppression obtainable by shimming over a small VOI (e.g., 1–18 mL for proton MRS). Moreover, the relatively short measuring times (e.g., 1–10 min) are clearly beneficial to a reliable quantification of metabolite concentration as they allow the use of long repetition times for fully relaxed acquisitions (i.e. TR $\geq$ 6000 ms for proton MRS).

In general, however, the selection of a technique should only depend on the question to be answered. While details about localization techniques are necessary to properly apply such methods, spectroscopic results used for metabolic research and medical decision-making must of course be independent of the chosen technique.

2.2 STEAM Localization Sequences

To accomplish the aforementioned goals the present investigations have been performed using fully relaxed single-voxel measurements. The two candidate localization techniques for proton MRS are PRESS[1,2] (point resolved spectroscopy) and STEAM[3–5] (stimulated echo acquisition mode). In this work, we have used a STEAM sequence in the form shown in Figure 1 as it provides optimum conditions for acquisitions at short echo times.

STEAM sequences comprise three successive 90° radiofrequency (rf) pulses that are commonly applied in the presence of orthogonal gradients. Thus, the stimulated echo (STE) signal represents magnetizations that exclusively originate from a VOI defined by the intersection of three perpendicular sections. During the course of the STEAM sequence transverse magnetizations excited by the first rf pulse transform into longitudinal magnetizations by application of the second pulse. Corresponding components are subject to T1 relaxation during the middle interval TM and refocus at TE/2 after application of the third pulse as a stimulated echo. During both TE/2 intervals transverse components are subject to T2 relaxation. To reduce T2 losses as well as to avoid J-modulation and multiple-quantum interferences of spin-coupled resonances, the echo time TE should be chosen as short as possible. For applications to proton MRS the sequence must be preceeded by one or more chemical-shift-selective (CHESS) water suppression pulses plus associated spoiler gradients.

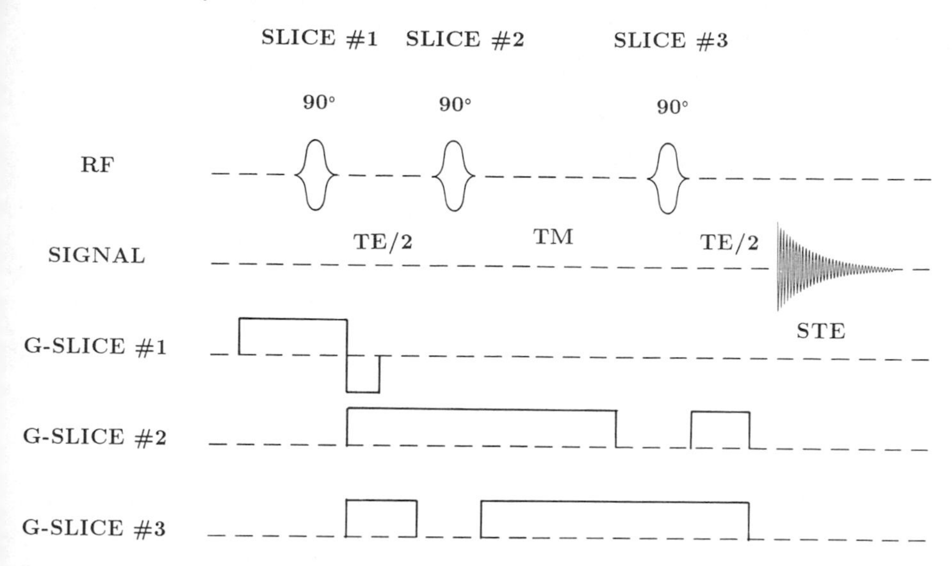

Figure 1 *Short-echo time STEAM localization sequence: acquisition of a stimulated echo (STE) signal from a volume-of-interest defined by the intersection of three orthogonal slices*

3 PROTON MRS OF HUMAN BRAIN: PHYSIOLOGY

3.1 Identification and Quantification of Metabolites

Biochemical data that may be derived from *in vivo* NMR measurements are metabolite concentrations or rate constants and fluxes, if detectable pools are changing in response to physiologic stimuli, pharmacologic treatment, or pathologic conditions. Prerequisites for such studies are a profound understanding of the proton (phosphorus, carbon, etc.) brain spectrum as well as an objective method for spectral evaluation unbiased by user interference.

Resonance assignments of the proton spectra shown in Figure 2 have been based on biochemical data, high-resolution NMR spectra of tissue extracts, and model metabolite spectra of individual compounds that were acquired under identical experimental conditions as chosen for human studies. This particularly applies to the same localization sequence and parameter settings. Without complications due to spectral overlap the model spectra further allow a better understanding of residual J-modulation effects at the short echo time of TE = 20 ms and of the influence of strong spin-spin coupling at the relatively low field strength of 2.0 T used here (Siemens Magnetom SP4000, Erlangen, Germany).

So far, a large number of metabolite resonances have been identified. In addition to N-acetylaspartate (NAA), N-acetylaspartylglutamate (NAAG), glutamate (Glu), creatine and phosphocreatine (Cr), choline-containing compounds (Cho), glucose (Glc), *myo*-inositol (*myo*-Ins) and *scyllo*-inositol (*scyllo*-Ins) indicated in Figure 2,

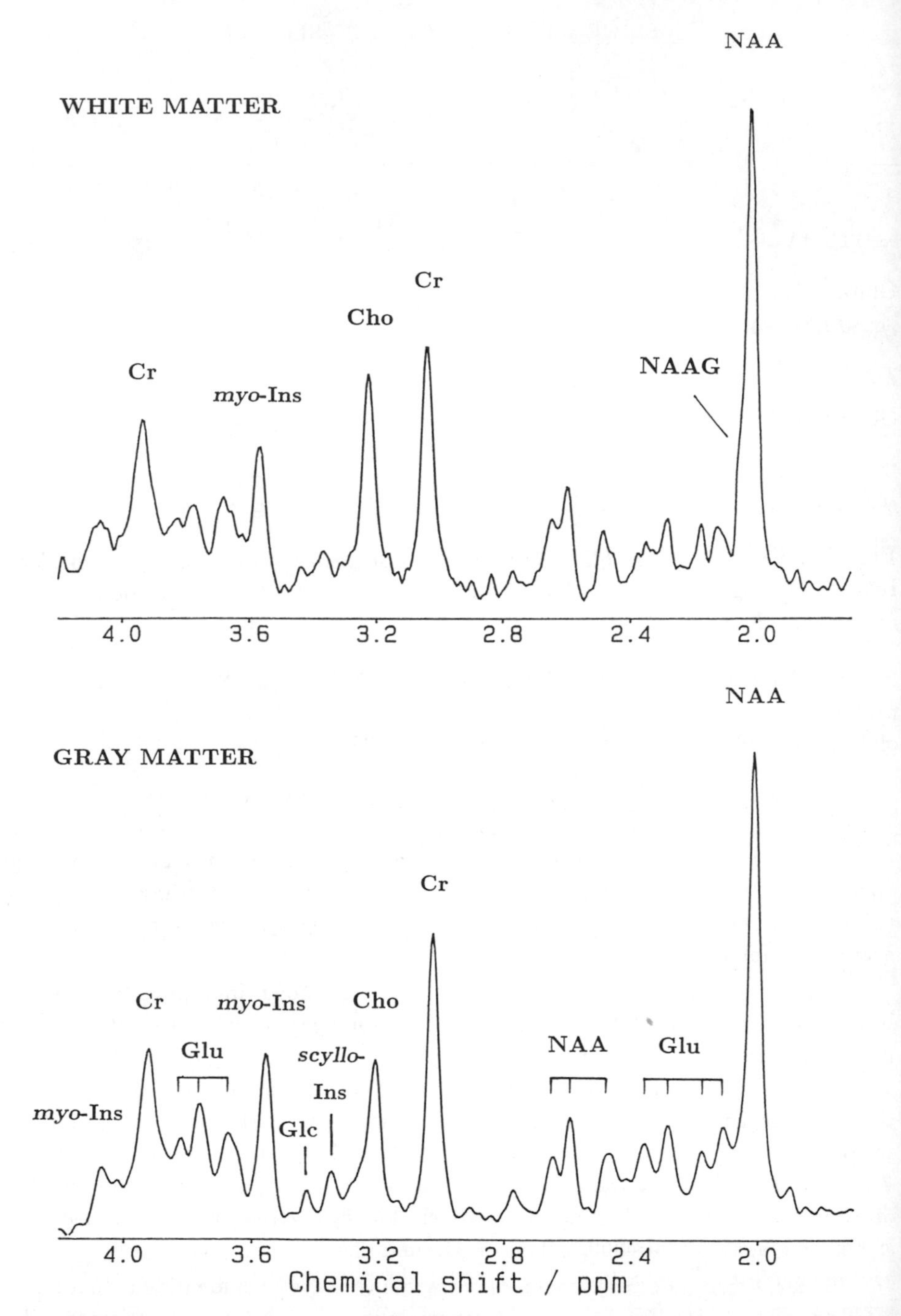

Figure 2 *Localized proton spectra of white matter (top: 12 mL VOI) and gray matter (bottom: 18 mL VOI) of a normal volunteer (STEAM, TR/TE/TM = 6000/20/30 ms, 64 acquisitions, for abbreviations see text)*

Table 1 *In vivo concentrations (mM) of major cerebral metabolites in parietal white (n=46) and gray matter (n=27) of young healthy adults (mean age 26 ± 4 years)*

Metabolite	*White Matter*	*Gray Matter*
N-Acetylaspartate	7.4 ± 1.3	7.0 ± 1.2
N-Acetylaspartylglutamate	1.6 ± 0.9	0.3 ± 0.4
Glutamate	5.0 ± 1.2	6.8 ± 1.4
Glutamine	2.0 ± 0.9	3.0 ± 0.8
Total Creatine	5.1 ± 0.9	5.5 ± 0.8
Choline-Containing Compounds	1.5 ± 0.3	1.0 ± 0.2
myo-Inositol	3.2 ± 0.9	3.7 ± 0.7
γ-Aminobutyrate	0.8 ± 0.9	1.0 ± 0.4
Glucose	0.3 ± 0.4	0.7 ± 0.4
Lactate	0.6 ± 0.5	0.5 ± 0.4

lactate, alanine, taurine, glutamine, γ-aminobutyrate, guanidinoacetate, cytosolic protein residues and mobile short-chain fatty acids may be observed in other (most often diseased) states of the brain.

For quantitation of metabolite concentrations a library of individual metabolite spectra of known concentration serves as a basis for fully automated spectral evaluation by LCModel.[6] Thereby this program takes advantage of prior knowledge in fitting *in vivo* time-domain data by a regularization method. In fact, since differences in rf coil sensitivity between different studies (phantom, small head, large head, etc.) may be easily compensated for by correcting with a reference pulse amplitude,[7] the procedure results in absolute metabolite concentrations largely independent of instrumental inadequacies such as eddy currents and with a minimum number of assumptions (e.g., no lineshape and baseline assumptions are necessary). Table 1 summarizes recent evaluations for gray and white matter.

3.2 Cerebral Uptake of Ethanol

A remarkable example of cerebral uptake after oral consumption of a substrate is the dramatic increase of brain ethanol shortly after drinking alcohol. As demonstrated in Figure 3 only 12 min after 1 mL ethanol per kilogram body weight (within 5 min) proton spectra from parietal gray matter revealed the presence of several millimole brain ethanol by the presence of the methyl triplet at 1.20 ppm originating from coupling to the methylene protons. This level of brain ethanol coincided with the occurrence of "classic" neuropsychologic symptoms such as alterations of mood and speech.

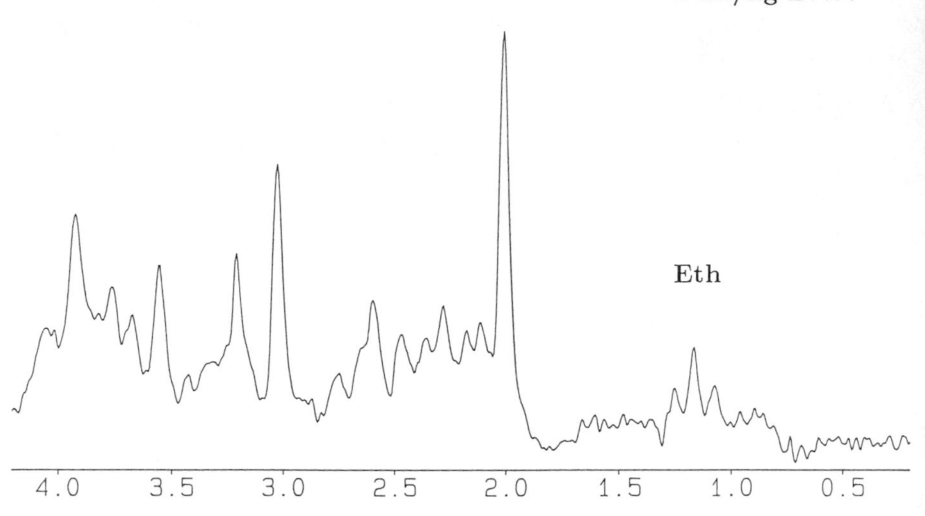

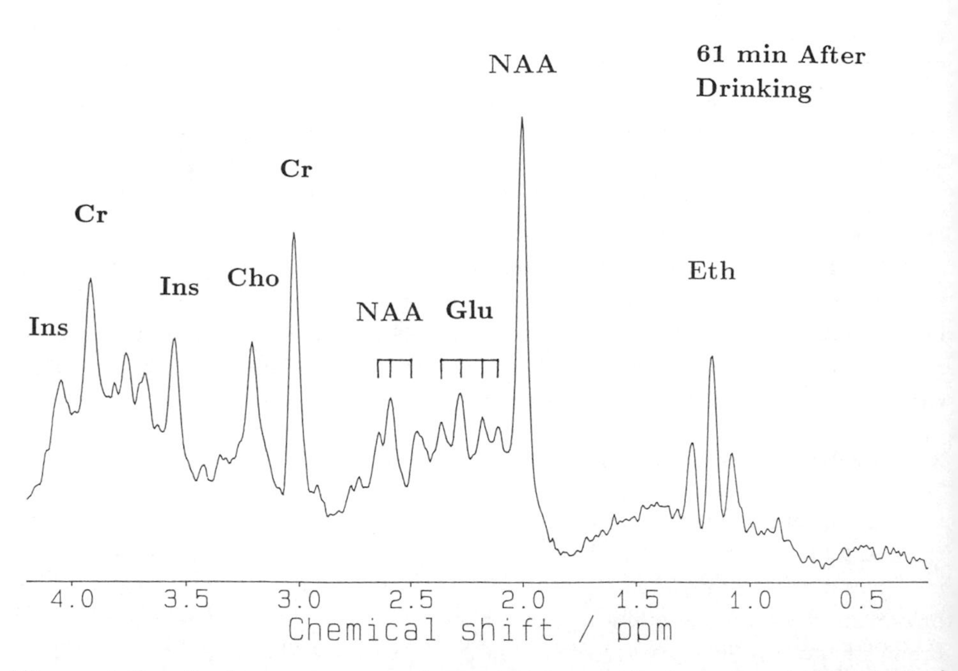

Figure 3 *Localized proton spectra demonstrating cerebral uptake of ethanol (Eth) in gray matter of a normal volunteer 12 min (top) and 61 min (bottom) after drinking (14.4 mL VOI, TR = 3000 ms, 128 acquisitions, other parameters as in Fig. 2)*

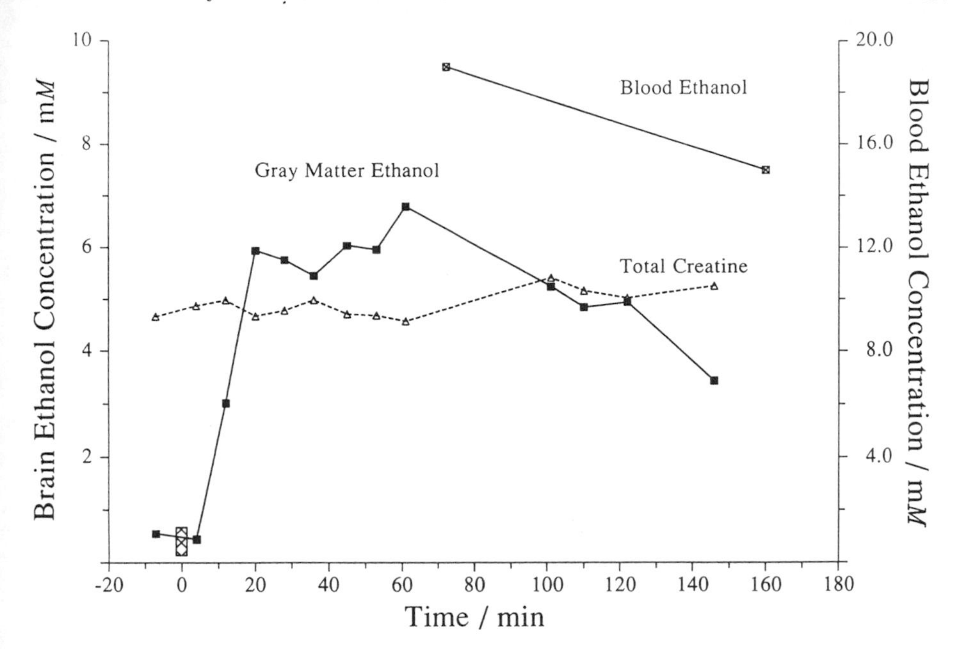

Figure 4 *Time course of brain ethanol concentration after drinking 1 mL/kg body weight ethanol within 5 min relative to total creatine (left scale) and in comparison to blood plasma levels (right scale)*

Figure 4 depicts the time course of gray matter ethanol concentrations (relative to total creatine and without corrections for mild T1 saturation effects at the repetition time of TR = 3000 ms. Typically, a rapid increase led into a plateau phase with maximal ethanol concentrations of about 7 m*M*. Such tissue concentrations were reached within 40–75 min for different volunteers and were followed by a slow decrease at a rate of about 1.5 m*M* per hour.

Interleaved acquisitions of gray and white matter spectra from the same volunteer (not shown) revealed regional differences with a gray/white matter concentration ratio of about 1.6. More important, there was an even more pronounced difference between gray matter and blood ethanol levels. In the plateau phase, i.e. 70 min after drinking in Figure 4, the plasma concentration was a factor of 3 higher than in gray matter. This discrepancy is not yet understood and subject of further investigations.

3.3 Homeostasis of Brain Glucose

When measuring arterial input and venous output parameters of the brain, its overall energy consumption is maintained by complete oxidation of glucose. Although the detailed interplay of glycolysis and oxidative phosphorylation as well as of glial and neuronal compartments is still a matter of ongoing research (see below), glucose and oxygen are essential to survival, so that their delivery via the vascular system needs to be carefully regulated.

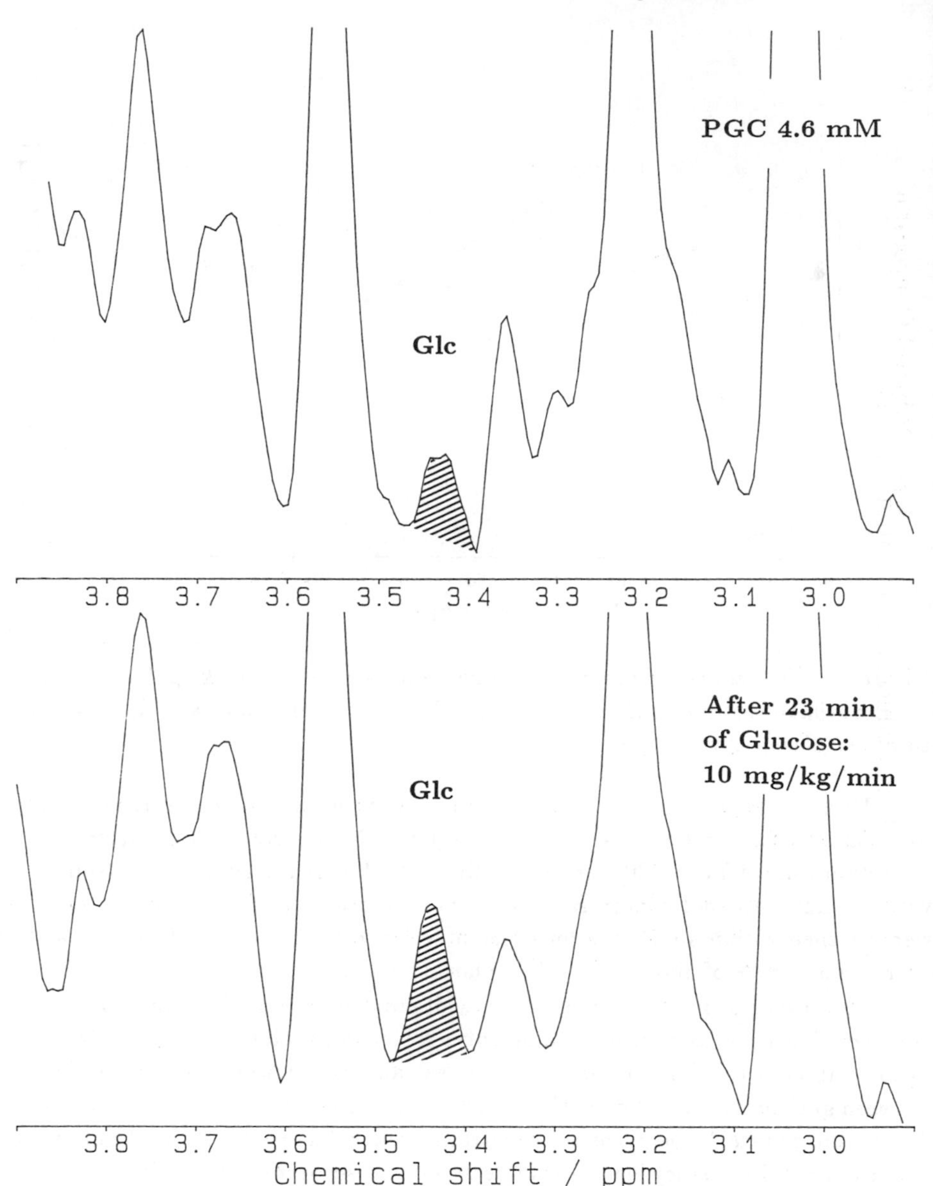

Figure 5 *Localized proton spectra demonstrating homeostasis of cerebral glucose in gray matter of a normal volunteer during normoglycemia (top: plasma glucose concentration 4.6 mM) and during transient hyperglycemia after 23 min of intravenous infusion of glucose (bottom: 10 mg/kg body weight/min, parameters as in Fig. 3)*

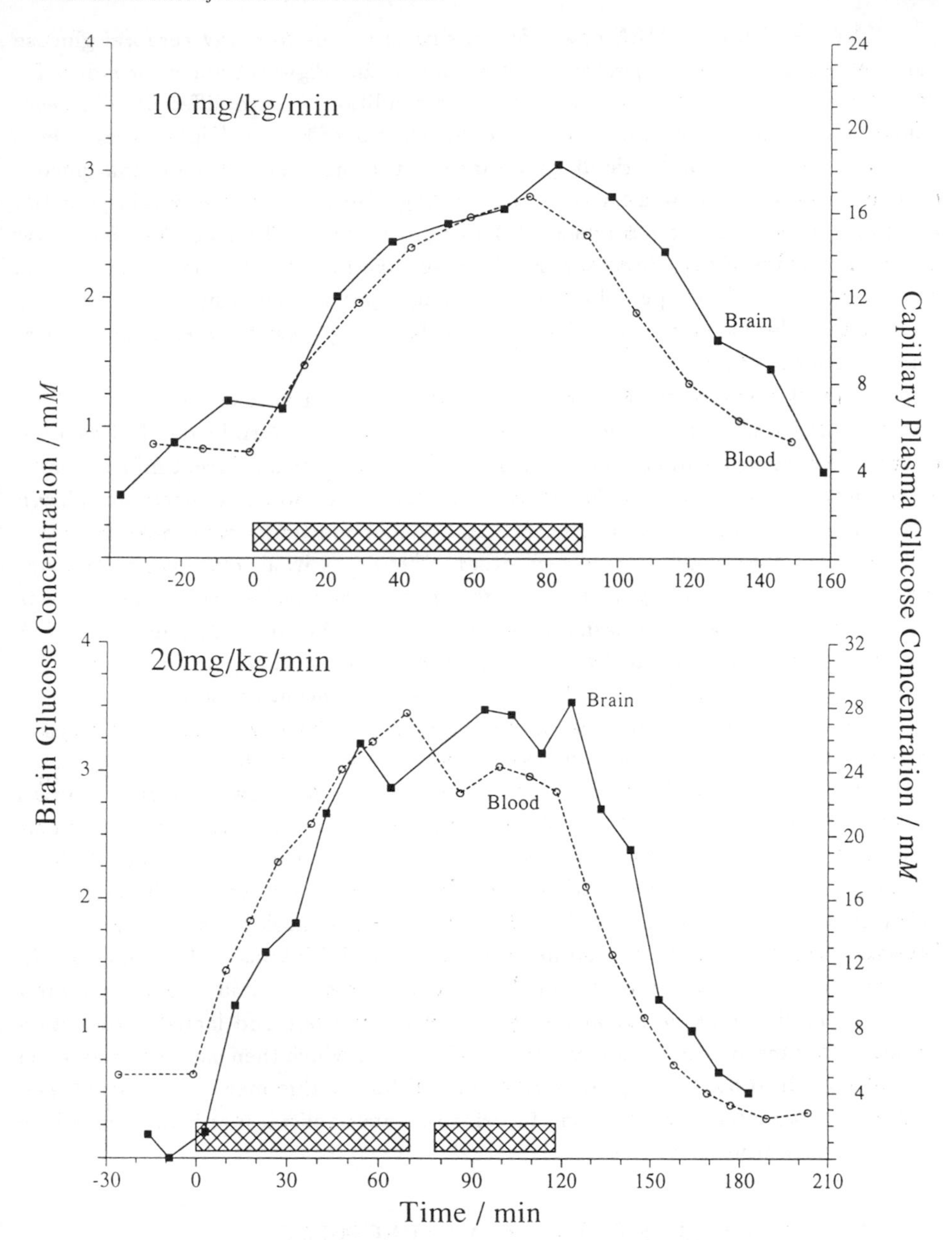

Figure 6 *Time course of brain glucose concentrations (left scale) in comparison to capillary blood plasma levels (right scale) during transient hyperglycemia in two healthy subjects following intraveneous infusion of glucose at 10 mg/kg body weight/min for 90 min (top) and at 20 mg/kg body weight/min for about 120 min (bottom)*

The use of proton MRS now offers a simple means to study cerebral glucose homeostasis (and transport) in the intact human brain. Figures 5 and 6 show data for normoglycemic and (transient) hyperglycemic conditions in two different volunteers following standard glucose infusion protocols. The upper trace in Figure 5 represents a proton spectrum from parietal gray matter of a subject with a plasma glucose concentration of 4.6 mM after overnight fasting. To improve the visibility of the strongly coupled glucose resonance at 3.43 ppm (shaded) the trace focuses on an expanded section of the aliphatic part of the spectrum. After 23 min of intravenous glucose infusion at 10 mg per kilogram body weight per min the resulting gray matter spectrum in the lower part of Figure 5 revealed a significant increase in cerebral glucose concentration.

Figure 6 compares the time courses of gray matter glucose concentration (filled squares) and capillary plasma glucose levels (open circles) determined in between the 6.5 min spectral acquisitions. The data were obtained for two different infusion rates in two subjects. At these rates there was no latency in the increase or decrease of brain levels after changing plasma concentrations. While the temporal profiles of tissue and plasma values closely follow each other, differences in absolute concentrations are by factors of 6–8 during acute hyperglycemia. Nevertheless, in normal subjects hyperglycemic gray matter/plasma ratios in of about 0.17 (10 mg/kg/min) and 0.13 (20 mg/kg/min) are close to the normoglycemic ratio of 0.14 calculated from the mean glucose concentration in Table 1 and a plasma concentration of about 5 mM.

Recently, time-resolved dynamic proton MRS studies of metabolic correlates of human brain function have demonstrated both a 40–50% decrease of brain glucose[8] in the calcarine cortex (primary visual cortex) and a *transient* 70% rise in brain lactate2–3 min after the onset of photic stimulation[9] (10 Hz red flickerlight, binocular matrix of 4×5 fiber optics). Interestingly, the observed increase in lactate from 0.4 mM to about 0.7 mM matched the observed excess glucose consumption of 0.07 mM min^{-1} when assuming that glucose utilization employs anerobic glycolysis (2 lactate molecules per glucose per minute) during the initial phase of stimulation. In the presence of enhanced rather than diminished oxygen availability this apparently paradoxic finding may be explained by anerobic glycolysis and lactate production within astrocytes followed by lactate uptake of neurons which then use lactate as a fuel for mitochondrial oxidative phosphorylation. Although this mechanism might lead to the observed latency in switching to oxidative metabolism, many open questions require further studies *in vivo*.

4 PROTON MRS OF HUMAN BRAIN: PATHOLOGY

4.1 Brain Glucose in Diabetes Mellitus

While measurements of glucose homeostasis and transport in normal subjects are of fundamental neurochemical interest, the possibility of monitoring tissue/plasma ratios may be of clinical importance as well. For example, Figure 7 shows proton spectra of gray matter in two chronic diabetics aged 60 and 69 years, respectively.[10]

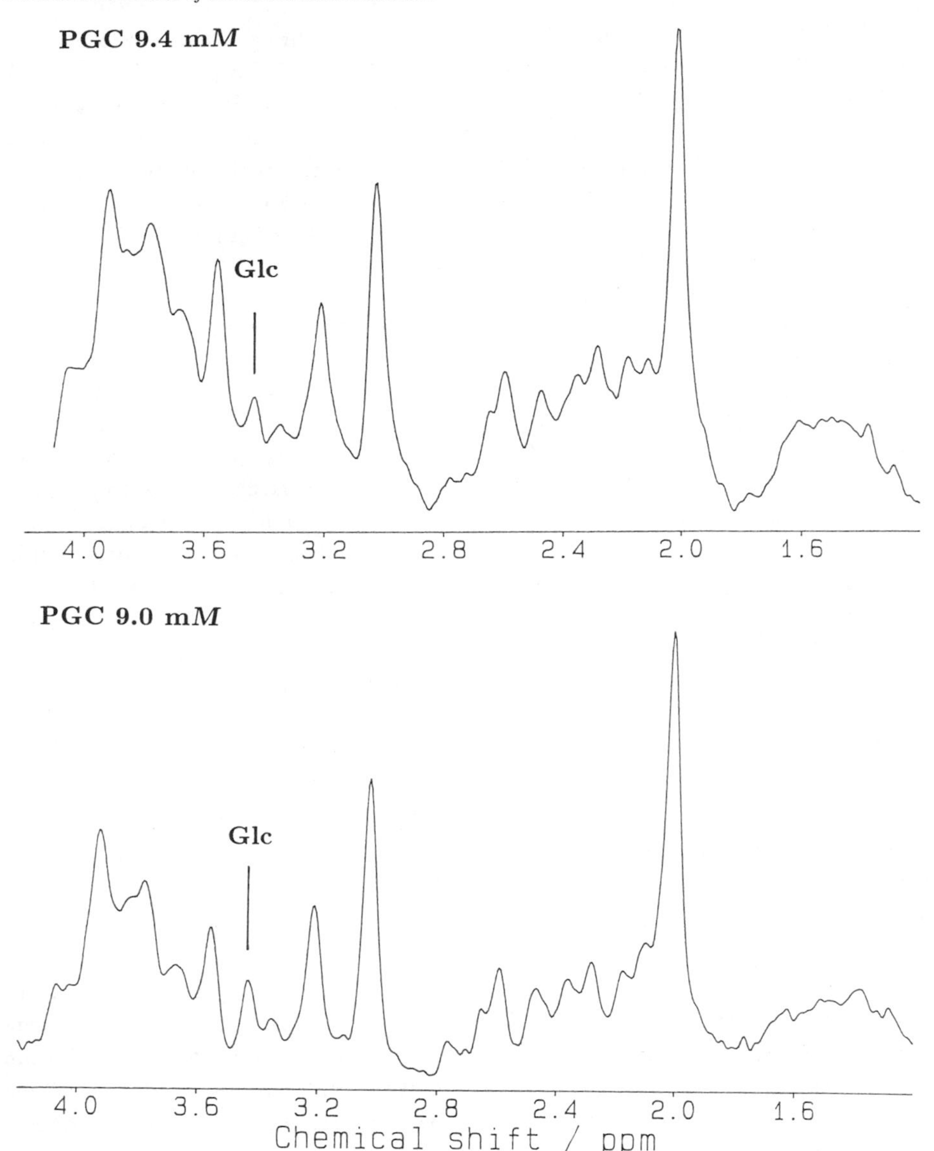

Figure 7 *Localized proton spectra demonstrating differences in brain glucose levels in parietal gray matter of two patients with diabetes mellitus type II that had similar plasma glucose concentrations of 9.4 mM (top: 72 years) and 9.0 mM (bottom: 69 years) at the time of the MRS investigation (18 mL VOI, other parameters as in Fig. 3). Evaluated brain concentrations are 3 mM (top) and 6 mM (bottom)*

The striking finding in these patients is the difference in brain glucose concentration despite a similarity of plasma levels at the time of the MRS examination. Such observations may explain why patients with insulin-dependent diabetes mellitus often present with warning symptoms of hypoglycemia while plasma values still indicate normoglycemia. Obviously, control of blood sugar does not ensure proper adjustment of brain tissue levels, if alterations of the affinity for blood glucose (e.g., membrane properties or tranporter systems) affect glucose transport kinetics via the blood brain barrier.

4.2 Choline-containing Compounds and Anorexia Nervosa

Malnourishment or even starvation may lead to cerebral metabolic abnormalities. For example, choline and lecithin are naturally occurring constituents of our daily food that have been orally used in several psychiatric disorders. Figure 8 shows white matter proton spectra of a 16-year-old girl with anorexia nervosa and an age-matched control. While no biochemical changes were noted in gray matter, there was a substantial and statistically significant decrease of choline-containing compounds in white matter in a cohort of anorectic girls. Although the choline concentration itself is almost negligible for the MRS-detected Cho resonance, it is an essential component of phosporylcholine, glycerophosphorylcholine, and the neurotransmitter acetylcholine as well as of membrane lipids such as phosphatidylcholine (lecithin) and spingomyelin incorporated into myelin that all contribute to the *N*-trimethylammonium signal in white matter. At this stage it may be hypothesized that the decrease of Cho is linked to the reversible (after weight gain) enlargement of CSF spaces and cerebellar atrophy seen in these patients. A pilot study of supplementation of cholines in anorectic patients has been started using proton MRS as a tool for monitoring its effect on cerebral white matter.

4.3 Creatine deficiency: a treatable inborn error of metabolism

Biosynthesis of creatine in the liver involves arginine and glycine, while delivery to the brain is accomplished via blood tranport. In a 22-months-old boy where localized proton MRS had diagnosed generalized and complete depletion of brain creatine, a hitherto unknown inborn error of metabolism, oral substitution of creatine over several months resulted in a complete recovery from his extrapyramidal movement disorder and epileptic seizures.[11]

Figure 9 shows representative proton spectra of gray matter at the age of 22 months as well as after 4 weeks of oral substitution of arginine and 8 months of oral substitution of creatine. During arginine treatment, the initial therapeutic trial, the lack of creatine was complemented by a substantial increase of guanidinoacetate (resonance G in Figure 9) as the immediate precursor of creatine (= α-methylguanidinoacetate). Since therefore creatine synthesis was blocked at the final step, subsequent direct substitution of creatine led to remarkable spectroscopic and clinical improvements. While NAA and cholines remained stable during treatment, guanidinoacetate returned to basal values and creatine increased from virtually

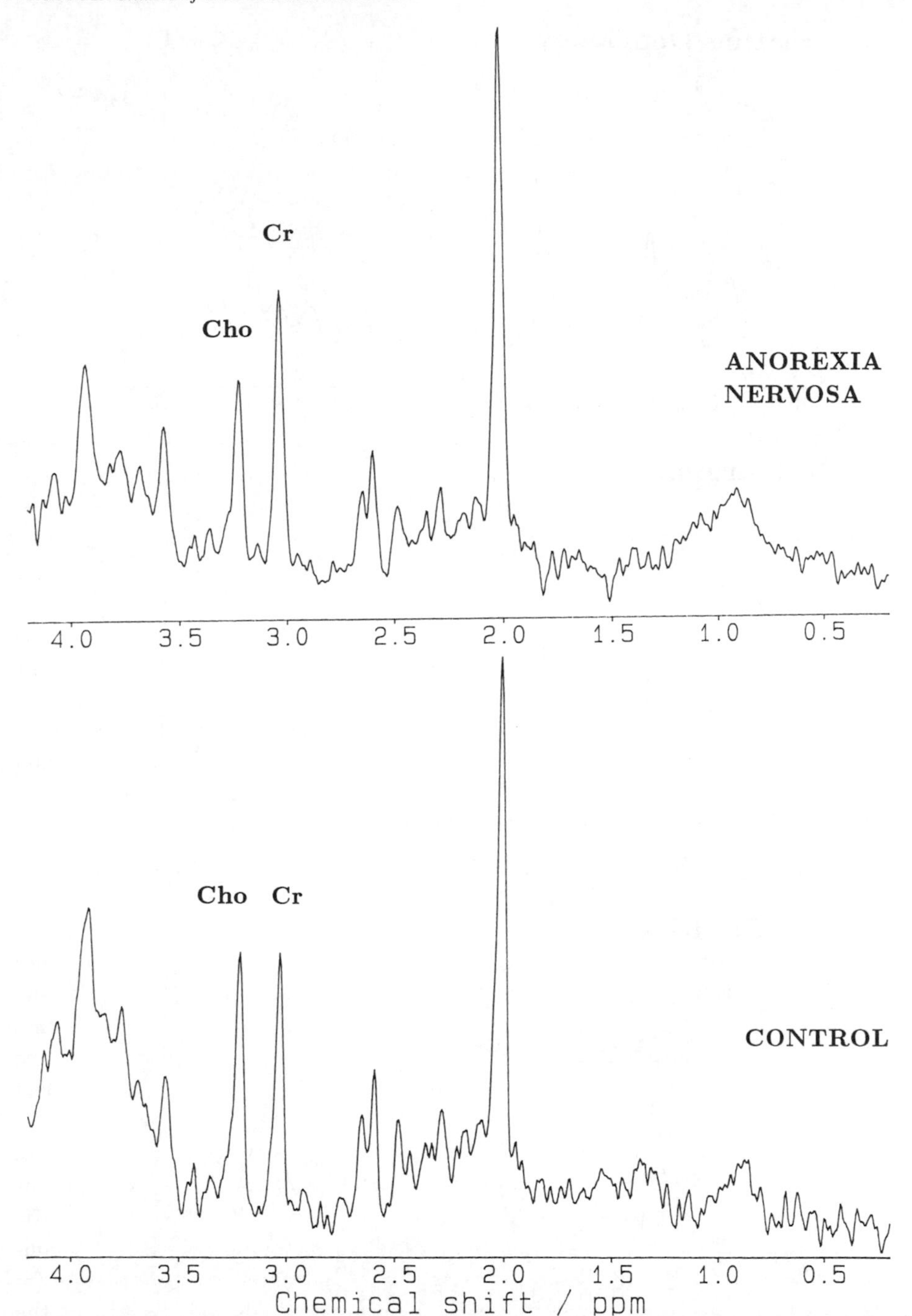

Figure 8 *Localized proton spectra demonstrating decreased cholines in white matter of a 16-year-old girl with anorexia nervosa (top: 12 mL VOI) in comparison to a control (bottom: 8 mL VOI, other parameters as in Fig. 2)*

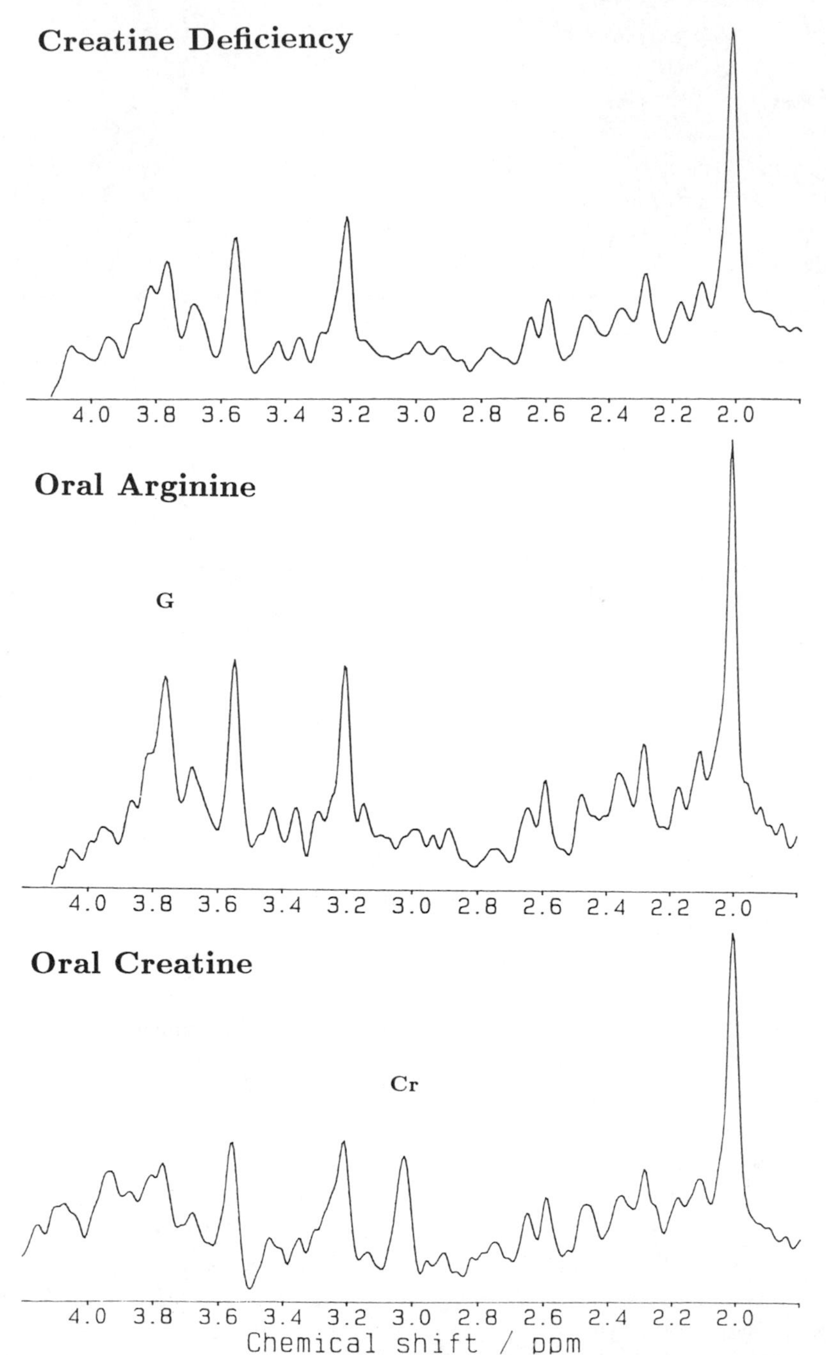

Figure 9 *Localized proton spectra of parietal gray matter in a patient with generalized deficiency of creatine at the age of 22 months (top), after 4 weeks of oral substitution of arginine (middle), and after 8 months of oral substitution of creatine (bottom, 8 mL VOI, other parameters as in Fig. 2)*

nothing to 2.5 m*M* or about 40% of young adult controls.

Phosphorus MRS (not shown) further revealed replacement of phosphorylated guanidinoacetate by PCr up to almost normal levels during creatine treatment. Moreover, pulsed saturation transfer experiments demonstrated a significant transfer from γ-ATP to the newly formed PCr clearly underlining that cerebral uptake of creatine is followed by phosphorylation and normal metabolic function, i.e. phosphate exchange with ATP.

A finding of general importance is the observation that refilling of the cerebral creatine pool by oral substitution seems to level off at about 40–45% of normal controls. This is in line with a previous study of the accessible creatine pool in rat brain,[12] where feeding a creatine analogue replaced only 40% of total creatine. It adds further evidence to the assumption of two different creatine pools in brain, only one of which is accessible via blood transport.

Combined proton and phosphorus MRS now further suggests that this pool – which has been speculated to be neuronal in nature almost entirely exists as PCr. This is also supported by the lack of creatine in primary glial tumors such as astrocytomas and gliomas.[13,14] Evidence accumulates that neurons perform oxidative phosphorylation, while astrocytes take up glucose from the blood, store glycogen, and produce lactate by anaerobic glycolysis – ideas that are very much related to the aforementioned investigation of energy metabolites during functional brain activation. No doubt, further impact of noninvasive NMR studies on human neuroscience is foreseeable.

References

1. R.J. Ordidge, M.R. Bendall, R.E. Gordon, A. Connelly, in: Magnetic Resonance in Biology and Medicine, Tata McGraw-Hill, New Delhi, 1985.
2. P.A. Bottomley, *Ann. N.Y. Acad. Sci.*, 1987, **508,** 333.
3. J. Frahm, K.D. Merboldt, W. Hänicke, *J. Magn. Reson.*, 1987, **72,** 502.
4. K.D. Merboldt, D. Chien, W. Hänicke, M.L. Gyngell, H. Bruhn, J. Frahm, *J. Magn. Reson.*, 1990, **89,** 343.
5. J. Frahm, T. Michaelis, K.D. Merboldt, H. Bruhn. M.L. Gyngell, W. Hänicke, *J. Magn. Reson.*, 1990, **90,** 464.
6. S.W. Provencher, *Magn. Reson. Med.*, 1993, **30,** 672.
7. T. Michaelis, K.D. Merboldt, H. Bruhn, W. Hänicke, J. Frahm, *Radiology*, 1993, **187,** 219.
8. K.D. Merboldt, H. Bruhn, W. Hänicke, T. Michaelis, J. Frahm, *Magn. Reson. Med.*, 1992, **25,** 187.
9. G. Krüger, Diploma Thesis, Universität Braunschweig, 1994.
10. H. Bruhn, T. Michaelis, K.D. Merboldt, W. Hänicke, M.L. Gyngell, J. Frahm, *Lancet*, 1991, **337,** 745.
11. S. Stöckler, *et al.*, *Pediatr. Res.*, 1994, **36,** 409.
12. D. Holtzman, *et al.*, *Brain Res.*, 1989, **483,** 68.
13. H. Bruhn, *et al.*, *Radiology*, 1989, **172,** 541.
14. J. Frahm, *et al.*, *J. Comput. Assist. Tomogr.*, 1991, **15,** 915.

NMR Study of Lipid Fluidity in Frozen Red Muscle of Atlantic Salmon (*Salmo salar*): Relation to Autoxidation of Lipids?

H. Grasdalen[1], M. Aursand[2], and L. Jørgensen[2]

[1] DEPARTMENT OF BIOTECHNOLOGY, THE NORWEGIAN INSTITUTE OF TECHNOLOGY, UNIVERSITY OF TRONDHEIM, N-7034 TRONDHEIM-NTH, NORWAY

[2] SINTEF, DIVISION OF APPLIED CHEMISTRY, GROUP OF AQUACULTURE, N-7034 TRONDHEIM-NTH, NORWAY

1 INTRODUCTION

Autoxidation of lipids in fish meat, leading to distinct lowering of product quality during long time cold storage, has been and is still a great problem in aquaculture product preservation. Normally the fish products are stored at low temperature, in the range from -20 to -60°C to prevent rancidity. However, rancidity and off-flavour still develope during storage. It is well known that hydrolysis and autoxidation of lipids play an important role, and the highly unsaturated ω-3 (omega-3) fatty acids are more reactive than less unsaturated fatty acids [1]. Lipid oxidation is a free radical process and it continues even at freezing temperatures [2]. The best precaution against oxidative deterioration is preventing oxygen from coming into contact with the product [2]. Therefore, in the mechanisms involved, effective diffusivity of different reactive species, among which oxygen is very essential, might play an important role.

Whereas liquid water in frozen tissue has been analysed and related to understanding of the mechanisms of protein denaturation and tissue degradation during frozen storage, the fluidity of lipids at low temperature and its significance for lipid oxidation has almost escaped attention.

For this reason we carried out a study on the state of lipids and water in a very fatty fish specimen, the red muscle of Atlantic salmon, as a function of temperature using NMR spectroscopy.

The extensive usefulness of NMR for studying lipids is attributed to major factors such as: NMR is rapid, direct and a noninvasive and nondestructive technique; the measurements can be performed on intact muscles. NMR in the study of lipids now ranges from determining the quantity and composition of lipids present in extracts and in living organisms using proton and carbon-13 [3,4] to determine the structure and dynamical characteristics of lipids in membranes by using deuterium [5] and phosphorus NMR [6]. In an NMR spectrum recorded at high magnetic field, the different lipids exhibit several individual resolved lines, whose integrated intensity yields their NMR-visible amounts present [7].

Quantitative information about the molecular mobility is achieved by relaxation time measurements. For ^{13}C NMR resonances the predominant contribution to spin relaxation is due to dipolar spin-spin interaction with the directly bonded protons. In the case of isotropic motion the dipolar spin-lattice relaxation time, T_1, is given by [8]:

$$1/T_1 = \frac{n_H(\mu_o h \gamma_H \gamma_C)^2}{640\pi^4 r^6} [1/(1+(\omega_H-\omega_C)^2\tau^2) + 3/(1+\omega_C{}^2\tau^2) + 6/(1+(\omega_H+\omega_C)^2\tau^2)]\tau \quad (1)$$

where n_H is the number of protons attached to the carbon in question, μ_o is the permeability constant, h is Planck's constant, γ_H and γ_C are the magnetogyric ratios of proton and ^{13}C, respectively, r is the carbon proton distance, ω_H and ω_C are the radial resonance frequencies for proton and ^{13}C, respectively, and τ is the correlation time, which is the time needed for a carbon-proton couple to rotate $\approx$ 1 radian. The temperature dependence of T_1 unambiguously discriminates between the two τ values fitting equation (1). The short correlation time representing the fast motion regime is applicable when T_1 decreases with temperature whereas the long correlation time typical for the slow motion regime is valid when T_1 increases by reducing the temperature. The T_1 minimum for ^{13}C occurs for $\omega_C\tau \approx 1$.

For lipids *in situ*, where they exist in a relatively viscous state, the molecular tumbling is too slow to average out the nuclear dipolar interaction completely and a resulting informative line broadening is clearly observed when the temperature is lowered. Below 0°C, selective stiffening within the triglycerides is reflected through selective line broadening, and signals from frozen lipids are broadened beyond detection. Lipids in partially immobilized anisotropic systems, like ordered membrane structure, produce severely broadened proton resonances that cannot be observed in high resolution spectra like those presented in this work, and their contribution to the ^{13}C-NMR lines is negligible.

We report here NMR spectra and spin-lattice relaxation times needed to characterize the NMR-visible lipid phase in the red muscle of Atlantic salmon in terms of acyl chain organization, composition and segmental motion in triglycerides in the temperature range from 40°C to -40°C.

Temperature dependent intensities of lipid NMR signals have earlier been observed in chicken pectoralis muscle [9] and bovin muscles [10].

2 EXPERIMENTAL

The fish, Atlantic salmon (*Salmo salar*) (2 days since killed and stored on ice) was obtained from the local fish market in Trondheim. The red muscle was cut into suitable pieces to fit inside a 5 mm NMR tube. The transfer to the NMR tube was performed carefully at $\approx$ 0°C in order to keep the tissue structure as intact as possible.

All ^{13}C NMR spectra were recorded with proton noise decoupling on a JEOL EX-400 NMR spectrometer operating at 100.4 MHz for ^{13}C. The NMR tube was spun at a rate of 15-20 Hz during accumulation of spectra. No field/frequency lock was applied. The magnetic field was homogenized (in advance) on a pure D_2O sample. Quantitative spectra were obtained by using a coaxial two-compartment 5mm NMR tube with hexamethyldisiloxane in $CDCl_3$ as an intensity standard in the central tube. A spectral width of 30 kHz, 65k data points, a 90° pulse angle, a pulse repetition time of 40s, and NOE suppressed, inverse-gated proton decoupling were used. Spin-lattice relaxation times (T_1) were measured by the inversion recovery method.

^1H NMR spectra were obtained at 400 MHz using 32k data points, a spectral width of 8 kHz, a 45° pulse angle and a pulse repetition time of 4 s on the same spectrometer.

NMR spectra and T_1 values were also obtained from commercial cod-liver oil enriched in n-3 fatty acids by P. Møller, Oslo, Norway.

3 RESULTS AND DISCUSSION

3.1 NMR-visible Content of Lipids.

Figure 1A and 1B show the ^{13}C NMR spectra of lipids in the dark muscle of Atlantic salmon obtained at two different temperatures, 20°C and -40°C, respectively. The spectrum is fairly well resolved at 20°C and the NMR-visible content agreed with that determined by traditional methods [7]. Signals are displayed in all spectral regions as expected from triglycerides containing n-3 and highly unsaturated fatty acids: carbonyl/carboxyl carbons (173 - 182 ppm), olefinic carbons (127 - 132 ppm), glycerol carbons (62 - 69 ppm) and the methylene and methyl carbons appearing in the region 14 - 35 ppm [7]. The carbonyl/carboxyl region far to the left in the spectra indicates whether free fatty acids, triglycerides or both are present. In the present case only one signal is observed, appearing at ≈ 172 ppm, which is typical of triglycerides. The presence of triglycerides is also evident from the relatively strong glycerol signals at 61.8 and 68.9 ppm.

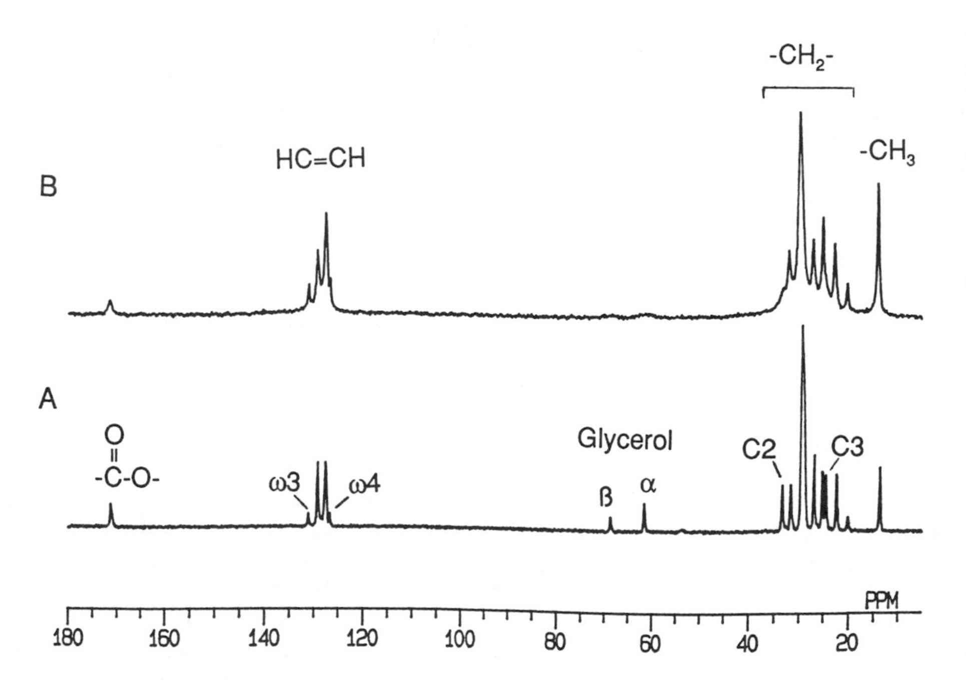

Figure 1. *^{13}C-NMR spectra of intact red muscle of Atlantic salmon (Salmo salar) obtained at (A) 20°C and (B) -40°C.*

In the methylene region, the dominating resonance at ≈ 30 ppm arises from long aliphatic chains, $-(-CH_2-)_n$, whereas the weaker lines represent CH_2 groups situated close to the chain ends or double bonds in the fatty acids. Two lines, important for the analysis, are displayed at ≈ 24.5 and ≈ 34 ppm, from carbon C3 next nearest to, and from C2 nearest to the carbonyl group, respectively. All fatty acids contribute at 34 ppm and all fatty acids except 22:6 contribute at 24.5 ppm.

The terminal methyl groups resonating at ≈ 14.1 ppm give a well separated line whose integrated area represents a measure of the total molar content of all fatty acids present. The spectral feature changed conciderably when the temperature was lowered to -40°C. As shown in Figure 1B, a pronounced selective broadening of lines occured. The resonances from the glycerol-carbons together with C2 and C3 close to the glycerol moiety have almost disappeared from the spectrum, whereas the methyl resonance is scarcely affected. These spectral changes with temperature were completely reversible. The reason for this diversity concerning dipolar line broadening is that the fish lipids exist exclusively in triglyceride form and the glycerol moiety, linking three chains together, constitutes the most rigid part in the lipid molecules. Therefore, a resulting dipolar broadening appears due to slow segmental motion in this region of the molecule. Similar line broadening was seen in the spectra of cod-liver oil when the temperature was lowered.

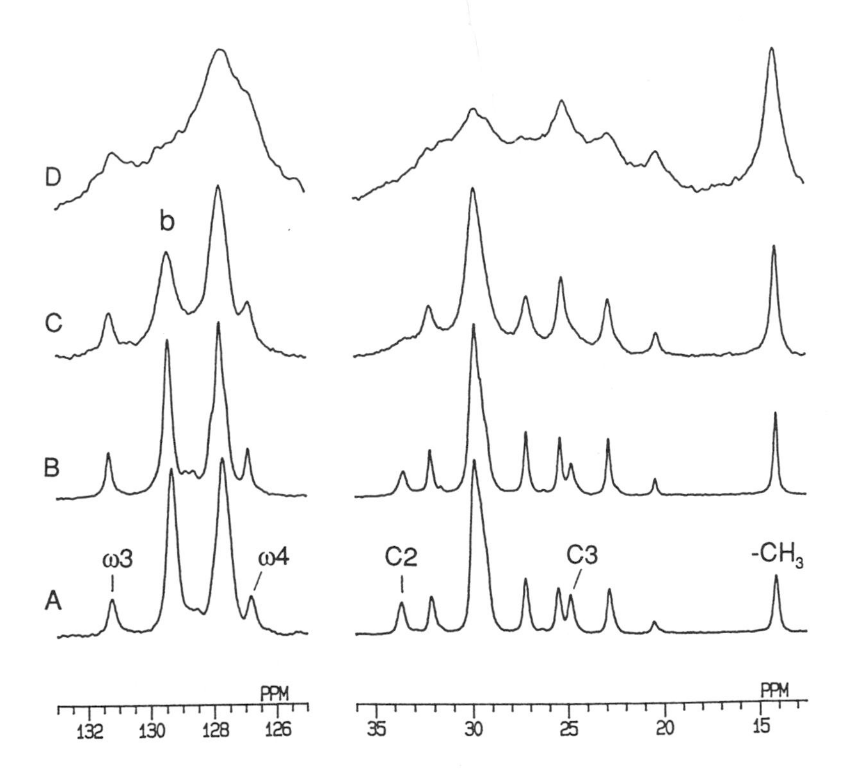

Figure 2. *Expansion of the olefinic, methylene, and methyl region in the ^{13}C NMR spectra of red muscle of Atlantic salmon (Salmo salar) obtained at (A) 0°C, (B) -20°C, (C) -40°C, and (D) -60°C.*

Figure 2 shows expansions of the olefinic, methylene, and methyl region in ^{13}C NMR spectra obtained from muscle at different temperatures ranging from 0°C to -60°C. The spectra appear almost unchanged down to -20°C, and, even at -60° the structural features in the line patterns can be recognised.

In the olefinic region, the two weak lines of equal intensity denoted by ω3 and ω4 at 131.5 ppm and 127.1 ppm, respectively, reveal the presence of n-3 fatty acids. They come only from unsaturated carbons ω3 and ω4 involved in the double bond close to the methyl end in all n-3 fatty acids. The contribution to the line at ≈ 130 ppm, denoted by b, arises from two carbons in mono- and polyunsaturated fatty acids other than n-3 fatty acids and one carbon in n-3 fatty acids. Thus, half the algebraic sum of its intensity plus that of ω3, accounts for one carbon in all unsaturated fatty acids. Hence, the NMR-visible amounts of total, unsaturated, saturated, and n-3 fatty acids can be evaluated by weighting their respective NMR intensities, I, against the intensity I_o of a calibrated standard, as follows:

Total molar content of acyl chains . . . $= I_{CH3}/I_o$ (2)
Molar content of n-3 chains . . . . $= I_{\omega3}/I_o$ (3)
Molar content of unsaturated chains . $= (I_{\omega3} + I_b)/2I_o$ (4)
Molar content of saturated chains . . $= I_{CH3}/I_o - (I_{\omega3} + I_b)/I_o$ (5)

The remaining olefinic resonance at 128 ppm comes only from polyunsaturated fatty acids, and the ratio of its intensity to that at 130 ppm reflects the relative amounts of polyunsaturated fatty acids.

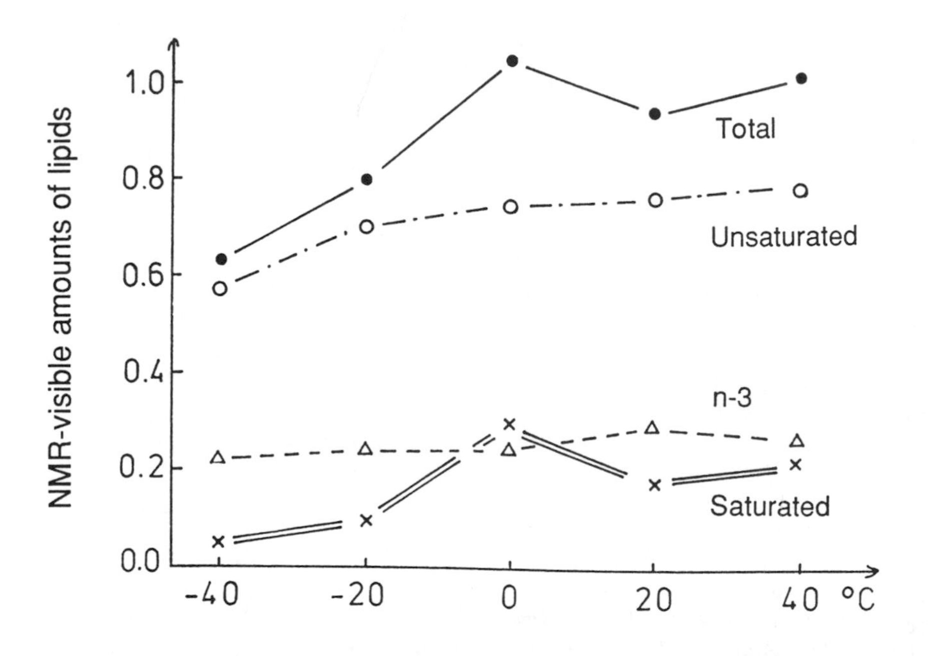

Figure 3. *Temperature dependence of the total amount of* ^{13}*C NMR-visible lipids (normalized to 1) and fatty acid classes in red muscle of Atlantic salmon (Salmo salar).*

As can be seen from Figure 3, the estimated fractions of the three different classes of NMR-visible fatty acids stay constant from 40°C to 0°C in proportions close to those determined by gas chromatographic analysis on lipid extracts [12]. Below 0 °C, the data indicated a gradual freezing out of both unsaturated and saturated acyl chains, whereas all n-3 fatty acids remained almost unaffected down to -40°C. At lower temperatures the total lipid phase became largely immobilized yielding completely unresolved ^{13}C spectra at ≈ -70°C.

It is well known that unsaturated fatty acids have a lower melting point than saturated ones, and that the melting point decreases with the number of double bonds and increases with the number of carbons in the fatty acid chain [13].

In the red muscle of Atlantic salmon, the content of saturated fatty acids is relatively small, about 20% of the lipid phase (mostly 16:0 (≈ 11%)), whereas the monounsaturated fatty acids account for ≈ 56% [12]. The amount of 20:1 and 22:1 as measured by GC reached a level of ≈ 15% each [12], and the plot showing the changing levels of saturated and unsaturated fatty acids by reducing the temperature (Figure 3), may be explained by assuming a gradual freezing out of these fairly long chained lipids together with the saturated ones. This explanation is supported by the selective decrease in the signal b (Figure 2), the only olefinic resonance gaining intensity from monounsaturated fatty acids as mentioned before.

The finding that the NMR-visible fraction of n-3 fatty acids remained constant independent of temperature down to -40°C is consistent with the remarkably low melting points of polyunsaturated lipids [13]. Since all acyl chains are integrated into triglycerides, this property of polyunsaturated fatty acids most probably plays an important role concerning the freezing pattern of triglycerides containing polyunsaturated acyl chains. A polyunsaturated acyl chain containing several *cis*-double bonds causes additional degrees of freedom between acyl chains in a triglyceride, and may function as a freeze-protection moiety by disrupting the packing of acyl chains. This is in accordance with the astonishingly large fraction of fluid lipids at - 40°C, approximately 60%, which is slightly more than twice the content of polyunsaturated fatty acids, ≈ 25%.

This property of n-3 and polyunsaturated fatty acids also has important biological significance when they are incorporated into cell membranes, because the segmental motion in the fatty acid chains affects the fluidity and dynamical features of the membrane. Many fish species and cold-blooded animals, live in habitats with low temperatures, and these would be expected to have a need for increased flexibility in the membrane lipid chains to secure proper functioning at lower temperature. However, in very recent studies, using deuterium NMR to monitor segmental motion of double bonded carbons in membrane lipids [14], questions have been posed as to a more biochemical role of the poly-unsaturated fatty acids concerning protein-lipid interactions.

3.2 Fluidity of lipids.

The chemical shifts of ω3, the glycerol line C_α and the CH_3 line, respectively, are similar for all fatty acids in question. Therefore, since they are well separated, these lines were chosen as markers for the mobility of the triglycerides.

The measured spin-lattice relaxation times shown in Figure 4, indicated a substantial motional restriction in the glycerolmoiety with a T_1 for C_α of 0.12s close to its T_1 minimum at ≈ 10°C. The approximately similar T_1 values for all CH_3 and ω3 carbons near the end of the chain in n-3 fatty acids are considerably longer and stayed in the fast motion regime decreasing from ≈ 3s to ≈ 0.6s by cooling from 20°C to -20°C. Practically similar T_1 values were measured for lipids in cod-liver oil, as shown in Figure 4 for comparison, except for a ≈ 50% longer T_1 value for the more mobile terminal CH_3 groups on n-3 acyl chains. In ^{13}C spectra of cod-liver oil the two CH_3 resonances were resolved down to -10°C.

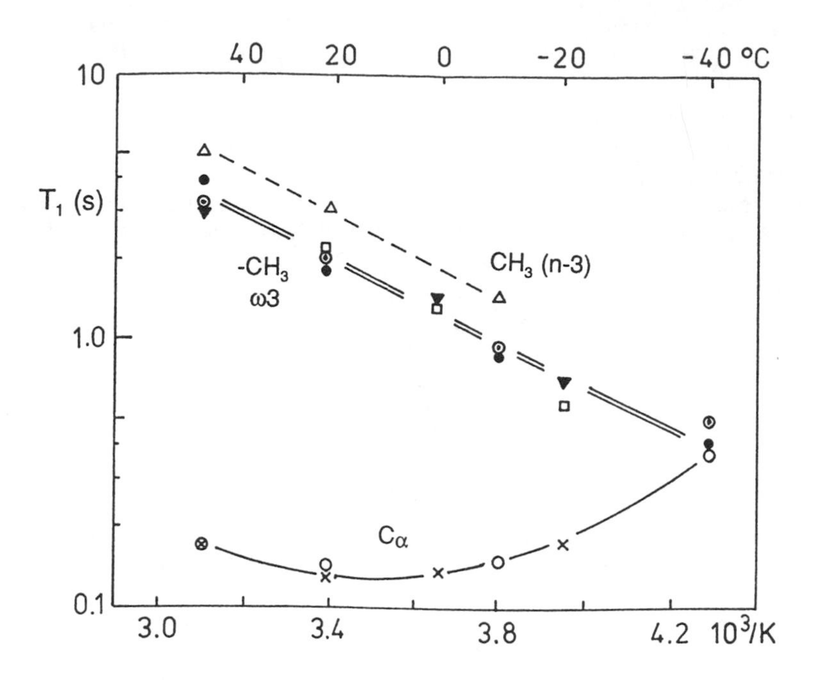

Figure 4. *Temperature dependence of the spin-lattice relaxation times, T_1, for carbons in the NMR-visible lipid phase in red muscle of Atlantic salmon (Salmo salar). Muscle: (□) ω3, (▼) CH_3, (x) C_α; Cod-liver oil: (●) ω3, (⊙) CH_3, (▲) CH_3 in n-3 fatty acids, (○) C_α*

Quantitative information about segmental motion in terms of deduced correlation times using equation (1), relies on measured T_1 values. Since the lipid phase constitutes a mixture of different lipids having somewhat different relaxation properties, the measured T_1 times represent averaged values for the mobile fraction. Nevertheless, no deviation from an exponential decay of nuclei magnetization was observed. Furthermore, the assumption that the segmental motions in the lipid chains are isotropic is probably not fulfilled. Hence, the evaluated correlation times serve only as a rough estimate of the lipid fluidity.

As expected, segmental motion is greatest at the end of the chains with a τ value of $\approx 8 \times 10^{-12}$s for the CH_3 carbons at 20°C. The anchoring glycerol moiety exhibited a noteworthy slower mobility, with a τ value of $\approx 10^{-9}$s for the C_α carbons which is about two orders of magnitude longer than that for the end groups. The double bonded ω3 carbon close to the ends in n-3 fatty acids exhibited a relatively slow segmental motion with $\tau \approx 2 \times 10^{-11}$s. This is understandable because two carbons are locked together via a stiff double bond and represent a relative heavy moiety. At -20°C, the evaluated correlation times had decreased to 2×10^{-11}s, 3×10^{-9}s, and 7×10^{-11}s for carbons CH_3, C_α, and ω3, respectively.

The mobility of the lipids *in situ* seems to be remarkably high down to -30°C which is attributable to their fairly well resolved ^{13}C spectra.

The fluid lipid phase in frozen tissue, comparable with that found in cod-liver oil, is compatible with an organization of lipids consisting of droplets both inside the cells and in the connective tissue.

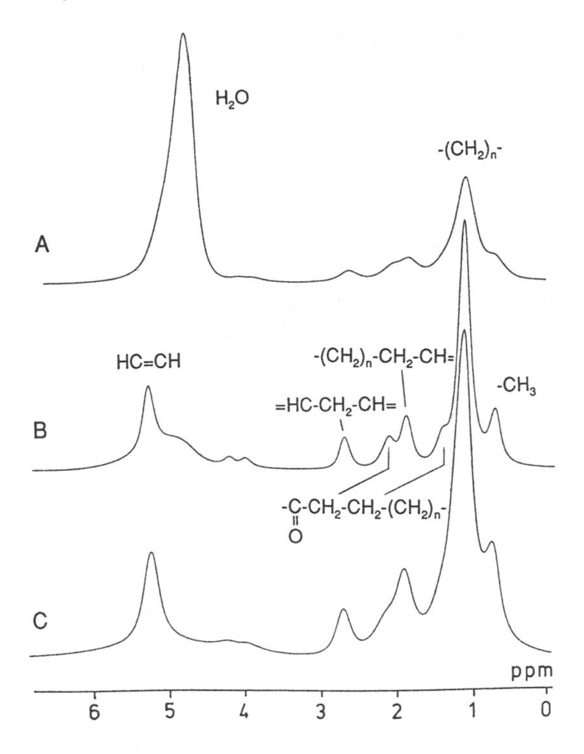

Figure 5. *^1H NMR spectra of intact muscle of Atlantic salmon (Salmo salar) obtained at (A) 0°C, (B) -10°C, and (C) -30°C.*

As shown in Figure 6, the ^1H NMR spectra are also remarkably well resolved and most conclusive information about lipid and water phases can be drawn by comparing relative intensities of denoted peaks. Firstly, at 0°C, the water resonance is dominating and completely masking the signal from protons bonded to unsaturated carbons at ≈ 5.4 ppm. At -10°C the water had partly frozen and its line has decreased to a broad shoulder on the right hand side of the olefinic protons line. At -30°C the line from the residual liquid water has broadened beyond detection. Secondly, the absence of the choline signal at ≈ 3.3 ppm, demonstrates that no phospholipids (polar lipids) were present in the mobile phase. Thirdly, characteristics for the NMR-visible lipids, like number averaged fatty acid chain length and average number of double bonds per chain, as calculated from line intensities in the spectra, agreed with those given by the composition of the lipid extract [12]. Hence, the majority of the extractable lipids also seems to show up in the ^1H NMR spectrum of intact red muscle supporting the organization of lipids within the tissue as mentioned above.

3.3 Significance for Lipid Autoxidation.

Lipid oxidation in fish is a very important factor responsible for deterioration during frozen storage. Even if freezing slows enzymes activity and bacterial growth, the development of rancidity by lipid oxidation proceeds. Hydrolysis of highly unsaturated fatty acids leading to accumulation of free fatty acids seem to be primary processes prior to autoxidation. The latter is a free radical process most effectively slowed by preventing contact with oxygen [2].

Therefore, whatever the mechanisms for the development of rancidity to proceed, some communication between the interior part of the flesh and the surface must occur by diffusive transport of either oxygen or lipids through water and/or lipids. About 7% (g/g wet weight of tissue) of the water was liquid (NMR-visible) at -10°C, and at -30°C any residual water signal could not be observed because of broadening and overlapping lipid resonances (Figure 5C). In postrigor-frozen tissue of cod containing more water than fatty salmon, about 7% by weight of liquid water has been found at -20°C [15]. When the temperature was lowered to -65°C all water was completely frozen [15]. However, later findings by differential thermal analysis showed that 24% of the water in cod muscle remained unfrozen down to -80°C [16]. The unfrozen water is associated with polar macromolecules and also exists partly as ice-salt solution mixtures. The mobile water is probably separated from the relatively large nonpolar lipid pool which appart from the saturated and the two long-chained monounsaturated fatty acids stayed partly fluid down to ≈ -70 - -80°C as evidenced by the NMR data in the present work. The fat in the red muscle of Atlantic salmon has been found to be located both incide the muscle cells and in the connective tissue surrounding them [12], and the distribution probably consists of lipid droplets as have been reported for red muscle tissue of other fishes [17]. This is also consistent with the relatively large content of fluid fat in these tissues.

As expected, lipid oxidation proceeds preferentially in skin during cold storage of fish. Yamaguchi and Toyomizu [18] suggested that movement of lipids took place from the connective tissue to the surface from the inner layer by disintegration of the connective tissue. However, different degrees of damage of muscle cells induced by slow and rapid freezing were found to have no effect on the oxidation rate by storage at -5°C [19]. Therefore, it is conceivable to believe that the transport of oxygen from the surface goes preferentially through the mobile water phase through the tissue. The mobility of lipids is probably an important factor within the cells to bring fatty acids in contact with the cell membrane surface, which is permeable for oxygen, and where the oxidative reaction might be initiated.

4 ACKNOWLEDGMENT

This work was carried out as a part of the SINTEF Strategic Technology Programme of Aquaculture and financially supported by The Norwegian Research Council (NTNF-project 26877 and NFR-project 104379/110).

References

1. J. I. Gray, *J. Am. Oil Chem. Soc.*, 1978, **55,** 539.
2. G. J. Flick, G.-P. Hong Jr., and G. M. Knobl, in *Lipid Oxidation in Food*, ACS Symp. Series 500, 1992.
3. L. F. Johnson, and J. N. Shoolery, *Anal. Chem.*, 1962, **34,** 1136.
4. M. Aursand, PhD Thesis, University of Trondheim, 1993.
5. I. C. P. Smith, , 'Biomembranes', Plenum Press, New York, (C. E Manson and M. Kates, eds.), 1984, Vol. 12, Chapter 4, p 133.

6. P. R. Cullis, and B. De Kruyff, *Biochim. Biophys. Acta* 1979, **559,** 399.
7. M. Aursand, J. R. Rainuzzo, and H. Grasdalen, *J. Am. Oil Chem. Soc.*, 1993,**70**,971.
8. F. W. Werhli, 'Topics in Carbon-13 NMR Spectroscopy', (Ed. G. C. Levy), Wiley, New York, 1976, Vol. 2, p. 343.
9. D. D. Doyle, M. Chalovich, and M. Barany, *FEBS Lett.*, 1981, **131**, 147.
10. P. Lundberg, H. J. Vogel, and H. Ruderus, *Meat Science*, 1986, **18**, 133.
11. M. Aursand, and H. Grasdalen, *Chem. Phys. Lipids*, 1992, **62,** 239.
12. M. Aursand, B. Bleivik, J. R. Rainuzzo, L. Jørgensen, and V. Mohr, *J. Sci. Food Agric.*, 1994, **64**, 239..
13. R. Brockmann, G. Demmering, U. Kreutzer, M. Lindemann, J. Plachenka, U. Steinberner, and H. KGaA, 'Ullmann's Encyclopedia of Industrial Chemistry', 5. ed., VCH Publishers, Deerfield Beach (USA), (Ed. Wolfgang Gerhartz), 1987, Vol. A10, p 245.
14. J. E. Baenziger, H. C. Jarrell, R. J. Hill, and I. C. P. Smith, *Biochemistry*, 1991, **30**, 894.
15. M. V. Sussman, and L. Chin, *Science*, 1966, **151**, 324.
16. R. B. Duckworth, *J. Food Technol.*, 1971, **6**, 317.
17. I. A. Johnston, *Comp. Biochem. Physiol.*, 1982, **73B**, 105.
18. K. Yamaguchi, and M. Toyomizu, *Bull. Jap. Soc. Sci. Fish.*, 1894, **50,** 2049.
19. M. C. Tomas, and M. C. Anon, *Int. J. Food Sci. Technol.*, 1990, **25**, 718.

Magnetic Resonance in the Study of Biopolymers and Complex Systems

Proton Magnetic Resonance and Relaxation in Dynamically Heterogeneous Systems

Robert G. Bryant, Denise P. Hinton, Xiaoqi Jiao, and Dawei Zhou
DEPARTMENT OF CHEMISTRY, UNIVERSITY OF VIRGINIA, CHARLOTTESVILLE, VA 22901, USA

1 Introduction

Foods pose an interesting and important problem for rapid, non-invasive analytical techniques that may provide detailed molecular information as well as characterize the components of very complex issues such as taste, mouth feel, transformation in storage, response to mechanical work, or temperature cycling. Magnetic resonance provides an efficient means for approaching many problems using a combination of high-resolution experiments on the liquid components, high resolution solids experiments on the rotationally immobilized systems, and relaxation spectroscopy in both cases. Magnetic relaxation is particularly interesting not only because understanding it is crucial to appropriate acquisition and analysis of routine data, but also because it provides observables that are directly tied to the molecular dynamics in the system studied.

Dynamically heterogeneous systems, i.e., systems where there is both a rotationally immobilized or solid spin system in contact with a liquid spin system such as water in most biological tissues, pose a slightly more difficult problem than the usual liquid or solid samples because rather efficient magnetic coupling between spin populations may make the observable response of either the liquid or solid spin system dependent on the properties or magnetic preparation of the other.

The definition of the liquid and solid is based on rotational and translational mobility rather than apparent bulk transport properties like viscosity. Thus, in a gel that does not pour from the vessel, the microdynamic behavior of the solvent molecules is often essentially identical with that of the bulk liquid solvent which therefore is dynamically still a liquid. Rotational mobility is the primary determinant of solid behavior magnetically. That is, if the rotational motion is so slow that the magnetic dipole-dipole couplings between the observed spins, which are generally protons, are not averaged, the system falls into the solid domain as far as the magnetic resonance experiment is concerned. Practically speaking, this means that if the vector connecting two protons reorients slowly compared with the dipole-dipole coupling strength, the system falls into the solid regime. Since the dipole-dipole coupling for protons is measured in the tens of kHz range, the time scale for the reorientation process for this distinction is in the range of tens of microseconds. The problems and opportunities raised by the effects of relaxation coupling are the focus of this discussion.

2 Relaxation Coupling Effects

Magnetic coupling of the relaxation response of spin pairs was treated early by Solomon who considered both hetero- and homo-nuclear effects on both longitudinal and transverse relaxation.[1] This fundamental study is the foundation of present applications of cross-relaxation effects or nuclear Overhauser effects in molecular structure determinations.[2,3] A sound qualitative picture of the relaxation coupling problem is summarized in Figure 1 which shows two reservoirs representing separately observable spin populations, the solid and the liquid, which are both coupled to the other degrees of freedom in the system by the standard spin-lattice relaxation pathways, but are also coupled to each other by a cross-relaxation path. Whenever the cross-relaxation or magnetization transfer rate is not negligible compared with the spin-lattice relaxation rates, the observable response of either liquid or the solid spin population will be affected by the coupling between the two. Considering only the z-component of magnetization, i.e., the population part of the problem, and denoting the z-component of the macroscopic magnetization of the immobile macromolecular protons as m_s and that of the water protons as m_w, the differential equations describing the coupled spin populations are

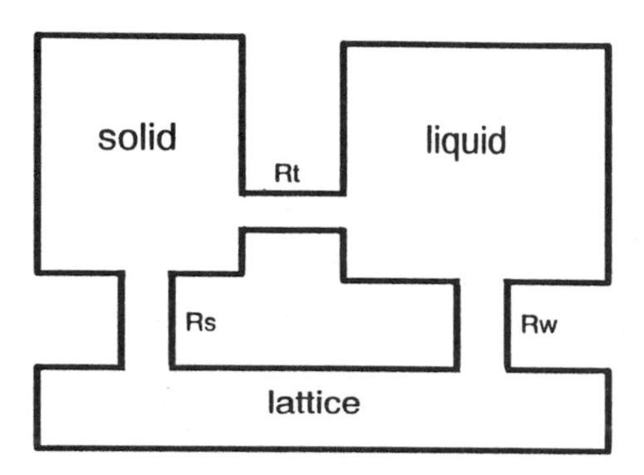

Figure 1. Schematic diagram of relaxation coupling between two spin populations.

$$\frac{dm_w}{dt} = -(R_{ws}+R_w)(m_w(t)-m_w(\infty)) + \frac{R_{ws}}{F}(m_s(t)-m_s(\infty))$$
$$\frac{dm_s}{dt} = -(\frac{R_{ws}}{F}+R_s)(m_s(t)-m_s(\infty)) + R_{ws}(m_w(t)-m_w(\infty)), \quad (1)$$

where R_w and R_s represent the water and protein proton spin-lattice relaxation rates in the absence of net magnetization transfer, and R_{ws} is the rate of transfer of magnetization from the water to the solid proton spin system which is assumed to be driven by dipole-dipole interactions between the solvent protons and the macromolecular protons. The variable, F, is equal to the ratio of the equilibrium magnetization of solid protons to that of water or liquid protons and enters from a consideration of detailed balance. A direct consequence of F, which is also apparent from inspection of Figure 1, is that the magnitude of the effects caused by the coupling and observed in one population, which is generally the liquid, is a function of the size of the other spin population. A second and crucial consequence is that the observable response to a perturbation applied to either spin population will not generally be an exponential growth or decay but at least a bi-exponential response where the time constants are the roots of the coupled equations which are themselves mixtures functions of the rate constants summarized in Figure 1.[4,5,6,7,8,9,10,11,12,13,14,15] Defining the reduced z-magnetization, M(t) as

$$M(t) = \frac{m(t) - m(\infty)}{m(\infty)}, \quad (2)$$

the solution may be written

$$M_{w,s}(t) = C^{+}_{w,s}\exp(-R_{+}t) + C^{-}_{w,s}\exp(-R_{-}t), \quad (3)$$

where the fast, R_{+}, and slow, R_{-}, decay rates are defined as

$$R_{\pm} = \frac{1}{2}[R_w + R_s + R_{ws}(1 + \frac{1}{F}) \pm [(R_s - R_w - R_{ws}(1 - \frac{1}{F}))^2 + \frac{4R_{ws}^2}{F}]^{\frac{1}{2}}] \quad (4)$$

The coefficients, $C_{w,s}^{\pm}$, are determined by the initial values of the magnetizations which are

$$C_w^{\pm} = \pm\frac{R_w - R_{\mp}}{R_+ - R_-}M_w(0) \pm \frac{R_{ws}}{R_+ - R_-}[M_w(0) - M_s(0)]$$
$$C_s^{\pm} = \pm\frac{R_s - R_{\mp}}{R_+ - R_-}M_s(0) \mp \frac{R_{ws}}{F(R_+ - R_-)}[M_w(0) - M_s(0)]. \quad (5)$$

It is clear that neglect of the relaxation coupling and interpretation of the slow component of the decay in the liquid magnetization, R_{-}, in terms of the usual models for an isotropic liquid relaxation, R_w, will possibly lead to major errors. The addition of other physical ideas to the relaxation equation such as a distribution of correlation times, although physically reasonable, may aggravate the difficulty and lead to vary unrealistic interpretations of the liquid component dynamics as pointed out by Shirley and Bryant for the case of a protein system.[4]

In spite of the difficulties, detailed consideration of the liquid relaxation in several systems has led to important conclusions about the dynamics of water at macromolecular interfaces. With a few exceptions to be noted below, the translational and rotational motions of water molecules at the surfaces of proteins, nucleic acids, and even glass are rapid and characterized by correlation times within a factor of ten of the bulk liquid values.[16 17 4 18] This conclusion is supported by high resolution measurements of water-proton-protein-proton nuclear Overhauser effect measurements, or their absence.[19 20 21 22] The time scale here is on the order of tens to hundreds of ps and the conclusions are not in conflict with x-ray or neutron diffraction measurements that often demonstrate water molecule positions are reproducibly occupied in crystals or proteins or nucleic acids. Thus, the concept of an ice-like shell of water surrounding proteins or carbohydrates in the presence of excess water is only useful in the sense of discussing the overall energy of the system. The dynamics of these water molecules, if they are routinely present in unique positions and orientations, is such that they apparently pose no kinetic barrier to a biochemical transformation. These very transient water-macromolecule interactions may, of course, alter the activity of the water, which may have a number of thermodynamic consequences.

3 Spectroscopic Applications of Relaxation Coupling: Z-Spectroscopy

The characteristics of the rotationally immobilized molecules generally found in bio logical materials are that the proton linewidth is typically on the order of 25-50 kHz, may be either Lorentzian or Gaussian in shape, and the spin-lattice relaxation times, which are both field and temperature dependent, are typically near a second and often nearly degenerate with those of the liquid components. The fact that the solid component linewidth is much broader than the liquid permits a class of experiments that has many parallels high resolution spectroscopy of liquids were one spin system, the solid spin system in this case, is selectively irradiated and the effects observed on the other population. Thus, irradiation of the solid proton signal far from the liquid resonance will partially saturate the solid spin system. Because of the magnetic coupling between the two spin systems, the population effects are transferred to the liquid, and shows up simply in the z-magnetization as a loss of polarization. The experiment is simply to apply a preparation pulse off-resonance from the solvent peak by many kHz, then detect the effects on the solvent or liquid magnetization with a second on-resonance pulse. An example of this experiment executed on a cross-linked bovine serum albumin gel is shown in Figure 2 where the frequency of the preparation pulse has been systematically changed from below to above the resonance frequencies of the solid and liquid components. The amplitude of the water resonance faithfully maps the lineshape of the rotationally immobilized protein resonance. There are many modifications in the detection scheme, including using an imaging or other two dimensional pulse sequence. The effects on the solvent resonance in the steady state limit have been discussed by several authors based on solutions of the coupled Bloch equations.[4, 5, 15 23 24 25 26] A useful formulation has been presented by Henkelman and co-workers which permits incorporation of the solid component lineshape function into the steady-state response for the liquid magnetization.[27]

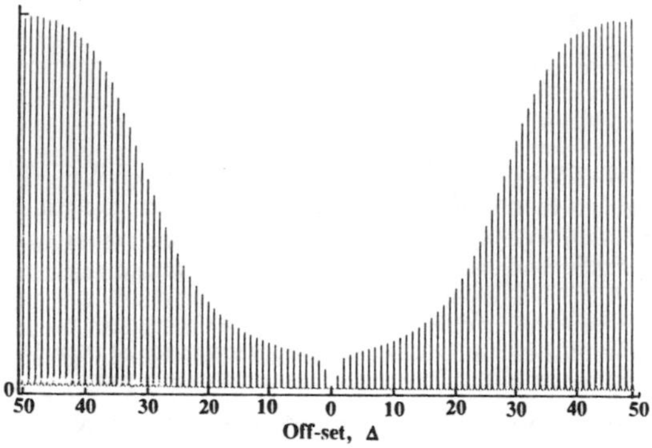

Figure 2. The relative water proton magnetization following a 3 second preparation pulse with 800 Hz amplitude as a function of the off-set from the water resonance frequency. The stacked plot presentation was made with no vertical off-set and the usual absyssa replaced by the value of the preparation pulse off-set rather than chemical shift of the water resonance which is constant throughtout the experiment.

$$\frac{M_w}{M_w^o}=\frac{R_sR_{ws}M_s^o+R_{rfs}R_w+R_wR_{ws}M_s^o}{(R_w+R_{rfw}+RM_s^o)(R_s+R_{rfs}+R)-RRM_s^o} \tag{6}$$

where

$$R_{rfw}=\frac{\omega_1^2T_{2w}^2}{1+(2\pi\Delta T_{2w})^2} \tag{7}$$

$$R_{rfs}=\frac{\omega_1^2T_{2s}^2}{1+(2\pi\Delta T_{2s})^2} \tag{8}$$

The line shape for many solids is best approximated by a Gaussian function, which, may be included in the above formalism by defining

$$R_{rbs}=\sqrt{\frac{\pi}{2}}\,\omega_1^2T_{2s}e^{-\frac{(2\pi\Delta T_{2s})^2}{2}} \tag{9}$$

This approach includes both the solid and water magnetization explicitly so that the parameter F is supressed and R_{ws} is now a second order rather than a pseudo first order rate constant. The crucial aspect of the water response is that it is determined predominantly by the relaxation parameters for the solid spin system, which is not observed directly. Thus, it is possible to characterize the solid component lineshape, and therefore local dynamics, by observation of only the liquid resonance, which is experimentally easier and less subject to experimental distortions than direct observation of the solid. The price for this gain is that, regardless of the detection scheme utilized, in the steady state limit, the experiment takes on the character of a continuous wave experiment. Thus, the solid-state lineshape detected indirectly is still subject to saturation or broadening effects caused by the magnitude of the preparation pulse.[4, 5, 15] However, the corrections for these effects are easy to make when necessary.[4, 5, 15]

4 Magnetic Field Dependence

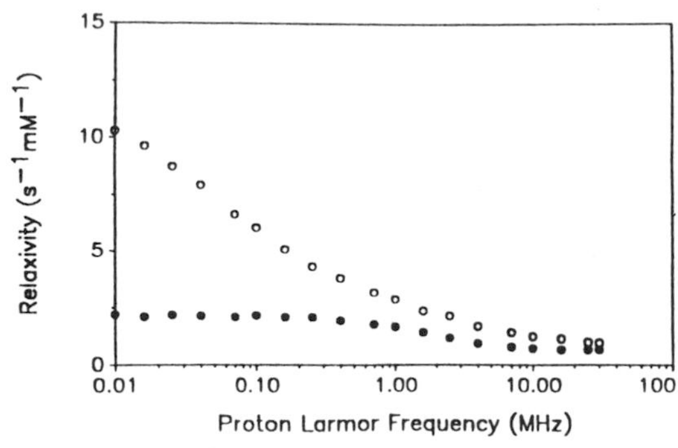

Figure 3. The water proton relaxation rate measured as a function of magnetic field strength reported as proton Larmor frequency at 298 K for 1.8 mM bovine serum albumin samples. The lower curve was obtained on the solution, the upper curve obtained after the solution was cross-linked with glutaraldehyde to immobilize the protein; i.e., to stop the rotational averaging of the proton-proton dipole-dipole coupling within the protein.

The magnetic field dependence of the spin-lattice relaxation rate provides fundamental test of the models used to interpret the relaxation data. The magnetic field dependence of the water spin-lattice relaxation rate in a protein system is shown in Figure 3 and is representative of a number of proton rich systems.[28] The lower curve is the relaxation rate profile for a protein solution. The magnetic field dependence is described approximately by a Lorentzian function, and the inflection point provides a report of the rotational frequency of the protein in the solution.[29] When the protein sample is cross-linked at constant composition the cross-linking reagent causes the protein rotational motion to stop. The resulting relaxation dispersion curve is fundamentally different in shape from that of the solution, but is representative of many heterogeneous systems including most animal tissues.

The shape of the relaxation dispersion profile for the protein gel is not described by any of the usual relaxation models for liquids. The dispersion curves have the following characteristics: 1) The relaxation profile follows a power-law where the rate is given by an equation of the form

$$Rate = A\,\nu^{-B} \tag{10}$$

where A is a constant, ν is the proton Larmor frequency, and B is generally 0.5 but in dry systems may be as large as 0.75.[30 31 32 33] 2) The magnitude of the relaxation rate at low magnetic field strengths is inversely proportional to the water content; i.e., the more water that is present, the lower the rate. 3) The relaxation rate profile shape is independent of the water content of the system. 4) The shape of the profile is the same whether water protons or methyl protons of a cosolvent are observed. 5) The relaxation rate for the water protons may be quantitatively described in terms of the cross-relaxation model; that is, the magnetic field dependence of the water proton relaxation rate is a scaled map of the relaxation rate profile for the rotationally immobilized spins in the heterogeneous system. Thus, the magnetic field dependence of the water proton spin relaxation does not carry any information directly about the water molecule motions in the heterogeneous systems. Why the magnetic field dependence of the proton relaxation rate in the protein or other solid material is this power law dependence is an interesting question in its own right but is beyond the scope of this discussion. Since the relaxation coupling between the liquid and solid spin systems determines the magnetic field dependence for the solvent spin-lattice relaxation in many dynamically heterogeneous systems, similar complications are expected in the interpretation of other relaxation rates such as $T_{1\rho}$.

5 Mechanisms of the Magnetization Transfer

The transfer of magnetization between a solid component and a liquid component may be more or less efficient depending on the mechanistic channels available. There are three main pathways operative: 1) Atom transfer involving labile protons in both the solid and liquid populations such as exchange of water protons with ionizable groups of the solid macromolecule such as alcohols, amines, or amides. 2) Exchange of the whole solvent molecule between a bound environment on solid phase and the liquid or mobile phase. 3) A transient dipole-dipole interaction that operates as the solvent diffuses in close proximity to the solid phase protons.

The third or transient diffusion pathway generally makes only a small contribution to the total relaxation rate because translational diffusion even at surfaces is rapid; thus, the correlation time for the transient dipole-dipole coupling between the solvent protons and the solid protons is on the order of that for the liquid itself and the cross relaxation efficiency low. Although translational contributions to relaxation in liquids are reasonably well understood, a complete theory for the present case is not available because the diffusing liquid spin may interact simultaneously with a number of solid spins that are in turn well coupled with each other. Nevertheless, this complication should not change the general magnitude of the overall coupling efficiency, and we conclude that this contribution is small and usually unimportant compared with the first two.

The first two contributions represent different types of chemical exchange coupling. In the first case, an atom transfer actually converts a solid spin proton into a liquid spin proton by an exchange of hydrogen between the two environments. The efficiency of this transfer is dependent on the participating functional groups in the solid phase, the acid-base properties of the solvent, the pH, and the temperature. Although the functional groups of a protein or carbohydrate have amines, amides, or alcohols where the proton exchange rates are not generally very high, the rates are usually catalyzed by hydroxide ion and therefore increase markedly with increasing pH. In addition, there are a relatively large number of exchangeable protons on the macromolecule which may participate in the exchange process. Therefore, even in the case where the exchange probability for a particular proton may not be high, there may be many protons so that in total, the pathway may make a significant contribution to the total transfer of magnetization between the liquid and solid phases.

The second chemical exchange pathway may be the only efficient one for solvents that do not have labile protons such as dimethyl sulfoxide or dimethylformamide. These solvents also show Z-spectra and magnetic field dependencies of the spin-lattice relaxation rates that are very similar to those observed in the water resonance.[4, 5, 15 28 34 35] The exchange processes in this case involves the exchange of whole solvent molecules with binding sites on the macromolecule; however, as pointed out above, the vast majority of what may be thought of as surface sites have solvent lifetimes that are very short, and therefore, make only small contributions via the third pathway above. However, x-ray and high resolution NMR have identified a few buried solvent molecule sites on proteins and nucleic acids.[19-22] In addition, the earliest explanation for the magnetic field dependence of the nuclear magnetic spin-lattice relaxation rates was based on the idea of a few thightly bound water molecules; however, the idea was not taken seriously because the numbers involved were so small compared with the total number of water molecules that have to interact with the total surface of a protein.[36] With the identification of specific sites by high resolution methods, and the observation of these magnetic coupling effects in a variety of solvents that do not have labile protons, this idea appears to provide the most likely explanation for a major but not atom-transfer pathway for the magnetic exchange. Recent experiments on mixed solvent have demonstrated that direct competition for some of these sites may be detected by observing the Z-spectrum of all each solvent in a high resolution experiment conducted on a protein gel.[35]

A general question is which of these pathways is dominant? The answer is clearly system, pH, and temperature dependent. A recent study of a protein system indicates that near neutral pH in a serum albumin gel, the atom transfer and molecular transfer paths make similar contributions to the total magnetization transfer rate.[37] However, in carbohydrate systems such as a Sephadex G-25 gel commonly used separations of proteins based on molecular size, the relaxation dispersion profile shape is very clearly pH dependent as shown in Figure 4. The data may be understood in terms of a pH dependent magnetization transfer rate where at neutral pH the rate is insufficient to keep the coupling between the solid and the liquid in the fast exchange regime over the whole range of magnetic field strengths. As a consequence, the exchange rate limits the low field relaxation rate observable and a plateau is reached giving the appearance of a more Lorentzian function. However, raising either the temperature or the pH raises the exchange rate of the OH protons, and moves the observed water proton relaxation dispersion profile towards the limiting fast exchange profile which again represents a scaled mapping of the solid spin magnetization response onto the solvent magnetization.[38]

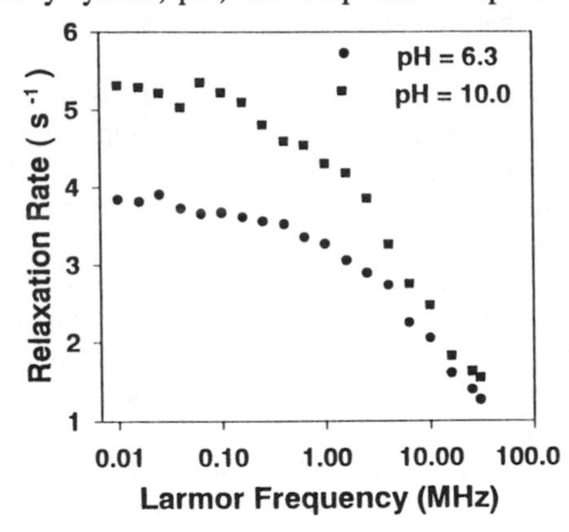

Figure 4. The water proton spin-lattice relaxation rate measured as a function of the magnetic field strength reported as the proton Larmor frequency at 298 K for a 27% by weight Sephadex G-25 gel at pH 7 and 10.

6 Effects of Domain Size

The formulation of the magnetization coupling summarized in Eqns. 1-4 depends on the assumption that all of the solid system spins communicate effectively with the liquid spin system. For a protein system, not all the protein protons are in Van der Waals contact with the water or other solvent molecules. Other solid systems may have similar difficulties of limited solvent contact. The communication between all of the solid protons and the water is facilitated by spin-diffusion processes in the solid spin system that keep the solid spins in equilibrium with themselves, i.e., at a constant spin temperature, even though some portion of the total solid protons are in more efficient communication with the liquid. Spin diffusion in solids is not infinitely efficient. There are difficulties in estimating what the spin diffusion constant is in a complicated structure like a protein, but estimates based on other solid systems suggests that there is an approximate limitation of the order of 50 Å as the depth or distance over which the spin diffusion will keep the total solid spin system in effective equilibrium if there is efficient relaxation at the surface.

Since not all of the interactions between the solvent and the protein are at the surface, most protein systems satisfy the constraints apparently imposed by this diffusion limitation. However, in other heterogeneous systems containing particulate solids of larger dimensions, generally one expects a size limitation on the efficiency of the coupling between the solid and the liquid. The formulation of Eqns. 1-5 may still be used, however, the meaning of the scaling parameter, F, must be changed. If we take F to be the ratio of the solid to the liquid spin populations, and define f as a fraction of the total solid spin population that is in efficient communication with the liquid because of the spin diffusion limitation, then the parameter F in the Eqns. 1 and 4 becomes f*F.

An obvious corollary of the solid spin system getting large is the case where there is not a significant solid proton population though there may be significant mass. An example of such a system would be microporous glass or the any of the microporous rock systems or aluminosilicates. Although there are often some surface OH groups present which may be magnetically in contact with the liquid, their population is very small compared with the systems discussed here. As a consequence, there is no detectable Z-spectrum, and the magnetic field dependence of the relaxation arises from a fundamentally different mechanism.[39]

7 Effects on Transverse Relaxation

The direct effects of dipolar coupling on the transverse relaxation rate are minimal as shown by Solomon's treatment.[1] However, both chemical exchange mechanisms may contribute to the transverse relaxation rate, and the magnitude of the contribution is a function of the dynamical state of the macromolecule system. An example of the effects is shown in Figure 5 where the transverse relaxation decays are shown for two protein systems: one in solution and one cross-linked at the same composition, temperature, and pH. The basis for the increase in the relaxation rate is that when the protein rotational motion is stopped, the transverse relaxation rate of the bound protons changes by several orders of magnitude. We may approximate the bound molecule or bound proton transverse relaxation time as that of the protein protons themselves or approximately 15 μs. The exchange process mixes this relaxation rate

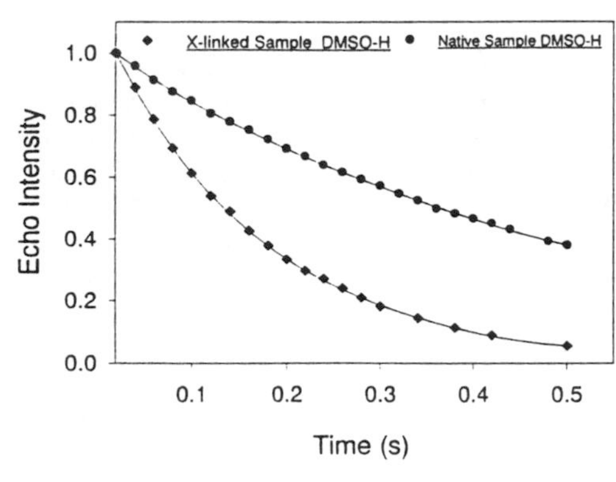

Figure 5. The decay of the transverse magnetization recorded by Carr-Purcell-Meiboom-Gill echo train methods for a 15% bovine serum albumin solution and a cross-linked gel at the same concentration at 298 K.

with that of the bulk solvent whether the protein is rotationally labile or not. In the solution case, the transverse relaxation times are much longer; generally permitting high resolution NMR experiments. If we estimate that the protein proton relaxation time is tens to hundreds of ms for the solution case, the increase in the size of the exchange contribution to the bulk solvent relaxation rate exceeds a factor of 10^3. The effects are observed for both chemical exchange pathways; i.e., the effects are observable in both water protons and methyl protons of co-solvents like dimethyl sulfoxide.[40]

8 Summary

Magnetic relaxation in dynamically heterogeneous system provides considerable information about the molecular dynamics in the system. The effects of magnetic relaxation coupling may generally superimpose the relaxation properties of the solid components of a heterogeneous system on the readily observable relaxation parameters of the liquid spins in the system which provides an complication in interpreting the magnitudes of the relaxation times measured, but also provides opportunities for indirect characterization of the solid spin components in the context of both spectroscopy and imaging experiments.

9 References

1. I. Solomon, *Phys. Rev.*, 1955, **99,** 559.
2. J. H. Noggle and R. E. Schirmer, "The Nuclear Overhauser Effect, Chemical Applications," Academic Press, New York, 1971.
3. K. Wuthrich, "NMR of Proteins and Nucleic Acids," Wiley, New York, 1986.
4. W. M. Shirley, R. G. Bryant, *J. Am. Chem. Soc.*, 1982, **104,** 2910.
5. J. Grad, R. G. Bryant, *J. Magn. Reson.*, 1990, **90,** 1.
6. R. G. Bryant, W. M. Shirley, in Water in Polymers, S. P. Rowland, Ed., ACS Symposium Series #127 (1980), pp.147-156.
7. R. G. Bryant, W. M. Shirley, *Biophys. J.*, 1980, **32,** 3.
8. R. G. Bryant, in Biophysics of Water, F. Franks and S. F. Mathias, Eds., Wiley, New York (1982), pp.125-126.
9. R. G. Bryant, in Mobility and Recognition in Cell Biology, H. Sunds and C. Veeger, Eds., Walter de Gruyter & Co., N.Y. (1983), pp.103-117.

10. S. D. Wolff, R. S. Balaban, *Magn. Reson. Med.*, 1989, **10,** 135.
11. H. T. Edzes, E. T. Samulski, *Nature,*, 1977, **265,** 521.
12. B. M. Fung, T. W. McGaughy, *J. Magn. Reson.*, 1980, **39,** 413.
13. J. C. Gore, M. S. Brown, J. Zhong, I. M. Armitage, *J. Magn. Reson.*, 1989, **83,** 246.
14. J. Zhong, J. C. Gore, I. M. Armitage, *Magn. Reson. Med.*, 1990, **13,** 192.
15. E. R. Andrew, D. J. Bryant, T. Z. Rigvi, *Chem. Phys. Let.*, 1983, **95,** 463.
16. C. F. Polnaszek and R. G. Bryant, *J. Amer. Chem. Soc.*, 1984, **106,** 428.
17. C. F. Polnaszek, R. G. Bryant, *J. Chem. Phys.*, 1984, **81,** 4038.
18. S.H. Koening, R.D. Brown III, *Magn. Reson. Med.*, 1984, **1,** 437.
19. G. Ottig, K.Wuthrich, *J. Am. Chem. Soc.*, 1989, **111,** 1871.
20. G. Ottig, E. Liepinsh, B.T. Farmer II, K. Wuthrich, *J. Biol. NMR,* 1991, **1,** 209.
21. G. Ottig, E. Liepinsh, K. Wuthrich, *Science*, 1991, **254,** 974.
22. G. Otting, E. Liepinsh, K. Wuthrich, *J. Am. Chem. Soc.*, 1991, **113,** 4363.
23. H. N. Yeung and S. D. Swanson, *J. Magn. Reson.*, 1992, **99**, 466.
24. X. Wu, *J. Magn. Reson.*, 1991, **94,** 186.
25. G. H. Caines, T. Schleich, J. M. Rydzewski, *J. Magn. Reson.*, 1991, **95,** 558.
26. S. D. Swanson, *J. Magn. Reson.*, 1991, **95,** 615.
27. R.M. Henkelman, X. Huang, Q.S. Xiang, G.J. Stanisz, S.D. Swanson, M.J. Bronskill, *Magn. Reson. Med.*, 1993, **29,** 759.
28. C.C. Lester, R.G. Bryant, *Magn. Reson. Med.*, 1991, **22,** 143.
29. K. Hallenga, S. H. Koenig, *Biochemistry*, 1976, **15,** 4255.
30. R. Kimmich, F. Winter, *Progr. Colloid & Polymer Sci.*, 1985, **71,** 66.
31. R. Kimmich, W. Doster, *J. Polymer Sci: Polymer Physics Edition.*, 1976, **14,** 1671.
32. R. Kimmich, F. Winter, W. Nusser, K.H. Spohn, *J. Magn. Reson.*, 1986, **68,** 263.
33. F. Winter, R. Kimmich, *Mol. Phys.*, 1982, **45,** 33.
34. R. G. Bryant, K. A. Mendelson, C. C. Lester, *Magn. Reson. Med.*, 1991, **21,** 117.
35. D. J. Hinton, R. G. Bryant, *J. Phys. Chem.*, 1994, **98**, 7939.
36. S.H. Koenig, W.S. Schillinger, *J. Biol. Chem.*, 1969, **244,** 3283.
37. D. Zhou, R. G. Bryant, *Magn. Reson. Med.* in press.
38. M. Whaley, L. Lawrence, R. G. Bryant, unpublished results.
39. R. G. Bryant, L. Lawrence, M. Whaley, J. P. Korb, unpublsihed.
40. X. Jiao, R. G. Bryant, unpublished.

Nuclear Cross Relaxation Spectroscopy and Single Point Imaging Measurements of Solids and Solidity in Foods

Thomas M. Eads[1] and David E. Axelson[2]

[1] DEPARTMENT OF FOOD SCIENCE AND WHISTLER CENTER FOR CARBOHYDRATE RESEARCH, PURDUE UNIVERSITY, WEST LAFAYETTE, IN 47907-1160, USA

[2] SINTEF UNIMED, OLAV KYRRESGT. 3, 7034 TRONDHEIM, NORWAY

1 INTRODUCTION: SOLIDS IN FOOD MATERIALS

In the hierarchy of structure and function in food materials, the role of structure-making molecular assemblies is paramount. Examples include polysaccharide crystallites and junction zones in gels, protein assemblies like myofilaments in muscle tissue and protein bodies in legumes, ice crystals in frozen foods, fat crystallites in semisolid fats, sugar crystallites in confections or condensed milk, protein/surfactant assemblies at droplet surfaces in emulsions, and so on. What these diverse assemblies have in common is that the bonds which hold their molecules together are of sufficient duration and strength that they form a recognizable microstructure, and the assemblies or their networks give a solid-like response to mechanical forces. At the molecular level, the meaning of the term "solid-like" is not so much the response to applied mechanical force as it is the dynamical state of molecules. Density in solid-like matter is sufficiently high that motion of a molecule or a chain segment is tightly coupled to motions of its neighbors. The activation energies for rotations of more than a few degrees or translations of more than a few angstroms are high. Motions are less probable, less frequent, and are greatly restricted in amplitude compared to molecules in liquid-like phases. The resulting differences in magnetic resonance signals have been exploited to quantify the solid/liquid ratio, to selectively observe one or the other, and in a few cases, to study molecular dynamics in each phase in intact food materials.

In the quest for non-destructive methods to probe microstructure of solid-like phases, and to measure molecular order and mobility within them, we have come to focus on two relatively new approaches: nuclear magnetic cross relaxation spectroscopy (CR),[1] which is useful for solids and solidity measurements on a broad range of complex food materials, and is illustrated here for a gel of starch crystallites; and single point imaging (SPI),[2,3,4] which can image solids directly, but instead of just visualizing "anatomy", it maps the spatial variation in solid mass fraction and internal dynamics of solid domains. Preliminary SPI experiments on the gel of starch crystallites are reported here.

1.1 NMR of Solids

In magnetic resonance, molecular motion permits partial to complete averaging of line-broadening magnetic interactions that are normally present in all materials. For example, within the crystallites of a retrograded starch gel, protons in the hydrogen atoms of glucosyl residues experience strong orientation-dependent magnetic fields due to neighboring protons (homonuclear dipolar interaction) or carbons (heteronuclear dipolar interaction), plus an orientation-dependent shielding due to anisotropic electron distribution in the molecular orbitals (chemical shift anisotropy). Thus the orientation of a glucosyl residue determines the NMR frequency of each of its hydrogen nuclei. In the solid echo[5]

^1H NMR spectrum of a starch sample, crystallites produce a very broad resonance about 35 kHz in width, in which resonances corresponding to all different crystallite orientations are superimposed (Figure 1).

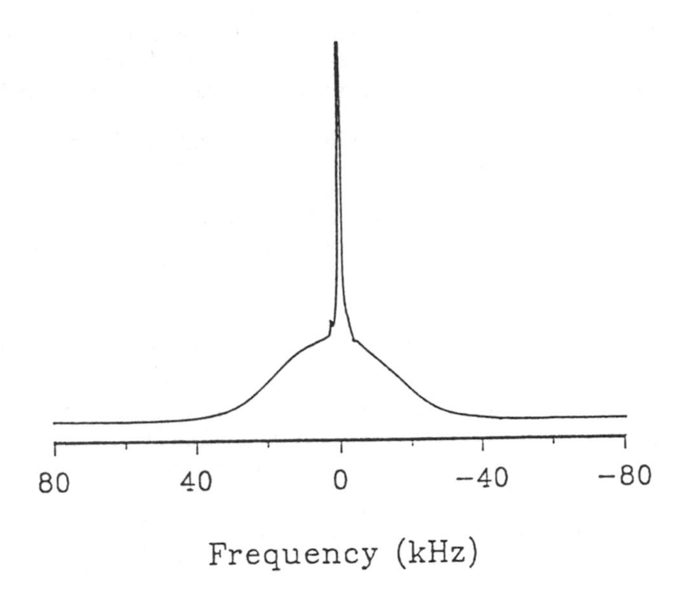

Figure 1 *The wide line spectrum of waxy maize starch granules obtained by the solid echo method.*[5] *The narrower resonance is due to water, which is present at about 14% by weight.*

Restricted thermal motions exchange orientations, hence frequencies, and if this motion is faster than the difference in frequencies between two positions, the average frequency is observed instead. Thus very rapid motions due to sample melting or dissolution finally collapse the broad spectrum to a very narrow one, called the high resolution spectrum, in which chemical shifts become evident.

Obtaining information about solid-like components in food materials by NMR normally requires application of pulsed wide line methods (either time domain or Fourier transform) or high resolution solids spectroscopy.[6] In pulsed wide line NMR, the entire solid resonance, which is of the order of 10 to 100 kilohertz in width, must be excited all at once by using very short pulses (microseconds) of high intensity. Such methods have been applied to starch.[7] In high resolution solids carbon-13 NMR spectroscopy, the strong heteronuclear dipolar interaction is eliminated by applying strong radiofrequency irradiation to the protons, and the chemical shift anisotropy is eliminated by spinning the sample at the magic angle. The result is a spectrum with narrow resonances. This method has been applied extensively to starch and to some other food materials.[6] In high resolution solids proton NMR spectroscopy, the strong homonuclear dipolar interaction is eliminated by application of a series of closely-spaced high-power pulses to the protons, and chemical shift anisotropy is removed by magic angle spinning. This technique, called combined rotation and multiple pulse spectroscopy (CRAMPS) (see, for example, ref. 8), should be useful for obtaining chemical information about solid phases in food materials but has received little use so far.

2 NUCLEAR CROSS RELAXATION SPECTROSCOPY OF SOLID PHASES

2.1 Principles of Cross Relaxation Spectroscopy

Nuclear magnetic cross relaxation spectroscopy[1] may be used to study solid phases in heterogeneous materials when the solid phase is in intimate physical contact with a liquid phase. We have used this approach previously to study starch retrogradation.[7,9] The principles of the method are now briefly reviewed.

2.1.1 Magnetic Coupling. Nuclear spins in solid and liquid are coupled by several possible mechanisms: spin state exchange, during which spin transitions in one phase are communicated to the other through space; chemical exchange, in which nuclei bearing a nuclear spin in a certain state are exchanged, as atoms or chemical groups, between solid and liquid (e.g. exchange of water protons with starch hydroxyl protons); and molecular exchange, in which liquid molecules more or less trapped within solid phase cavities exchange with the bulk liquid phase.

2.1.2 Selective Saturation. Due to the extremely large difference in resonance widths, solid may be irradiated separately from liquid by applying a radiofrequncy pulse of sufficient strength to partially saturate the solid magnetization (strength is expressed in terms of the proton precession frequency $\omega_1 = \gamma B_{sat}/2\pi$), of sufficient duration that a steady state partial saturation is achieved, and at any frequency within the broad solid resonance except for about +/- 500 Hz from the liquid resonance. The excitation bandwidth of such a pulse is very narrow.

2.1.3 Saturation Transfer or Cross Relaxation. During irradiation of the solid, the liquid also becomes partially saturated due to magnetic coupling, and this is easily detected as a drop in intensity in a single pulse liquids experiment (application of a short pulse and collection of the free induction signal). A simple cross relaxation pulse sequence is shown in Figure 2.

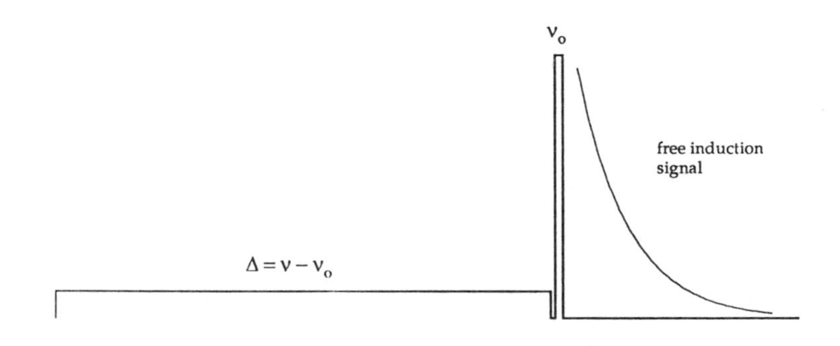

Figure 2 *The pulse sequence for the cross relaxation experiment achieves selective steady state partial saturation of solid by a long duration, low power radiofrequency pulse at a variable offset frequency, and detects the partially saturated liquid magnetization by a short, high power pulse on resonance with liquid.*

2.1.4 Offset Frequency Dependence of Saturation. As the solid irradiation frequency, i.e. the offset frequency, is varied, the intensity of the liquid signal (the first point of the free induction signal, or the area of the liquid peak in the NMR spectrum) is observed to change in intensity. The plot of liquid saturation versus offset frequency, called the nuclear cross relaxation spectrum (Figure 3), is reminiscent of the solid wide line spectrum (Figure 1).

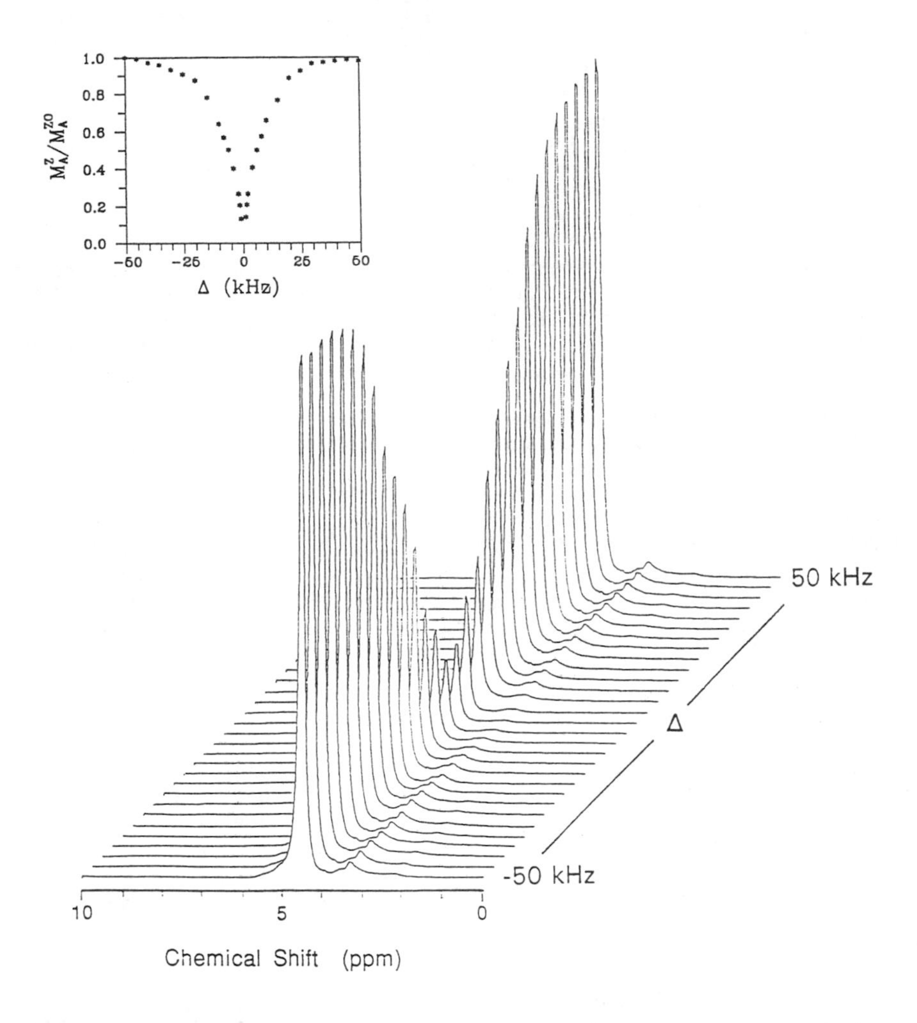

Figure 3 *Construction of the cross relaxation spectrum from the offset dependence of liquid saturation. (Reproduced from ref. 7 with permission of the American Chemical Society).*

2.1.5 Cross Relaxation Theory. The cross relaxation spectrum can be modeled. Several approaches have been taken. In the first, the Bloch equations are modified by adding a cross relaxation term and solved under steady state conditions.[1] Other terms may be added, corresponding to off-resonance irradiation effects on the liquid signal,[10,11] or accommodation for the decidedly non-Lorentzian shape of the spectrum of the solid spins.[12,13] An analytical solution has been obtained in an alternative approach which removes the restriction of the line shape function assumed used to represent the solid component.[14]

The Grad and Bryant result gives the steady state z-magnetization of the liquid phase as

$$\frac{M_A^{SS}}{M_A^{o}} = 1 - \frac{2\alpha}{\beta + 4\pi^2\Delta^2\gamma} \quad (1)$$

$$\alpha = \frac{fR_{BA}T_{2B}\omega_1^2}{2R_AR_B},$$

$$\beta = \frac{R_{BA}}{R_B} + f(\frac{R_{BA}}{R_A + 1})(\frac{T_{2B}\omega_1^2}{R_B + 1}),$$

$$\gamma = T_{2B}^2[\frac{R_{BA}}{R_B} + f(\frac{R_{BA}}{R_A} + 1)].$$

Here M_A^{SS} and M_A^o are the z-magnetization of the liquid (A spins) in the presence and absence, respectively, of steady state saturation of the solid (B spins), R_B is the spin-lattice or longitudinal relaxation rate of the solid proton population in the absence of magnetization transfer, R_A is the corresponding rate for the liquid, T_{2B} is the solid transverse relaxation time, Δ is the offset of the saturation radiofrequency field from the center of the liquid resonance (ν_1-ν_0), ω_1 is proton precession frequency in the rf field of the saturation pulse, and f denotes the ratio of the number of solid protons effective in cross relaxation to the number of liquid protons effective in cross relaxation.

2.1.6 Fitting Data to Theoretical Expressions. By fitting cross relaxation data to such theoretical expressions it is possible to extract physically meaningful parameters. The number of these is five or more, which is more than can be obtained uniquely by fitting. Thus it is necessary to obtain two or more NMR parameters in independent measurements. For example, we used the inversion recovery saturation (IRSAT) method[15] to measure R_A and R_{BA}. Other procedures have been proposed, such as varying the strength of the solid saturation pulse in CR measurements, measuring R_A as a function of Δ, and fitting all data simultaneously.[13,16]

2.1.7 Useful Parameters. In studies of food materials, the parameters of greatest interest are: f, a measure of solid in contact with liquid; T_{2B}, a measure of internal mobility of solid-like domains; and R_{BA}, a measure of concentration-dependent accessibility of the solid to the liquid. The parameters R_A (liquid) and R_B (solid) are usually less useful because they are more difficult to interpret unambiguously. Alternative representations of the theory permit extraction of the concentration-independent cross relaxation rate constant.[13] Observed values of these parameters may be rationalized in terms of the molecular processes underlying them, but this is beyond the scope of this chapter.

2.2 Some practical aspects of cross relaxation spectroscopy

The method can, in principle, be executed on a variety of NMR instruments. We have used a high field, high resolution NMR spectrometer and a magic angle spinning (MAS) solids probe.[9] While spinning is certainly not necessary for the cross relaxation experiment, it removes broadening of liquid phase resonances due to magnetic susceptibility inhomogeneity, which is common in heterogeneous materials like foods. Instead of simply taking the first point of the free induction signal following the detection pulse, the FID is Fourier-transformed, resulting in a high resolution ^{1}H NMR spectrum. A cross relaxation spectrum may then be plotted for each resolved liquid phase component in contact with solid-like domains. Note, for example, the Δ-dependent intensity of resonances from highly mobile starch between 2.5 and 4 ppm in Figure 3.

In the high resolution CR method, following probe tuning, the smallest water line width and shape is achieved by small adjustments to room temperature shims. The water resonance frequency (in Hz) in a single pulse experiment is noted, and a list of offset frequencies Δ is generated to cover the range from -60 kHz to +60 kHz in 5 kHz increments. The rf channel used for transmitting the saturation pulse on our spectrometer is known as the "decoupler" channel, although that term has no significance in the CR experiment. The proper length of the saturation pulse is determined by selecting the offset frequency to be 10 kHz, selecting a power level expected to be useful in a subsequent CR experiments, then executing the CR pulse sequence with increasing saturation pulse lengths until there is no further significant decrease in the water signal intensity. This time corresponds to achievement of steady state saturation of the solid spin system. It is also long enough that water diffusion to solid surfaces will not limit the cross relaxation efficiency. To determine the exact magnitude of the saturation pulse B_{sat}, which enters as ω_1 (Hz) in the theoretical expressions, single pulse experiments are performed with the "decoupler" channel to find the 90° pulse length for the chosen "decoupler" power level. Then B_{sat} is calculated from the Larmor relation. The CR experiment is now executed for the list of offset frequencies. The cross relaxation signal intensities $M^{SS}{}_{A}(\Delta)$ expressed equivalently as FID intensities, peak areas, or peak heights obtained by interpolation, are then normalized by $M^{o}{}_{A}$. The value of $M^{o}{}_{A}$ is the liquid signal intensity without the saturation pulse, but may also be taken as the cross relaxation signal intensity at offsets far from the liquid resonance, e.g. $M^{SS}{}_{A}(\Delta > 60\text{ kHz})$.

2.3 Cross Relaxation Analysis of Gel of Starch Crystallites

We have previously examined concentrated retrograding waxy maize starch gels from 10 to 45% solids by nuclear cross relaxation spectroscopy.[7,9] During these studies the method of fitting cross relaxation data to theoretical expressions had not yet been developed in our lab. Data were parameterized instead by fitting the CR spectrum to the sum of a broader Gaussian and a narrower Lorentzian function. The area and width of the Gaussian were then used to monitor evolution of solid. However, it was not possible by this method to tell whether increases in area and width were due to increased solid content, increased rigidity within solid, more efficient cross-relaxation, or some combination of these. To distinguish among the possibilities it is necessary to compare data to theory.

As an example we have chosen a particle gel of starch crystallites. Used as a fat replacing ingredient, the material is known commercially as Stellar (A.E. Staley Manufacturing Company, Decatur, IL, USA). The hydrated material contains ~15% of its total mass as oligomers (degree of polymerization > 13) in a crystalline state presenting an "A" type X-ray diffraction pattern, ~5% of mass as soluble glucose, maltose and short oligomers, and about 80% water. The rheology is much like hydrogenated vegetable shortening, having a yield stress, high viscosity, and some viscoelasticity. The material has an opaque white appearance.

The gel of starch crystallites presents a cross relaxation spectrum (Figure 4) which is typical of a sample in which the solid phases present are internally quite rigid: i.e. the shape is Gaussian-like. The saturation signal in the center of the spectrum is fairly narrow, and is due primarily to the off-resonance saturation effect on the water. Thus for the gel of starch crystallites there is no evidence for a second solid-like component. In contrast, the CR spectra of gels of native waxy maize starch, the parent material of Stellar, are more complex, suggesting that several kinds solid-like domains co-exist.[9]

Equation 1 is a Lorentzian function of Δ. However, any line shape may be encountered, depending primarily on the strength of the saturation pulse and the amount and rigidity of the solid-like domains. For example, Gaussian-like shapes are observed for rigid solid particles when ω_1 is of the order of 500 Hz or greater. In addition, the equation does not predict the narrow component of the cross relaxation spectrum due to off-resonance irradiation effects on the liquid (Caines et al., 1991; Wu, 1991). We made the adjustment for solid line shape according to Swanson (Henkelman et al., 1993; Swanson, 1992), and handled the narrow component by simply fitting it to a narrow Lorentzian whose width and intensity are obtained in the fit but are of no further interest. The fitting expression is:

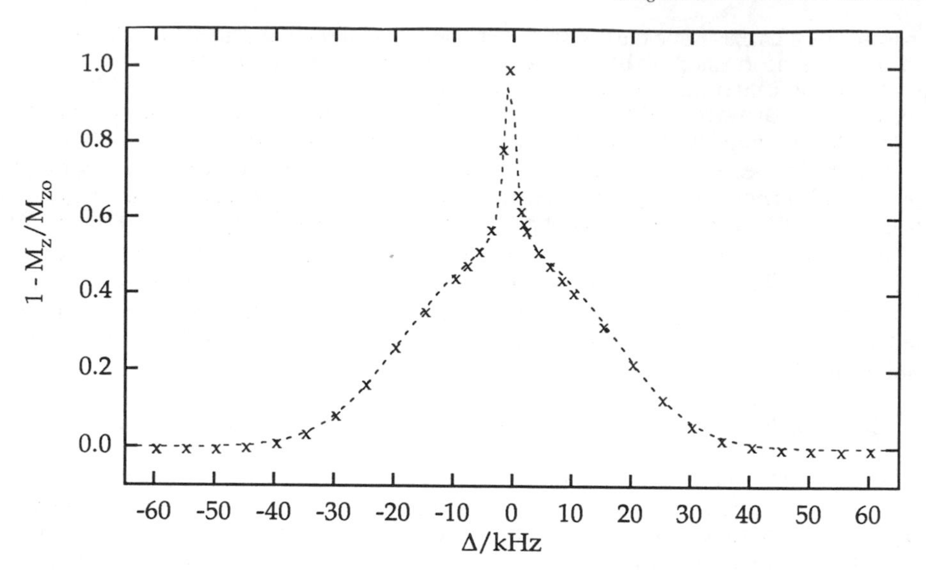

Figure 4 *Nuclear magnetic cross relaxation spectrum of a particle gel of starch crystallites. Solid circles: data. Dotted lines: best fit of the data to Equation 2.*

$$1 - \frac{M^{SS}}{M^{o}} = \frac{2R'}{\frac{R_A R_{BA}}{f} + (R_{BA} + R_A)(R_B + R')} + \frac{C}{1 + (2\Delta/\Delta\nu_{1/2})^2} \qquad (2)$$

where

$$R' = T_{2B}\omega_1{}^2(\frac{\pi}{2})^{1/2} \exp[-\frac{(2\pi\Delta T_{2B})^2}{2}],$$

C is the intensity parameter for the narrow Lorentzian, and $\Delta\nu_{1/2}$ is its width at half height in Hz.

There are eight parameters in Equation 2: the value of ω_1 was determined during experimental setup, and Δ is known. Since the data usually have about four degrees of freedom (i.e. they can be fit to a sum of two simpler line shape functions, each having an intensity and a width), it is necessary to measure at least two parameters independently. In the current approach, the water longitudinal relaxation rate R_A and the cross relaxation rate R_{BA}, were obtained by fitting inversion-recovery saturation (IRSAT) data (Figure 5) to the following expression:[15]

$$M'(t) = \frac{M_\infty - M(t)}{2M_\infty}$$
$$= [M'(t{=}0) - \frac{R_{BA}}{2(R_{BA} + R_A)}]\exp[-(R_{BA} + R_A)t] + \frac{R_{BA}}{2(R_{BA} + R_A)}, \qquad (3)$$

where $M(t)$ is a normalized magnetization. M_∞ is measured without saturation at a value of t that is very much greater than the water T_1.

The IRSAT method exploits the fact that cross relaxation alters longitudinal relaxation in both liquid and solid spin systems.

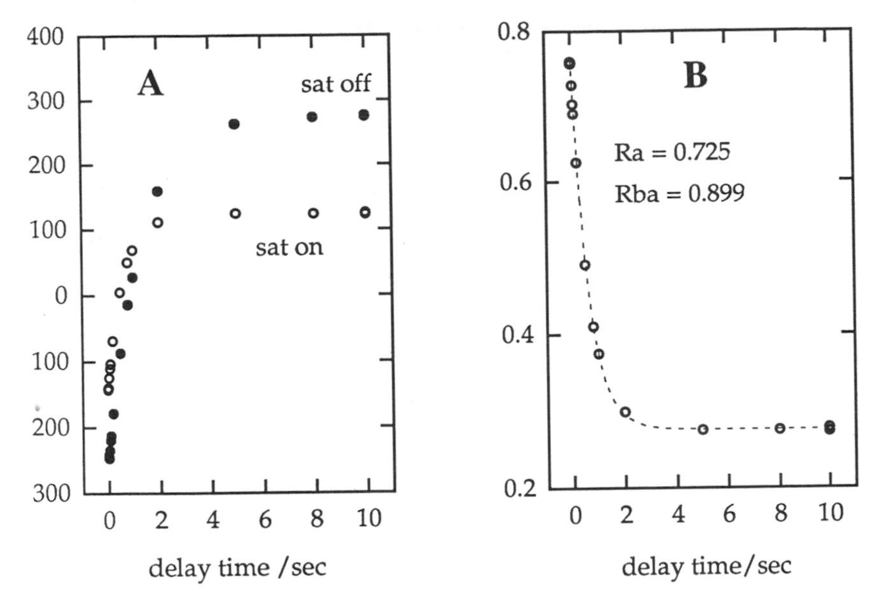

Figure 5 *Inversion-recovery saturation (IRSAT) method for determination of water longitudinal relaxation rate $R_A(s^{-1})$ and cross relaxation rate $R_{BA}(s^{-1})$in a particle gel of starch crystallites. A: Data obtained by the usual inverstion-recovery pulse sequence (D1 - $180°_x$ - t - $90°_x$ - acquire), except that during both D1 (fixed) and t (variable) a saturation pulse of strength equal to that used in the cross relaxation experiment is applied at an offset frequency Δ = 10 kHz. Ordinate: signal intensity in arbitrary units. B: Fit (dotted line) of normalized data to Equation 3.*

Although the significance of single values of R_A and R_{BA} obtained at a single offset frequency has been argued,[13] we believe at present that the method is a practical alternative to measuring Δ dependence of R_A and saturation field strength dependence of CR spectra in the absence of instrumentation for rapid CR data acquisition (e.g. broad band cross relaxation[17]).With R_A (i.e. $1/T_{1A}$) and R_{BA} fixed at their measured values, the four parameter fit of the cross relaxation spectrum (Figure 4) to Equation 2 becomes straightforward. The value of f so obtained, f_{obs} = 0.087, is somewhat smaller than that calculated with the assumption that all starch polymer is solid and participating in cross relaxation, $f_{stoichiometric}$ = 0.104. This suggests that not all starch protons within the solid particles are available for cross relaxation with liquid. Simple arguments based on spin diffusion within solid particles suggest that f_{obs} will be less than $f_{stoichiometric}$ when particle diameters exceed about 10 nm. This situation is an opportunity for using CR to follow changes in particle size associated with processes like microparticulation, aggregation, dissolution, and so on. The value of T_{2B} obtained, 11.6 μs, is typical of crystalline organic solids, but appears to be longer than the value of 10.6 μs for T_2^* estimated from the line width ($T_2^* = 1/\Delta\nu_{1/2}$) in the pulse-wide line NMR spectrum of a retrograded starch gel made with D_2O,[7] the value of 9.9 μs for T_{2B} obtained by a cross relaxation technique on starch granules within unripe banana tissue,[18] and the value of 8.3 μs estimated from the line width of the spectrum of starch granules in Figure 1.

3 SINGLE POINT IMAGING OF SOLID PHASES

3.1 Principles of Single Point Imaging (SPI)

By applying a saturation pulse at a suitable offset frequency, the water signal observed in an imaging experiment can be made to reflect the concentration and internal dynamics of local solid-like domains via the cross relaxation effect.[19,20] The method, usually referred to as magnetization transfer contrast (MTC), may thus be considered as an *indirect* method for imaging solids, and is not discussed further here. However, *direct* methods also exist (see, for example, reference 21).

The method of single point or constant time imaging (SPI)[2,3,22] obtains the NMR signal directly from the solid nuclei. Solid domain proton T_2^*s (the time constant for decay of the free induction signal) will normally be of the order of tens to hundreds of microseconds, in sharp contrast to molecules in liquid-like domains whose T_2^*s are normally of the order of hundreds to milliseconds to seconds. Direct detection of the solid signal requires very short acoustic ring-down times in the radiofrequency coil to permit start of digitization long before the signal has decayed. A fast digitizer is also required.

3.1.1 Basic Imaging Protocol. In the SPI protocol (Figure 6), the gradients used to encode position as frequency are switched on and allowed to stabilize. A small angle (broad bandwidth) excitation pulse is applied and at some fixed time after this, i.e. at the *detection time*, a point in the free induction signal (FID) is sampled. The gradients are successively ramped from their minimum to maximum values in equal steps, thus selecting different parts of the sample, and creating an image for the given detection time.

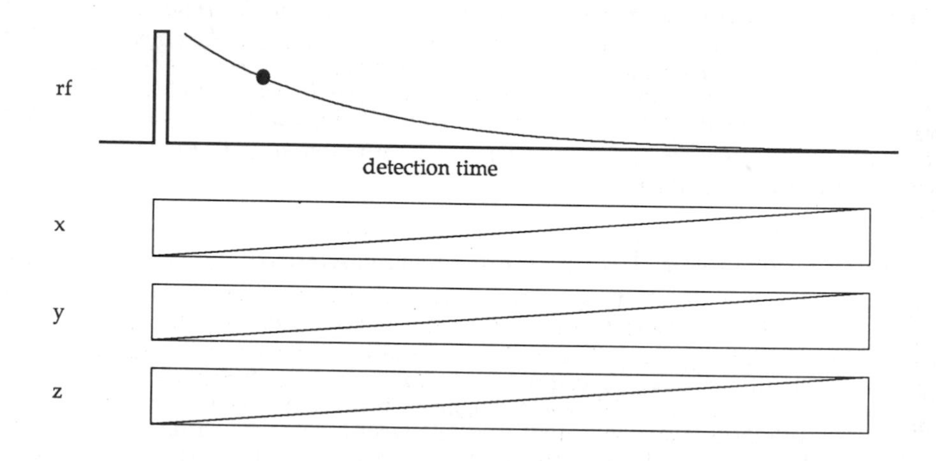

Figure 6 *Basic protocol for single point imaging. Top panel shows radiofrequency excitation pulse, free induction signal, and the single point at which the signal is sampled. Bottom panels signify phase- and frequency-encoding gradients for acquisition of a rhree-dimensional image.*

3.1.2 Variation of Detection Time. An image is obtained at each of a series of detection times encompassing the free induction decay of solid-like protons. Thus for each pixel position, a FID-like curve exists, which upon analysis yields the amplitudes and relaxation times of its multiple decay components.

3.1.3 Mapping Mass Fraction and Time Constants. Each amplitude and each relaxation time may be plotted separately, thus creating maps of the spatial distribution of mass fraction and rigidity of several solid-like components with microscopic or near-microscopic resolution (see Figure 7, below).

3.1.4 Plotting Pixel Intensity Histograms. Instead of being used to generate a map, which for visually homogeneous materials does not contain patterns recognizable to the eye, pixel parameter values may be plotted in a histogram. The underlying distribution function presumably reflects the distribution in solid states in the sample.

3.2 Some Practical Aspects of Single Point Imaging

The presence of solid is manifested by a rapid decay in the first tens of microseconds of the free induction signal. As solid mass fraction becomes smaller than about 20%, the rapid decay becomes less distinguishable from the liquid signal. For this reason, we used D_2O instead of H_2O as diluent in our initial SPI experiments on 20% starch gels. The rapid decay of the FID from such samples is due almost entirely to protons at non-exchangeable hydrogen positions on glucosyl residues.

Data were acquired with a Bruker Biospec 24/30 system, operating at 100 MHz. Typical experimental conditions were as follows. A 15 or 35 mm ^{1}H probe set within a Bruker microimaging (G060) gradient set was used. The maximum gradient strength of this set was 175 Gauss/cm. The parameters are: field of view 4 x 4 cm^2, acquisition matrix size 30 x 30 mm, reconstruction matrix size 128 x 128, spectral width 500 kHz, pulse length 1 μs, number of averages 4, detection time variable from 12.5 μs to 10,000 μs, repetition time 100 ms, gradient strength up to 38.7% of maximum values, total experimental time six minutes per image with 16 images/data set, gradient power supply duty cycle <2%. In-plane resolution is approximately 330 μm. Slice thickness was equal to sample thickness (approximately 10 mm).

Features of this protocol are different in many respects from others commonly used in imaging. These features are explained in greater detail in a related paper.[4]

Amplitudes and relaxation times were analyzed assuming various models for the decay rates in order to find the most appropriate one. From the starch data we consistently obtained best fits to a bi-exponential decay $A_1\exp(-t/T_{2,1}^*) + A_2\exp(-t/T_{2,2}^*)$, yielding two decay constants and two relative mass fractions, each of which could be mapped. The asterisk is a reminder that the decay constants are not necessarily the same as transverse relaxation times.

3.3 Results of Single Point Imaging of Gel of Starch Crystallites

The particle gel of starch crystallites was imaged at a series of detection times and decay curves at each pixel were analyzed. The component with shorter time constant greatly dominated the decay curve (mass fraction 0.87, standard deviation 0.06). The map of the shorter decay component, T_2^* is shown in Figure 7A, which reveals the spatial distribution of *mobility* of polymer in rigid solid domains. That is, longer T_2^* means greater mobility. We are not aware of a previous report of a mobility map for solid domains in such a dilute system.

Each point in a plot corresponds to a measurement on a volume element (voxel) whose dimensions are 330 x 330 x 10,000 μm, i.e a volume of 1 μl. This is a macroscopic element, of the order of thousands of times larger than individual crystallites observed by electron microscopy and tens of times larger than the aggregates observed by light microscopy (unpublished results). Therefore it seemed surprising to find that the average mobility varied from point to point. This kind of heterogeneity was also evident in the surface plot of mass fraction of the shorter decay component.

A histogram for shorter relaxation time (Figure 7B) was constructed from the same MRI data. The histogram shape indicates that mobility in the rigid domains may be characterized by a fairly narrow distribution function, in turn suggesting that solid domains occur as a single type with internal rigidity, consistent with other observations on such

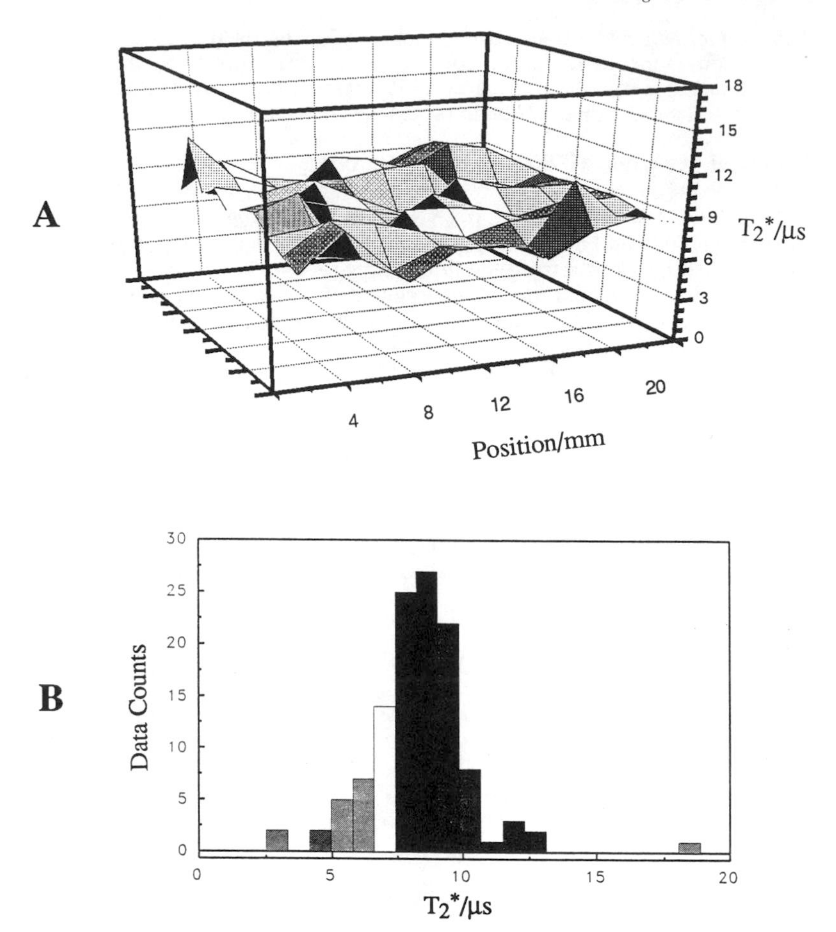

Figure 7 *Spatial distribution of mobility in a particle gel of starch crystallites obtained by variable detection time single point imaging. A: surface plot of* T_2^* *(rigid component). B: Histogram of* T_2^* *values (same data as used in panel A).*

gels. The arithmetic mean was 8.4 μs, standard deviation 1.9 μs, 95% confidence interval 8.0-8.7 μs, n = 119 points. The imaging experiment produced reasonable results, as shown by comparison with the arithmetic mean of the histogram and the value of 11.6 μs for T_{2B} measured by cross relaxation spectroscopy in the previous section. The SPI value is likely to suffer from a systematic error: when the earliest detection time is 12 μs, the Gaussian-like shoulder occurring at shorter times would be missing, the data after 12 μs would appear to be steeper, and $T_{2,1}^*$ would be artifactually shorter.

SPI results for the gel of starch crystallites were compared to those for a retrograded (partially recrystallized) waxy maize starch gel at the same solids level. That comparison will be reported separately (Axelson, Fisher and Eads, manuscript in preparation).

4 SOME NMR STRATEGIES FOR COMPLEX FOOD MATERIALS

Many kinds of information can be obtained from a single food sample by using a versatile NMR spectrometer, particularly one equipped for NMR microscopy. The choice among possible measurements should be guided by the technical or scientific demand. A broad characterization by NMR should precede more detailed study. In all cases, good NMR experiment design requires some prior knowledge of composition and structure. Interpretation of NMR results requires that they be integrated with results from complementary physical and chemical measurements. Speculation on the molecular origin of functionalities will then be more sound.[23]

For dilute materials like vegetable tissue, cross relaxation spectroscopy is currently the only feasible method for investigating solids and solidity. For more concentrated materials, pulse wide line NMR (e.g. ^{1}H or ^{2}H) may yield similar information, and high resolution solids ^{1}H or ^{13}C NMR may provide not only chemical information about solid domains, but also the geometry and rates of restricted molecular motion within them. Isotopic labeling may be worth the expense and effort in such studies. Spin diffusion experiments can yield a rough idea of solid domain size. Magnetization transfer experiments can provide an idea of the intimacy of contact of solid phases. Magic angle spinning single pulse ^{1}H and ^{13}C experiments can be used to quantify organic liquid phase components via their chemical shifts and intensities, while line shapes and relaxation times reveal small molecule rotational diffusion constants and chemical exchange rates. Compartmentation, exchange, and mobility of inorganic species including K^{+}, Na^{+}, inorganic phosphate, and so on, can also be measured by high resolution methods. Pulsed gradient spin echo (PGSE) measurements provide translational diffusion constants of mobile species. The high resolution version of PGSE gets this information for each resolvable species. A rough idea of the microstructural space "visited" by diffusive species can be obtained from pulse-spacing dependence of CPMG T_2 experiments. Finally a more refined picture of anatomy, diffusion, flow and spatial distribution of abundant liquid species, and distribution of solids and solidity may be obtained in a carefully designed suite of NMR imaging experiments,[22] the parameters for which may conveniently be deduced from less expensive, simpler NMR spectroscopy experiments.

Following are a few descriptions of recent NMR studies in this laboratory of well-known food materials for which the technical issues are defined.

Cross relaxation spectra have been obtained as part of a study of the molecular basis for functional effects of control of cell wall solidity in tomato pericarp by genetic modification. Intact pericarp is ~4% solids present as cell wall polysaccharides and proteins. Genetic type determines the evolution in amount and solidity of cell wall matter during ripening both on and off the vine. This study included high resolution MAS ^{1}H NMR for non-destructive analysis of liquid phase composition in the same tissue samples. Three sugars, several organic acids, liquid-like fatty acyl compounds, and water were determined quantitatively, and the evolution in composition during ripening was measured (Ni, Handa, and Eads, submitted for publication).

Intact banana tissue at various stages of ripeness, and having from 4% to 20% solids as starch and cell wall polysaccharides, was examined by cross relaxation spectroscopy and high resolution NMR, showing nearly quantitative conversion of starch to soluble sugars, and evolution of sugar composition during ripening.[18]

Licorice, a low moisture, chewy confection with ~40% of its mass present as partially hydrated starch and protein, and ~40% as partially hydrated sugars in a viscous phase, was examined by cross relaxation spectroscopy and by pulse-Fourier transform wide line ^{1}H NMR to characterize macromolecular solids, high resolution MAS ^{1}H NMR to characterize mobile liquid-phase components (water and fatty acyl lipids), and high resolution solids ^{13}C NMR which detected three sugars present in crystalline form. There was no previous evidence for crystalline sugars in licorice. The NMR results depend on formulation. NMR was combined with other physical measurements to propose that molecular events occurring during ageing are responsible for various sensory defects (Ni, Lu and Eads, manuscript in preparation).

In a study of wheat crackers, which may be considered as an organic glass, the moisture dependence of solid mobility was measured by pulse-time domain and pulse-Fourier transform wide line ^{1}H NMR, and by efficiency of cross-polarization in high resolution solids ^{13}C NMR. High resolution MAS ^{1}H NMR detected the mobile water fraction as well as the liquid phase fatty acyl compounds. CPMG T_2 measurements revealed that water that is highly immobilized due to strong interaction with the hydrophilic macromolecular matrix, and that triacylglycerols in the liquid phase of ingredient shortening are highly mobile regardless of moisture content. These results were related to technical issues, namely the rate of spontaneous breakage and the breaking strength of crackers (Ni, Grenus, Okos and Eads, manuscript in preparation).

4 ACKNOWLEDGMENTS

This is paper No. 14480 of the Purdue Agricultural Experiment Station. This research was supported by grants from: A.E. Staley Manufacturing Co., Decatur, IL, U.S.A; Hershey Foods Corp., Hershey, PA, U.S.A; Nabisco Brands, Inc., E. Hanover, NJ, U.S.A; and by sustaining members of the Whistler Center for Carbohydrate Research at Purdue University. Cross relaxation data and computer programs for fitting cross relaxation data to theoretical expressions were developed by Mr. Joseph P. Smith of TME's laboratory. Starch samples were kindly provided by Dr. Gary A. Day at the Staley research laboratory in Decatur, IL, U.S.A. Access to the imaging equipment at the Petroleum Recovery Institute, Calgary, Alberta, Canada is gratefully acknowledged.

Use of product trade names for materials used as examples in this work does not constitute endorsement by Purdue University of the products or their manufacturers.

References

1. J. Grad and R. G. Bryant, *J. Magn. Reson.*, 1990, **90**, 1.
2. D. G. Cory, *Adv. Magn. Reson.*, 1992, **24**, 87.
3. S. Gravina and D. G. Cory, *J. Magn. Reson.*, 1994, **B104**, 53.
4. D. E. Axelson, A. Kantzas and T. M. Eads, *Can. J. Appl. Spectr.*, 1994 (in press).
5. P. Mansfield, *Phys. Rev.*, 1965, **137**, A961.
6. M. J. Gidley, *Trends in Food Sci. Technol.*, 1992, **3**, 231.
7. J. Y. Wu, R. G. Bryant and T. M. Eads, *J. Agric. Food Chem.*, 1992, **40**, 449.
8. B. C. Gerstein, *Phil. Trans. R. Soc. Lond.* 1981, **A 299**, 521.
9. J. Y. Wu and T. M. Eads, *Carbohydrate Polymers*, 1993, **20**, 51.
10. G. H. Caines, T. Schleich and J. M. Rydzewski, *J. Magn. Reson.*, 1991, **95**, 558.
11. X. Wu, *J. Magn. Reson.*, 1991, **94**, 186.
12. S. D. Swanson, in 'Book of Abstracts, Annual Meeting, Society for Magnetic Resonance in Medicine', 1992, 255.
13. R. M. Henkelman, X. Huang, Q.-S. Xiang, G. J. Stanisz, S. D. Swanson and M. J. Bronskill, *Magn. Reson. Medicine*, 1993, **29**, 759.
14. H. N. Yeung, R. S. Adler and S. D. Swanson, *J. Magn. Reson.*, 1994, **A106**, 37.
15. J. Grad, D. Mendelson, F. Hyder and R. G. Bryant, *J. Magn. Reson.*, 1990, **86**, 416.
16. H. N. Yeung and S. D. Swanson, *J. Magn. Reson.*, 1992, **99**, 466.
17. S. D. Swanson, *J. Magn. Reson.*, 1991, **95**, 615.
18. Q. X. Ni and T. M. Eads, *J. Agric. Food Chem.*, 1993, **41**, 1035.
19. S. D. Wolff and R. S. Balaban, *Magn. Reson. Med.*, 1989, **10**, 135.
20. J. Eng, T. L. Ceckler and R. S. Balaban, *Magn. Reson. Med.*, 1991, **17**, 304.
21. D. G. Cory, *Annual Reports on NMR Spectroscopy*, 1992, 87.
22. P. T. Callaghan, 'Principles of Nuclear Magnetic Resonance Microscopy', Oxford University Press, Oxford, 1992.
23. T. M. Eads, *Trends in Food Technology*, 1994, **5**, 147.

Water–Macromolecular Interactions in Chocolate

C. Jackson, R. Weiler*, M. Smart, and B. Campbell

KRAFT GENERAL FOODS, TECHNOLOGY CENTER, 801 WAUKEGAN ROAD, GLENVIEW, IL, 60025, USA

1 INTRODUCTION

Water-macromolecular interactions in chocolate are not well understood; however, their importance in chocolate manufacturing is well known. Several methods of using water to provide heat resistance for chocolate have been patented[1-4]. Water has been added to chocolate in order to incorporate fruit, liquor and creamy flavors[5-8]. Water addition to chocolate has also been applied for antiblooming effects[9]. Other applications of water addition to chocolate include improved shelf life[10], calorie reduction[11], and texture improvement[12].

Water addition to chocolate has been accomplished by several methods. Water has been added directly to chocolate[13,14], in the form of water-in-oil[15] and oil-in-water emulsions[16], as foams[17], as hydrated emulsifiers[18], as polyol/humectant syrups[19], as microencapsulants[20], and as amorphous/hydrophilic ingredients[21]. This work will examine the effects of direct addition of water to a chocolate mass.

Water addition to chocolate can have several deleterious effects on processing. Large viscosity increases are observed upon addition of water to chocolate making processing difficult or impossible[22]. Also, water addition to chocolate can lead to demolding problems[23].

Several hypothesis have been proposed to explain the effects of water addition to a chocolate mass. One common hypothesis proposes that water displaces emulsifiers coating carbohydrate and protein surfaces at the fat interface thereby exposing these surfaces. These surfaces are then free to interact with other hydrophilic surfaces resulting in heavy internal friction in the chocolate resulting in resistance to mass deformation. The idea that moisture in chocolate can induce carbohydrate and protein interactions leading to increased viscosity has been suggested in the patent literature as early as 1956[24]. However, the body of scientific literature directed at supporting this hypothesis with data is meager. Also a few articles describe NMR applications to chocolate, but most have been limited to quantitative studies of solid/liquid fat ratios. The aim of this work was to acquire a better understanding of appropriate applications of NMR and microscopy techniques to test hypothesis, as well as to investigate ingredient effects on molecular dynamics in chocolate. In particular, large quantities of water were added to normally rather anhydrous chocolate systems to magnify water-macromolecular interaction dynamics, in order that the mechanisms causing processing problems in chocolate could be elucidated.

Like many foods, chocolate is chemically complex containing lipids, carbohydrates, proteins, emulsifiers, cocoa particles and small amounts of water. The majority of the lipids, mainly triglycerides, form a continuous fat phase in which is imbedded a heterogeneous disperse phase. This inherent complexity of chocolate necessitates an array of investigative techniques of which two are described below.

2 MATERIALS AND METHODS

Samples A and B were commercially obtained American and European milk chocolate tablets respectively. Both samples contained sugar, milk powder, cocoa butter, water, emulsifier and flavoring but differed in the ratio of the listed ingredients. Water added samples were prepared in duplicate by addition of the appropriate amount of water to chocolate that had been heated at 50°C for 0.5 hours followed by hand mixing for approximately 1 minute. The samples were allowed to cool at room temperature for 18 hours prior to NMR measurements. Water contents of the samples were measured in triplicate by Karl Fischer assay before and after each NMR experiment[25].
Proton FT high powered wide-line and ^{13}C cross-polarization[26] magic angle spinning (CPMAS) experiments[27] were performed at 400.13 MHz and 100.613 MHz respectively using a 9.4 Tesla Oxford magnet and a Bruker MSL400 spectrometer equipped with a Doty Scientific Incorporated solids probe and an Aspect 3000 computer operated using DISMSL software version 911101. Sample rotors with air tight macor o-ring caps were used for all samples measured by both wide line and CPMAS experiments. ^{13}C 90° pulse experiments with WALTZ proton decoupling were performed on a Bruker MSL400 NMR spectrometer at 100.613 MHz using a standard 10 mm Bruker broad band probe unit with standard 10 millimeter Wilmad glass NMR sample tubes sealed with Saran Wrap and Parafilm. Solid/liquid ratios were measured as a function of temperature with a Bruker Minispec 120 spectrometer equipped with 110 PROM, a Bruker PC PH 20/10 VTs probe unit and a Haake F3-CH Temperature Controlled Water Circulator.

Chocolate samples for microscopy were frozen to -20°C and cryo-sectioned at a nominal thichness of 20 micrometers. Sections for slides were stained by flooding them with either 1 % 1-anilino-8-naphthalene sulfonic acid (ANS) for protein or 1 % aqueous Nile Blue for fat. The slides were observed by epifluorescence optics on a Zeiss Axiophot microscope using filter set 02 for ANS and filter set 09 for Nile blue magnifications.

3 RESULTS AND DISCUSSION

The proton FT high powered wide line spectum shown in Figure 1(1), as well as solid/liquid ratio measurements, shown in

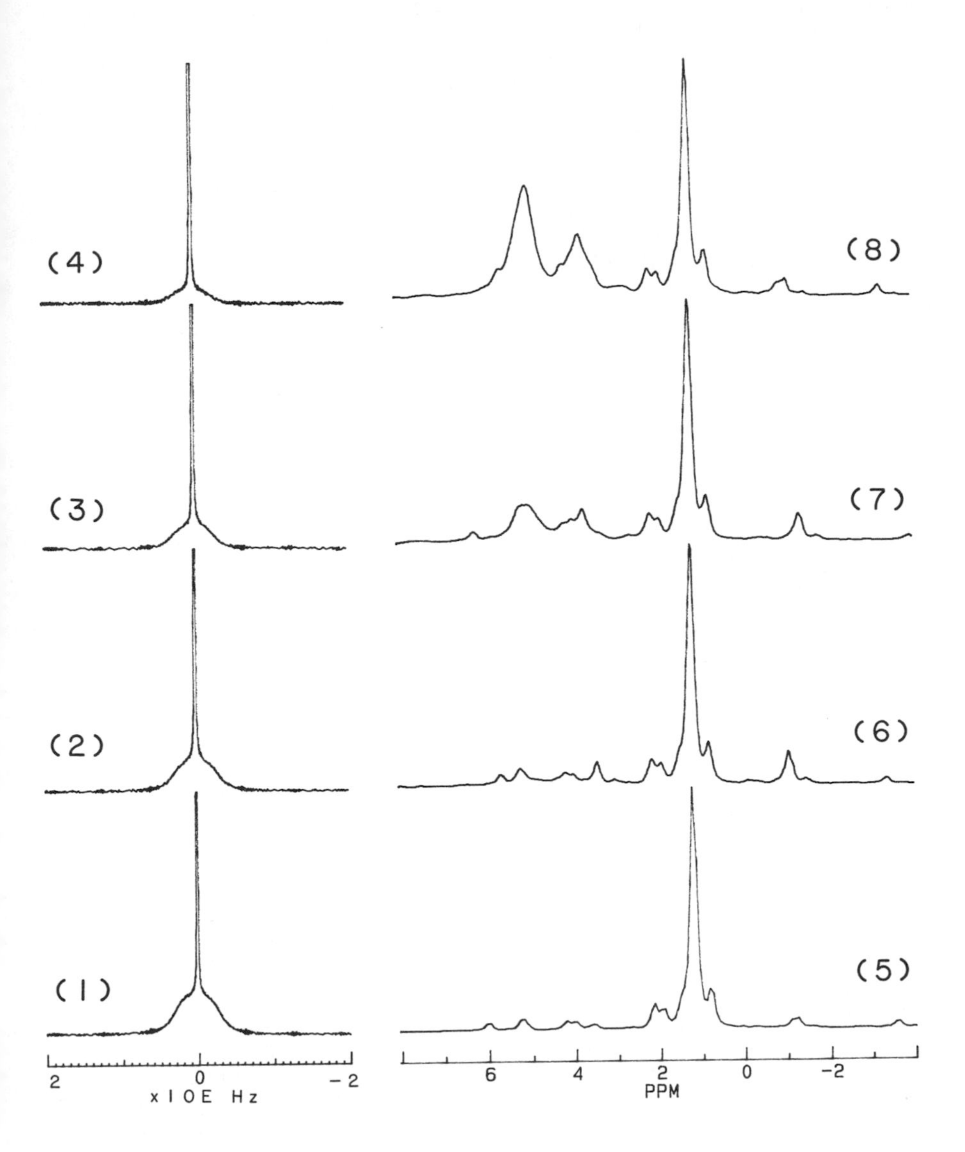

Figure 1 Proton FT High Powered Wide Line of Sample B as a function of percent (%) (w/w) water at 25°C with 100 transients, and a sweep width of 500,000 Hz. Spectra 1(1-4) were obtained for static samples using 123 Hz line broadening and a 2 microsecond pulse length and % water (w/w) of (1) 1.17 %, (2) 2.15 % (3) 5.07 %, (4) 10.38 % respectively. Spectra 1(5-8) were obtained under MAS conditions using 30 Hz line broadening, rotor frequency of 900 Hz and % water (w/w) of (5) 1.17 %, (6) 2.15 %, (7) 5.07 %, (8) 10.38 % respectively.

Figure 8, indicate that about 30 percent of a chocolate mass are liquid at room temperature (25°C). Solid/liquid ratios obtained from measuring peak areas of the narrow and broad peaks in the proton FT high powered wide line spectra using the Bruker SIMFIT program were within 1 percent of the values found using a Bruker Minispec 120 spectrometer 110 PROM. Proton FT high powered wide line experiments performed with magic angle spinning (MAS) show that the liquid component is comprised almost entirely of triglyceride (Figure 1(5)). If highly mobile fractions of proteins or carbohydrates exist in a chocolate at 25°C prior to water addition, they are not observable in this spectrum.

The dynamic response of the componenents in Sample B to water addition is shown in Figures 1(1-4). These spectra obtained from proton FT high powered wide line experiments indicate that the solid/liquid ratio of a chocolate mass decreases as the amount of added water increases as would be expected. As the water concentration is increased, as shown in the spectra in Figure 1(1-4) and 1(5-8), a few peaks, not visible at water contents below 5 percent, begin to be apparent in the 2.5 - 5.5 ppm region of the ^{1}H FT high powered wide line spectra shown in Figure 1(7-8). The appearance of these peaks suggests that solubilization of carbohydrates and protein side groups in the sample may be occurring. The addition of water to the chocolate does not appear to affect the liquid fat present in the sample as indicated by the lack of change in the fat signal as a function of added water level.

The dynamics of the solid components in Sample B were observed by monitoring the ^{13}C signal during CPMAS NMR experiments. The ^{13}C CPMAS spectrum of a milk chocolate tablet obtained at 25°C is shown in Figure 2(1). A comparison of this spectrum with the ^{13}C CPMAS spectra shown in Figure 2(2-3) indicate that the solid signal is dominated by sucrose and cocoa butter signals. Protein signals are not visible in the spectrum.

^{13}C CPMAS spectra obtained for Sample B as a function of added water are shown in Figure 3(1-4). The signal-to-noise decreases with increasing water concentration again indicating solubilization of the solid components. Whereas only three peaks are observed in the spectrum of the zero-added water sample in the region between 70-75 ppm, four peaks are observed in this region of the spectum for all samples with water contents greater than 2 percent suggesting that water is interacting directly with the sucrose in the sample. The narrow lines widths of the sucrose resonances also indicate that the sucrose in chocolate exists in a crystalline form rather than in an amorphous state. The solid fat signals observed between 10-35 ppm are also changed by the addition of water indicating that the fat environment or structure is changed in the presence of added water. After increasing the percent water in the sample by only 0.2 percent (w/w) water above the normal amount present in the sample, the amount of solid fat relative to that of solid sucrose is reduced sigificantly.

The hydration response of the mobile components in chocolate was monitored during a ^{13}C 90°-pulse WALTZ decoupled NMR experiment and is shown in Figure 4(1-4). The spectrum in Figure 4(1) for Sample B with no added water again indicates

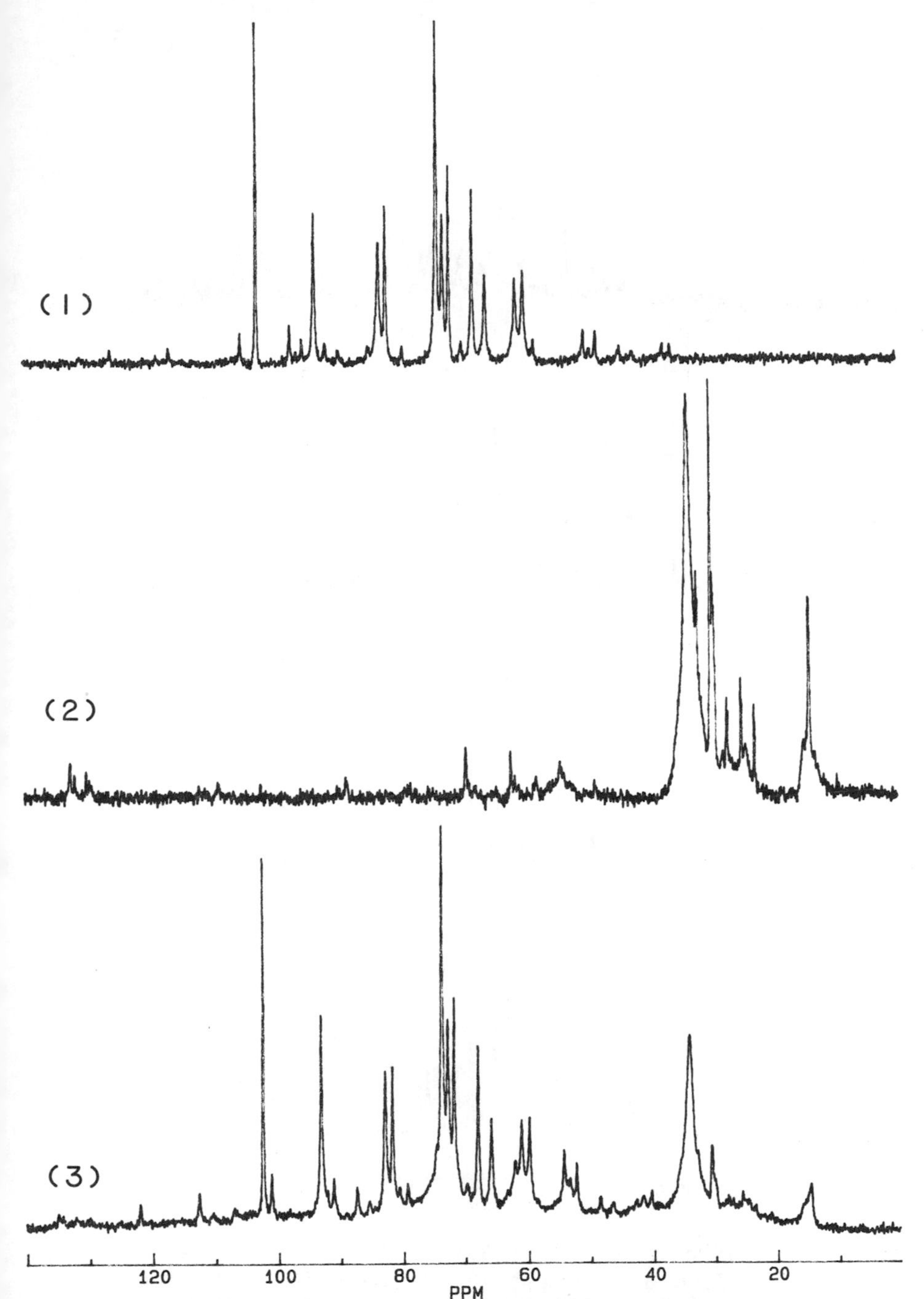

Figure 2 ^{13}C CPMAS NMR spectra obtained at 25°C with 1000 transients, line broadening of 3.6 Hz, 3000 microsecond contact time and a rotor frequency of 2.0 KHz for (1) crystalline sucrose (2) cocoa butter and (3) Sample B.

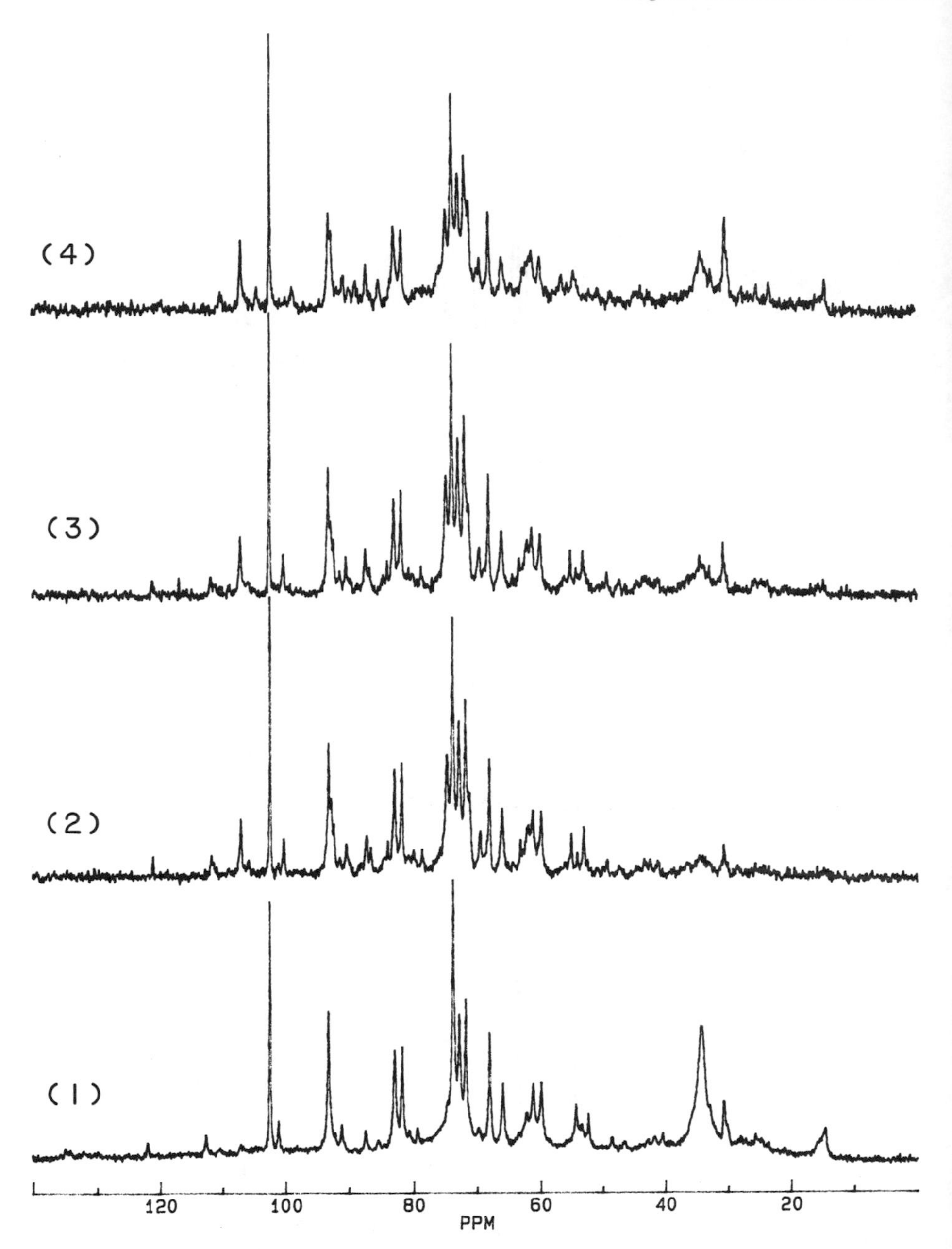

Figure 3 ^{13}C CPMAS spectra of Sample B at 25°C collected as a function of percent (%) (w/w) water with 1000 transients, 5000 microsecond contact time, 3.6 Hz line broadening and rotor frequencies of (1) 2000 Hz, (2) 1880 Hz, (3) 2000 Hz, (4) 1750 Hz and % (w/w) of (1) 1.05 %, (2) 1.78 %, (3) 2.81 %, (4) 6.35 % respectively.

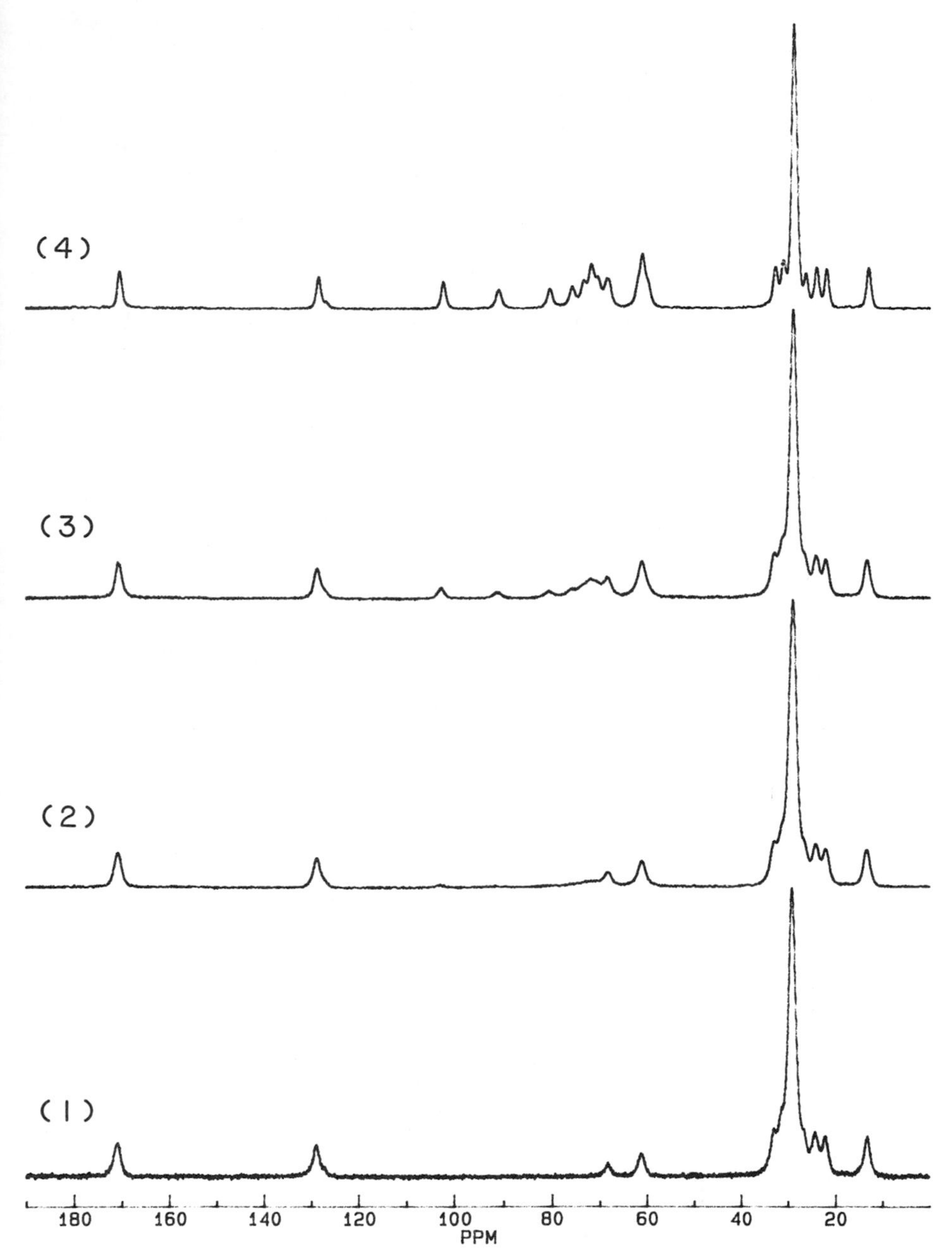

Figure 4 ^{13}C 90°-pulse spectra acquired with WALTZ decoupling at 25°C as a function of percent (%) water (w/w) of Sample B using 3.05 Hz line broadening. (1) 6400 transients, 1.05 % water (w/w); (2) 8384 transients, 2.08 % water (w/w); (3) 8240 transients, 4.77 % water (w/w); (4) 10,000 transients, 7.81% water (w/w) respectively.

that the mobile components in the chocolate mass prior to water addition are triglycerides. As the percent water in the chocolate samples is increased, new peaks appear in the spectra as observed in Figure 4(2-4). By 8 percent added water the anomeric carbon resonances are apparent in the spectrum. The other sucrose resonance carbons are present but are broad and not resolved indicating that the mobility of the mobilized sucrose has not reached the solution limit. Again detectable protein carbon resonances are not observed in the spectra. The lack of protein signal can be explained by several factors. First the signal-to-noise of the protein signals would be expected to be less than that for sucrose as the sucrose:protein ratio in the samples is approximately 2:1. Secondly the protein signal would be distributed over a greater number of resonances due to the greater number of chemically different carbons in a protein versus a sucrose molecule. Thirdly the whole protein may not be entirely solubilized and free to rotate isotropically leading to broader lines which may be buried in the baseline noise. The first two factors could also contribute to the lack of observed protein resonances in the ^{13}C CPMAS spectra. Since the milk proteins in chocolate are added as amorphous powders, the possibly large distribution in local environments for any particular protein carbon within the sample would result in a large chemical shift anisotropy. The chemical shift distribution would further broaden the protein signal making it even more difficult to observe under the fat and carbohydrate signal in the CPMAS spectra.

The hydration response of Sample A was also observed via ^{13}C 90° pulse with WALTZ decoupling NMR experiments and is shown in Figure 5(1-4). The sucrose was again beginning to be solubilized by 2 percent (w/w) added water and was fully solubilized by 10 percent (w/w) added water. The milk:cocoa butter ratio of the Sample A is greater than that for the Sample B; however, this particular ingredient difference does not appear to change the dynamic effects that water has upon the carbohydrate in Sample A when compared to Sample B. In fact, no dramatic differences are observed in the spectra of the two samples of two different ingredient ratios.

Although the protein signal was not observable in the NMR experiments described above, the protein response to hydration was monitored by fluorescence microscopy. Micrographs of Sample B stained for protein and obtained as a function of percent (w/w) water as shown in Figure 6(1-2) indicate that by 2 percent added water, the boundaries of the protein aggregates in chocolate have become diffuse. By 10 percent added water, the proteins have formed a continuous diffuse network. Microscopy data of Sample B stained for fat using Nile Blue show complete separation of fat and protein by 10 percent added water as shown in Figure 7(1-2). Micrographs of Sample B stained for fat also show that protein aggregates contain fat. Neither of the dyes used stained for carbohydrate, thus sucrose crystals could not be observed in either set of fluorescence micrographs. Sucrose and lactose crystals were observed, however, by birefringence[28]. Solid/liquid ratios were measured for Sample B as a function of percent water (w/w) and as a function of temperature. The data is shown in Figure 8. The data indicate that the percent solid decreases faster than the

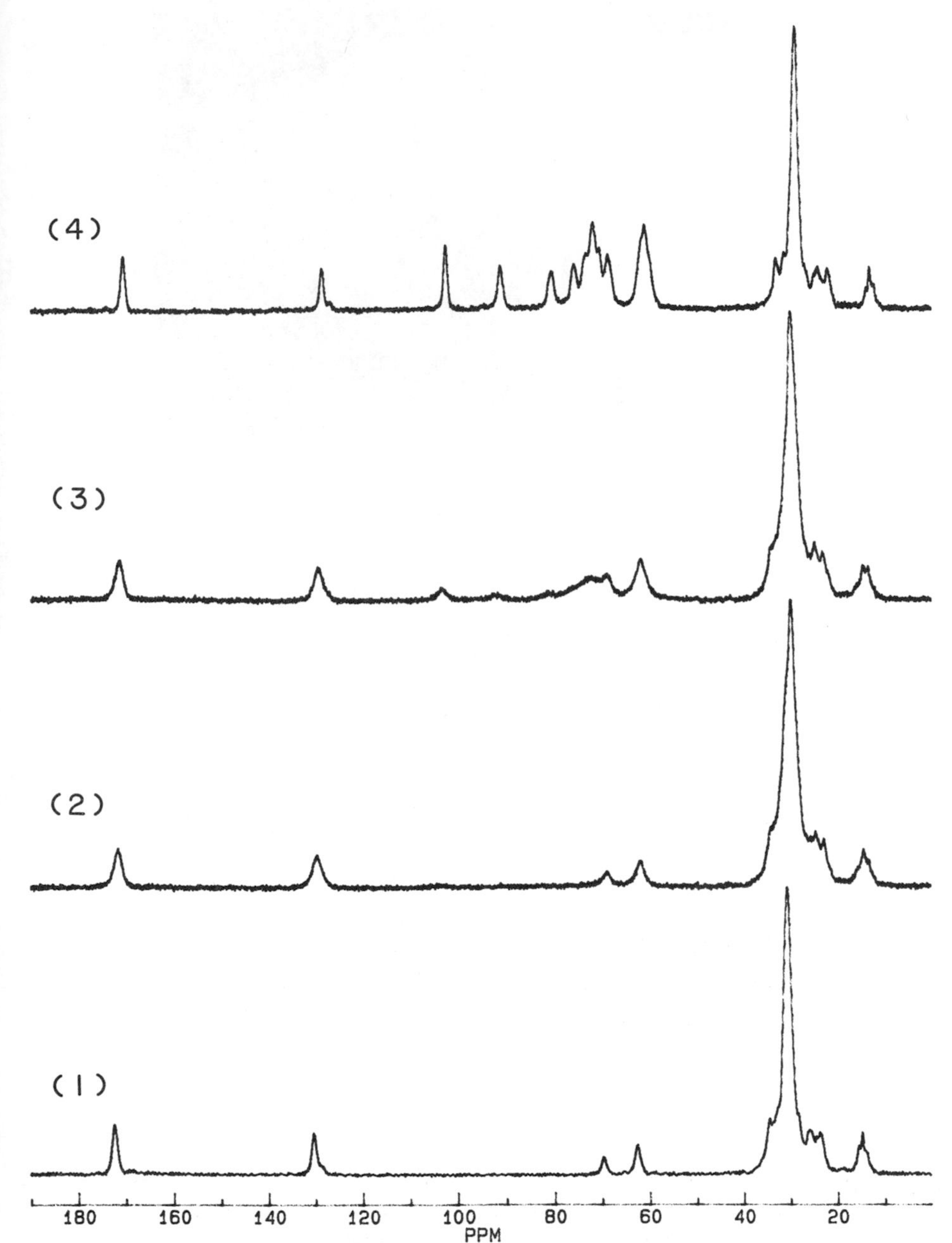

Figure 5 ^{13}C 90°-pulse spectra acquired with WALTZ decoupling at 25°C and 10,000 transients as a function of percent (%) water (w/w) for Sample A using 3.05 Hz line broadening. (1) 1.26 % water (w/w), (2) 2.72 % water (w/w), (3) 5.38 % water (w/w), (4) 9.68 % water (w/w) respectively.

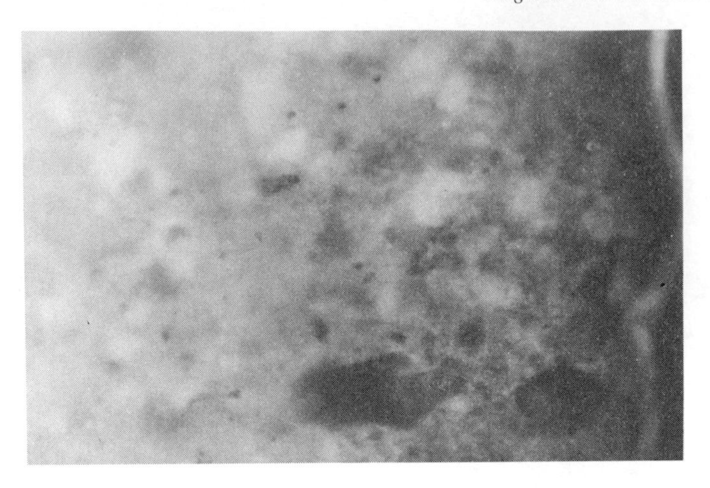

(2)

(1)

Figure 6 (1) Fluorescence micrograph of Sample B hydrated to 2% (w/w) water then stained for protein with ANS. (2) Fluorescence micrograph of Sample B hydrated to 8% (w/w) water then stained for protein with ANS.

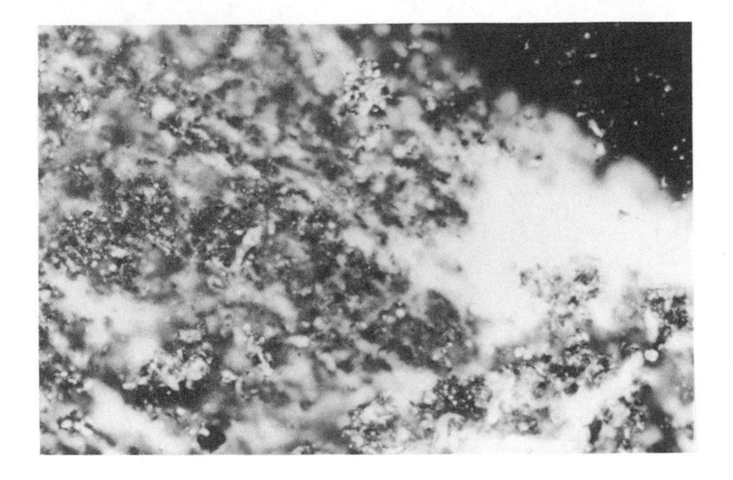

(2)

(1)

Figure 7 (1) Fluorescence migrograph of Sample B hydrated to 2% (w/w) water then stained for fat with Nile blue. (2) Fluorescence micrograph of Sample B hydrated to 8% (w/w) water then stained for fat with Nile blue.

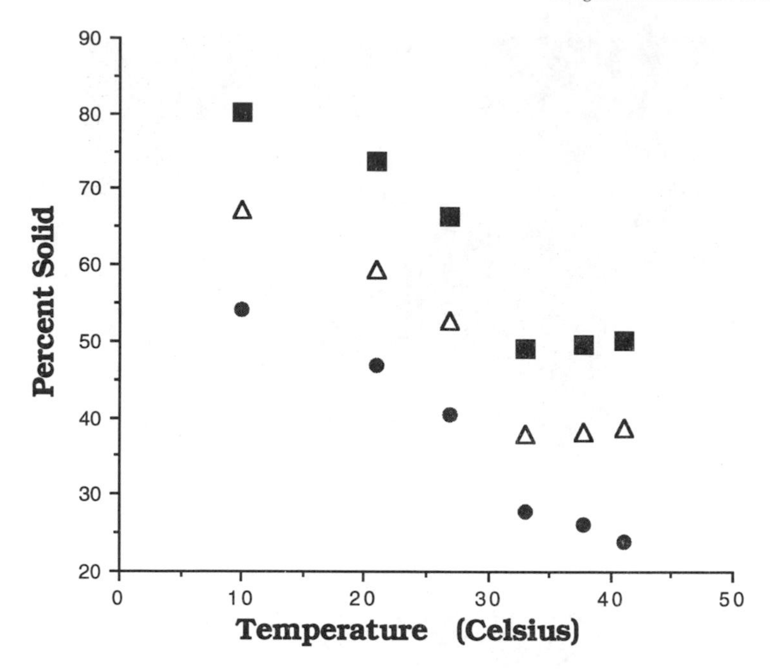

Figure 8 Solid/liquid ratio of Sample B as a function of temperature and percent (%) water (w/w). (■) 1.055 % water (w/w), (Δ) 5.35 % water (w/w), (●) 9.68 % water (w/w).

percent water is increased in the sample again indicating solubilization of the carbohydrates and proteins in the chocolate mass. The decrease in the percent solid between 10°C and 33°C decreases with increasing water content suggesting that a portion of the solid fat becomes mobilized upon water addition. That the shape of the fat melting profile remains relatively constant with water content suggests that the added water is not changing the form of the solid fat in the sample.

NMR and microscopy data could not be correlated directly with viscosity measurements as the viscosity of chocolate even at levels of 1 percent added water is so high that accurate measurement using any standard viscometry is not possible. Attempts to measure yield stress values using a squeeze flow technique are in progress. Although our initial squeeze flow data had a high degree of variability, reproducible increases were observed for samples with increasing water content up to 10 percent added water after devising appropriate sample preparation protocols[29].

4 CONCLUSIONS

The NMR and microscopy data together support the hypothesis that water added to a chocolate mass solubilizes the carbohydrate and some portion of the proteins present in a chocolate mass. Solubilization of the carbohydrates is evident from both the ^{13}C CPMAS and ^{13}C 90°-pulse WALTZ decoupled spectra. The observed solubilization could lead to increased interactions between carbohydrate and protein. Increasing the amount of water also appears to lead to networking and swelling with subsequent fat separation as supported by the fluorescence micrographs. Both events could contribute to the huge viscosity increases observed in chocolate upon water addition. Also with separation of the fat from the protein aggregates, the continuous fat phase may be too disrupted to allow for the proper contraction of the chocolate necessary for demolding. Neither the microscopy nor the NMR data as described could be used to determine the hydration response of emulsifiers in chocolate.

Acknowledgements

The authors would like to thank Dr. Michael McCarthy for useful discussions and Ms. Vikki Nicholson for completion of the Karl Fischer assays and assisting with the manuscript.

References

1. U.S. Patent 2,760,867, N .Kempf and P.Downey, 1956.
2. U.S Patent 5,149,560, Kirk Kealey, 1992.
3. U.S. Patent 4,081,559, M. Jeffrey et al, 1978.
4. U.S. Patent 2,904,438, Mars, Inc., 1959.
5. Japanese Patent 4,228,647, Lotte, 1991.
6. European Patent 0,440,203 A, Toshio Takemori, 1991.
7. European Patent 0 427 544 A2, Fuji Oil Co., 1991.
8. European Patent 0,401,427 A, Takemori et al, 1990.
9. European Patent 0,459,777 A1, Frank Kincs, 1991.
10. European Patent 0, 438,597 A, Fuji Oil Co. 1991.
11. European Patent 0,397,247 A2, Padley Frederick, 1990.
12. Japanese Patent 59,156,246, Fuji Oil Co., 1984.
13. Swiss Patent 662,041, Jacobs Suchard, 1987.
14. German Patent 389,127 DRP, Padley Frederick, Unilever, 1990.
15. European Patent 0,393,327 A, Toshio Takemori, 1990.
16. Japanese Patent 3,091,443, Fuji Oil Co., 1991.
17. European Patent 0,297,054 A, C. Giddey and G. Dove, 1988.
18. U.S Patent 2,760,867, General Foods, 1956.
19. European Patent 5,063,080, Christof Kruger, 1990.
20. Swiss Patent 519,858, Interfoods.
21. Swiss Patent 399,691.
22. U.S. Patent 2,760,867, N. Kempf and P. Downey, 1956.
23. U.S.Patent 5,149,560, Kealey et. al., 1992.
24. U.S.Patent 2,760,867, N. Kempf and P. Downey, 1956.
25. AOAC Method 977.10, Official Methods of Analysis of the Association of Official Analytical Chemists, 15th Edition, p. 763, 1990.

26. B. C. Gerstein and C. R. Dybowski, "Transient Techniques in NMR of Solids", Academic Press; Orlando, Florida, 1973.
27. E. R. Andrews and R. G. Eades, Proc. R. Soc., London 1953 **a218**, 537.
28. M. Smart, unpublished data.
29. Bruce Campbell, unpublished data.

Conformation and Dynamics of Polysaccharide Gels as Studied by High Resolution Solid State NMR

Hazime Saitô, Hiroyuki Shimizu, Takako Sakagami, Satoru Tuzi, and Akira Naito

DEPARTMENT OF LIFE SCIENCE, HIMEJI INSTITUTE OF TECHNOLOGY, HARIMA SCIENCE GARDEN CITY, KAMIGORI, HYOGO, JAPAN 678-12

1 INTRODUCTION

Gel-forming ability may be the most important property of several types of polysaccharides such as curdlan, agarose, starch, carrageenans, etc. in relation to their biological functions as well as food, industrial or other applications. In many instances, gelation of these polysaccharides[1,2] occurs as a result of physical association of individual chains adopting an ordered conformation through either formation of cross-links or junction-zones by multiple-stranded helices or aggregation of single or multiple-stranded helices, or both. It is essential to analyze conformation of the respective polysaccharide chain in order to clarify the network structure and gelation mechanism as well. In practice, this task is not always straightforward, because the network structure is highly heterogeneous as viewed from conformational and dynamic point of view.[3] The existence of solid-like domains from cross-linked region is characteristic of a gel-formation and can be analyzed by X-ray diffraction if crystalline portion is available but tends to prevent one performing a detailed study of conformation, because it is not always guaranteed that conformations of such solid-like domains are well retained in liquid-like domains.

Instead, it is emphasized that high-resolution ^{13}C NMR spectroscopy is a more suitable means for conformational characterization of polysaccharides both in the solid and gel states, because the above-mentioned solid- and liquid-like domains can be analyzed separately by means of cross-polarization (CP) and dipolar decoupling (DD) experiments as combined by magic angle spinning (MAS), respectively. In addition, it has been demonstrated that ^{13}C chemical shifts of backbone carbons in polysaccharides are substantially displaced (up to 8 ppm), depending on their respective conformations as defined by a set of two torsion angles (ϕ,φ) about the glycosidic linkages, and can be used as a diagnostic means for conformational elucidation.[4,5]

In the present article, we demonstrate how conformation and dynamics of the solid- and liquid-like domains of gel networks consisting of (1->3)-β-D-glucans, amylose and starch and agarose are revealed by the solid-state NMR technique with emphasis on clarifying their gelation mechanism .

2 CONFORMATION-DEPENDENT ^{13}C CHEMICAL SHIFTS AS A MEANS FOR CONFORMATIONAL CHARACTERIZATION

It appears that (1->3)-β-D- and (1->4)-α-D-glucans have a common feature to exhibit gel-forming ability by heating. It is thus essential to clarify the conformation-dependent displacements of ^{13}C NMR peaks of these polysaccharides by examination of their polymorphs, to be used as reference data for conformational characterization in the gel state.

A number of linear and branched (1->3)-β-D-glucans were isolated from a variety of cell walls of plants, bacteria, fungi or reserved polysaccharides.[6] These glucans are known to take the following three different conformations, single chain, single helix and triple helix, in the solid state which are mutually converted by a variety of physical treatments (Figure 1).[3,7] In particular, it was demonstrated[3] that in hydrated state curdlan (a linear glucan) takes the single helix form, whereas a number of branched glucans (lentinan, schizophyllan, HA-β-glucan, etc.) take the triple helix form. This distinction arises from how easily the individual polymer chains are hydrated. In fact, curdlan polymer is very hydrophobic and is not sufficiently hydrated untill its aqueous suspension is heated at a temperature above 150°C.[9] The branched glucans, on the other hand, are well hydrated immediately after they are exposed to aqueous media. In addition, it should be also taken into account that the triple-helical curdlan is formed with expense of thermal depolymerization of chain from DP (degree of polymerization) 3980 to 106, after heating at 180° C for 10 min.[10] A similar conversion diagram was obtained for (1->3)-β-D-xylan,[11] but the pathway from the single helix to the triple helix form in this case is not yet established probably because (1->3)-β-D-linked xylan is more hydrophobic than curdlan.

These three forms are readily distinguished by their characteristic ^{13}C NMR spectra,[3] as manifested from those of the three kinds of curdlan samples (Figure 2). The C-3 ^{13}C peak of anhydrous sample (single chain form) (89.8 ppm) is displaced downfield by 2.5-3.3 ppm from that of the

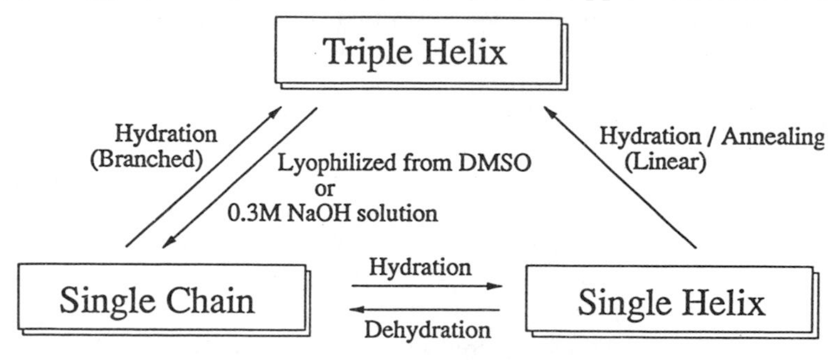

Figure 1. Conversion diagram for (1->3)-β-D-glucans by a variety of physical treatments (based on Refs. 7 and 8).

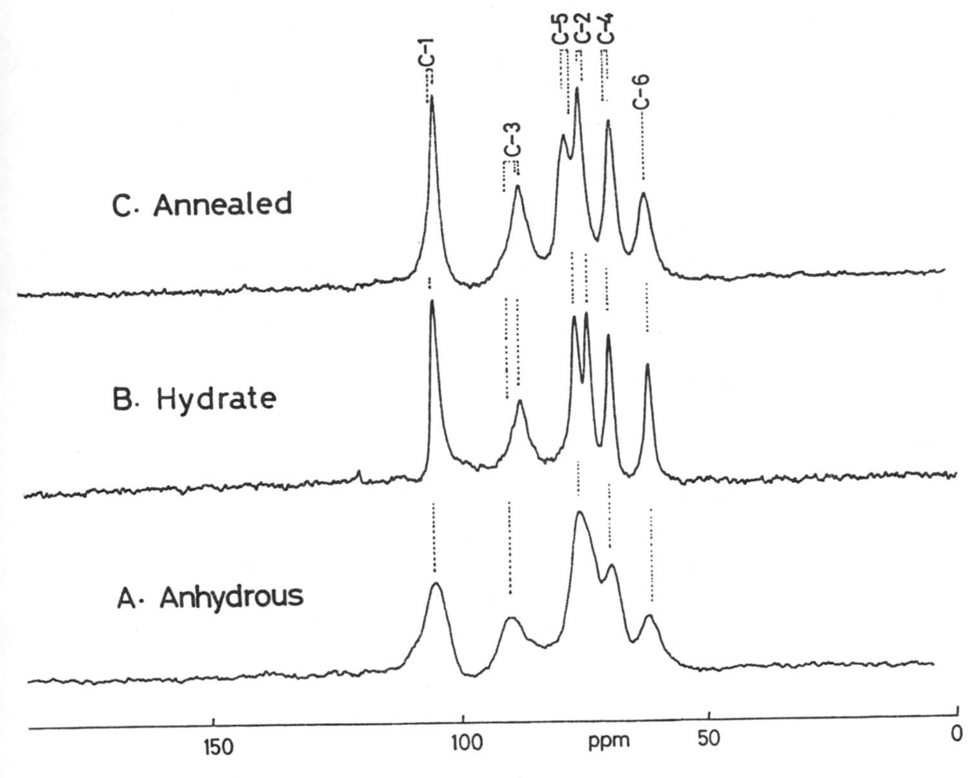

Figure 2. ^{13}C NMR spectra of curdlan in anhydrous (A), hydrate (B) and annealed state (C) (Ref. 7).

hydrated sample (single helix) (87.3 ppm) or annealed sample (triple-helix) (86.5 ppm). A slightly narrowed but almost the same spectral pattern was observed when curdlan was lyophilized from DMSO solution. At first glance, it appears that the ^{13}C chemical shifts of the hydrated sample are similar to those of the annealed curdlan. Distinction of these two spectra is straightforward on closer examination of the C-5 chemical shifts (75.8 for the former and 77.5 ppm for the latter, respectively) and the peak-separation between the C-5 and C-2 carbons (2.0 and 3.2 ppm, respectively).[7] The reason why the C-3 ^{13}C chemical shifts are not significantly different between the single and triple helix forms could be ascribed to the fact that the torsion angles between the two types of helices are very similar as manifested from the lengths of the c-axis (fiber axis) being 18.78 Å (triple helix)[12] and 22.8 Å (single helix).[13] In addition, random coil conformation is readily distinguished from the above-mentioned three forms on the basis of the conformation-dependent displacement of peaks.[14] Thus, it is now possible to characterize unequivocally the conformations of the liquid- and solid-like domains of gels consisting of (1->3)-β-D-glucans, as will be described later in more detail.

It has been shown that amylose and starch exhibit the following polymorphs, as analyzed by X-ray diffraction study:[15,16] V-, A, B, and C- forms. The V-form exists as complexes with small organic molecules and has in common a left-handed, single six residue helix, whereas the A- and B-forms are found for cereal and tuber starches, respectively. The latter two forms are readily distinguished by ^{13}C NMR spectra, because the C-1 peaks are split into a triplet and a doublet for the A- and B-forms, respectively.[17-19] On the contrary, the C-1 and C-4 signals of the V-forms give rise to single lines and are displaced downfield from those of the B form (4-5 ppm for the C-4 peak) (Figure 3).[20] It is interesting to note that the ^{13}C NMR spectra of both A- and B-forms of starch and amylose are substantially distorted by drying to give a spectral profile of amorphous form.[21-23] In contrast, it was shown that hydration of amorphous amylose of low molecular weight (DP 17) results in complete conversion to the B-type form.[20] In addition, Senti and Witnauer previously demonstrated that the B-form of amylose can be obtained from the V-form by hydration.[24] Consistent with this view, we noticed that the V-form amylose of high molecular weight (DP 1000) complexed with DMSO was converted to the B-form by humidification by 96% relative humidity for 12 hr, as shown in Figures 3A and 3B. A similar result was obtained for amylose of low molecular weight (DP 17). These findings are summarized in Figure 4. It turned out, however, that this sort of conversion is incomplete (50%) for amylose of intermediate molecular weight (DP 100).

The B-form was initially considered as a single helical conformation,[25] since the conversion of V- to B-amylose takes place on humidification.[24] Later, the structure was refined as a right-handed double-helix in an antiparallel fashon.[26] The handedness of the double-helix, however, was recently revised as a left-handed one.[27] Nevertheless, it is hardly likely that simple humidification of amylose in a desiccator causes such an unfolding/folding process leading from the single stranded helix (V-form) to double stranded helix (B-form). In this connection, it was shown that complete dissolution in aqueous solution is an essential requirement for the conversion of the single helix to the triple helix as found for (1->3)-β-D-glucans.[7] This means that unfolding of the polymer chain followed by refolding in aqueous media is a necessary condition for such conversion.

There remains an essential question how the secondary structure of the "B-form" converted from the V-form can be distinguished from that of the native B-form so far studied. Further extensive studies might be required to fill the gap between the two different views.

3 NETWORK STRUCTURE AND GELATION MECHANISM

Figure 5 illustrates the ^{13}C NMR spectra of curdlan gel recorded by a variety of NMR methods including broad-band decoupling by a conventional NMR spectrometer (the liquid-like domain; B), DD-MAS (intermediate domain; C), and CP-MAS (the solid-like domain; D).[28] The present result indicates that the single helix form is dominant for all of the motionally

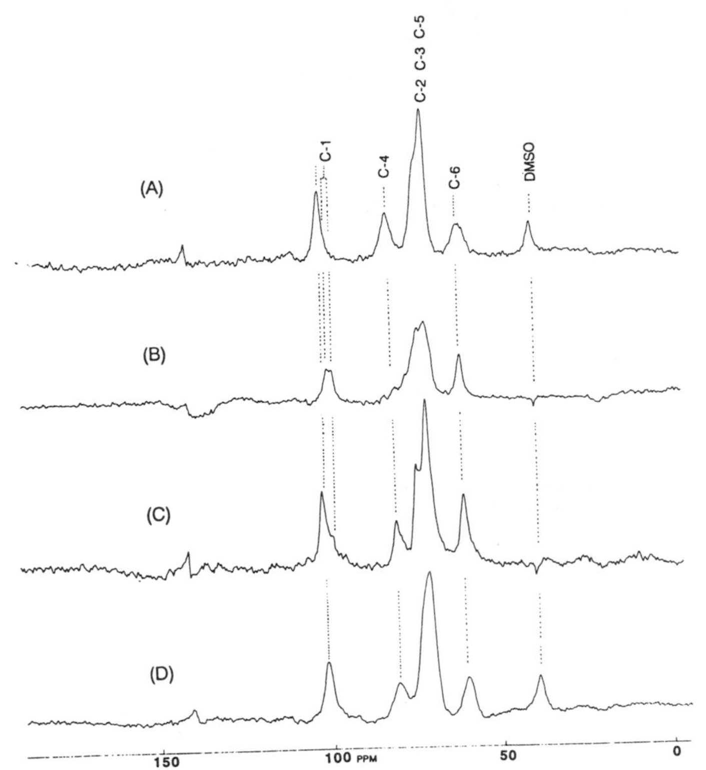

Figure 3. ^{13}C NMR spectra of amylose film (DP 1000). (A) anhydrous, (B) hydrated, (C) hydrated iodine complex, and (D) anhydrous iodine complex (Ref. 20).

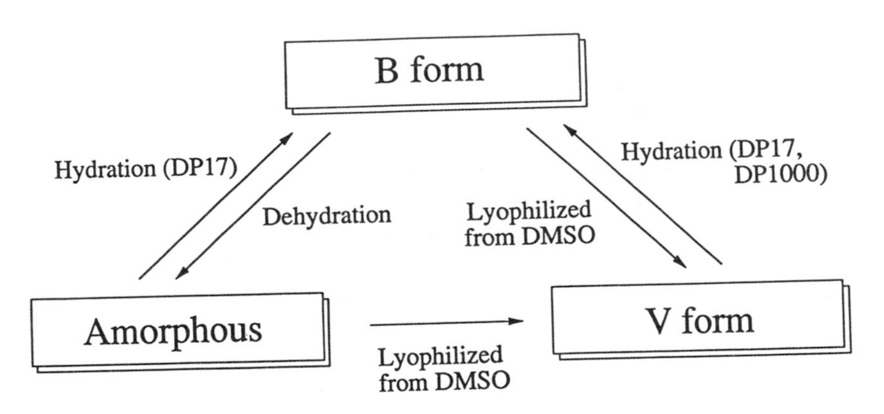

Figure 4. Conversion diagram of amylose by various physical treatments (based on Ref. 20)

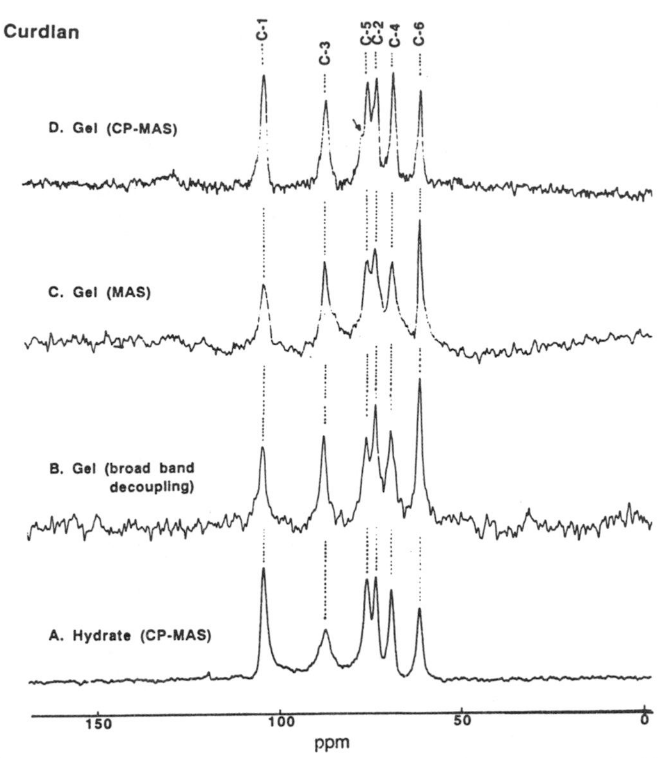

Figure 5. ^{13}C NMR spectra of curdlan gel recorded by a variety of NMR methods. A and D, CP-MAS NMR technique; b, conventional NMR using broad-band decoupling; C, DD-MAS (Ref. 28).

Table 1 Conformational feature of network structure as examined by NMR

	Chemical structure	Liquid-like domain (by DD-MAS)	Solid-like domain (by CP-MAS)
Curdlan	linear (1->3)-β-D-glucan	single helix	single helix
Schizophyllan	branched (1->3)-β-D-glucan	–	triple helix
HA-β-glucan	branched (1->3)-β-D-glucan	–	triple helix
Amylose/starch	(1->4)-α–D-glucan	random coil	single helix?
Agarose		random coil	single helix?

different domains of curdlan gel as summarized in Table 1 and seems to account for an elastic property due to the presence of rather flexible single helical chains. The amount of the triple-helical chains is thus nominal, if any, as indicated by the arrow of Figure 5D, as far as heating temperature is kept below 80 °C (low-set gel). However, the increased gel strength by heating at a temperature between 80° and 120°C (high-set gel) is well explained in terms of the additional formation of cross-links due to hydrophobic association of the single helical chains as well as increased proportion of the triple-helical chains.[9] On the contrary, we found that the ^{13}C NMR signals characteristic of the triple helix form are dominant in the ^{13}C CP-MAS NMR spectra of the branched (1->3)-β-D-glucans including HA-β-glucan (see Figure 6), lentinan and schizophyllan and all of the NMR signals were completely suppressed as recorded by the conventional liquid-state NMR (Table 1).[3,28] This means that gelation of the branched glucans proceeds from partial association of the triple-helical chains. This network structure is far from formation of an elastic gel and seems to be consistent with formation of brittle gel structure.

It was shown that amylose gel contains two kinds of ^{13}C NMR signals: the B-type signals from motionally restricted regions as recorded by CP-MAS NMR technique and the signals identical to those found in aqueous solution.[29] Undoubtedly, the latter signals could be ascribed to a flexible molecular chains adopting random coil conformation (liquid-like domain). On the other hand, the former peaks are ascribed to the solid-like domain of cross-links, either double helical junction-zones[29] or aggregated species of single helical chains as discussed already (see Table 1).[28] So far two different views have been proposed for gelation mechanism of amylose gels. Miles et al.[30] have suggested that amylose gels are formed upon cooling molecularly entangled solutions as a result of phase separation of the polymer-rich phase, whereas Wu and Sarko[26] proposed that gelation occurs

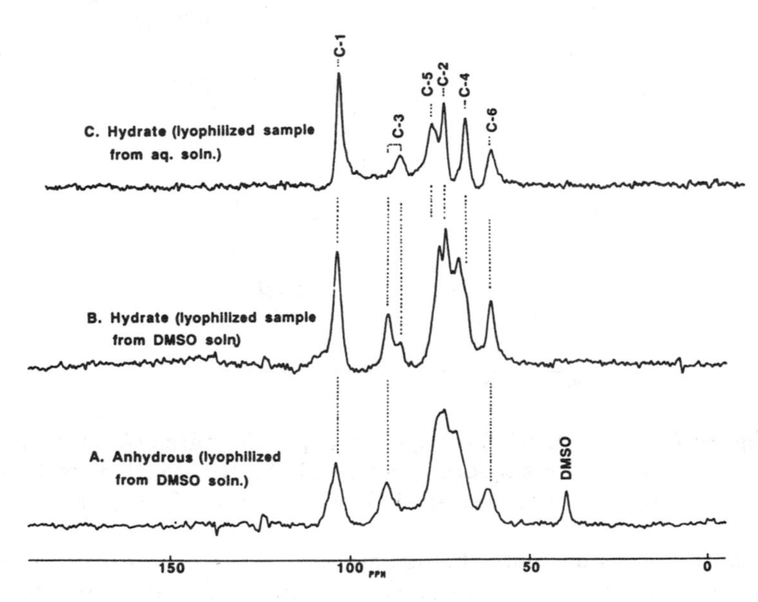

Figure 6 ^{13}C NMR spectra of HA-β-D-glucan at anhydrous (A) and hydrate state (B) and (C). (Ref. 28).

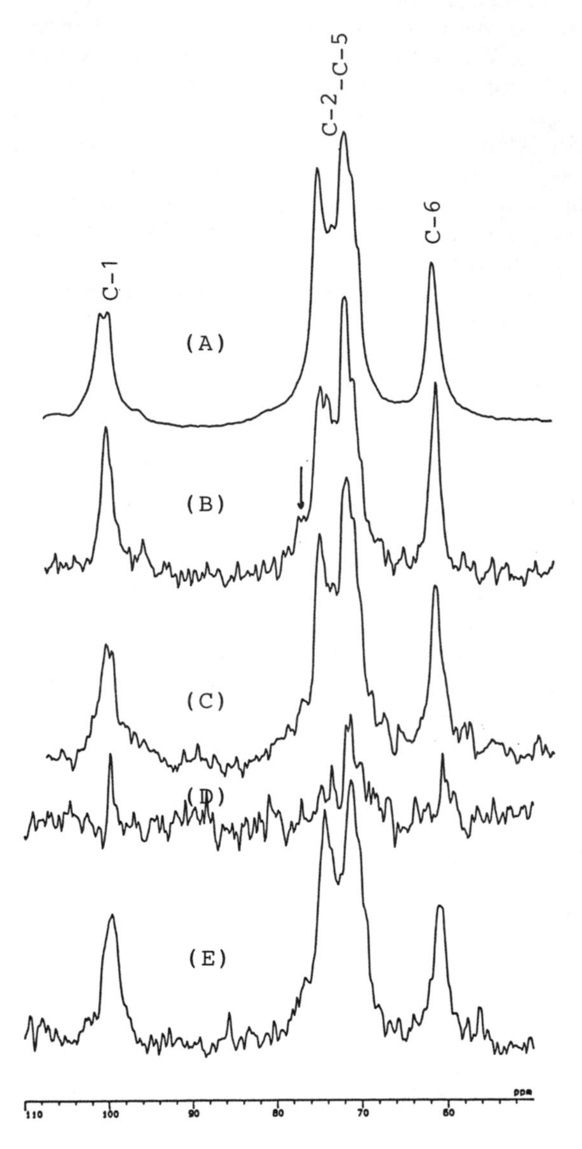

Figure 7 ^{13}C CP-MAS NMR spectra of potato starch (A) and its gel (B-E). Freshly prepared gel (33%; B), after 3 days (33%,C), freshly prepared gel (17%, D), and after 3 days (17%, E)

through cross-linking by double helical junction-zones. It is thus important to explore a new means to distinguish the single helical chains from the multiple helical chains as secondary structures of amylose. In this connection, we recorded ^{13}C DD-MAS and CP-MAS NMR spectra of freshly prepared starch gel and its retrograded samples after 3 days in a refrigerator. We recorded intense ^{13}C NMR signals taking random coil conformation by DD-MAS NMR for samples with and without retrogradation. On the contrary, ^{13}C NMR signals of starch gels recorded by the CP-MAS technique are ascribed to the presence of B-form chains whose peak-intensities were significantly increased by retrogradation (Figure 7). This finding is consistent with the previous finding that retrogradation results in crystallization as detected by X-ray diffraction.[30] Thus, it is concluded that retrogradation of starch gel is accompanied by a conformational change from random coil to B-form (single helix). Further discussion on this subject will be performed based on the data of relaxation parameters in the next section.

The well-documented network model of agarose gel[31] arises from the junction zones consisting of associated double helical chains. It is unlikely that this sort of rigid network structure gives rise to the liquid-like domain which provides signals visible by a conventional NMR spectrometer. Nevertheless, it was shown that intense ^{13}C NMR signals are visible either by a conventional spectrometer or DD-MAS experiment. Here, we illustrate ^{13}C NMR spectra of agarose gel (Nacalai Chemicals, LGT, 20%) recorded by DD-MAS (top; a broad peak at 110 ppm being from the probe assembly) and CP-MAS techniques (bottom), respectively (Figure 8). At first glance, it appears that the spectral positions of the two spectra are very similar. A detailed examination, however, shows the presence of a significant change in the relative peak-intensities at 77-78.5 ppm region. As demonstrated already, the peaks which exhibit the conformation-dependent displacements are C-3 and C-4 carbons for (1->3)- and (1->4)-linked galactosyl residues,

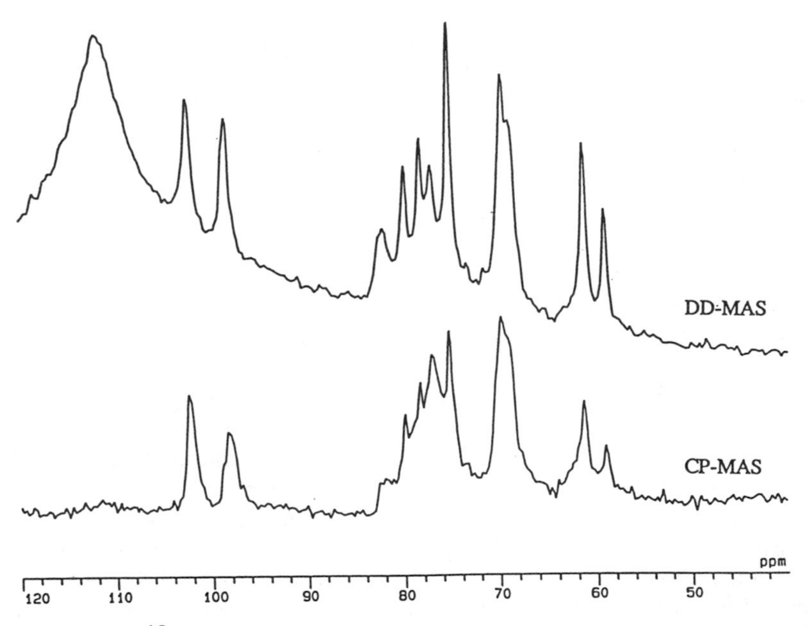

Figure 8 ^{13}C NMR spectra of agarose gel. liquid-like domain (top) and solid-lke domain (bottom)

respectively. Therefore, it is expected that the peaks which exhibit conformation-dependent change of agarose gel are C-3 peak of (1->3)-linked galactosyl and C-4 peak of (1->4)-linked 3,6-anhydro-α-L-galactosyl residues, respectively. Usov[32] showed that these peaks are resonated at 81.9 and 77.0 ppm in aqueous solution. Therefore, the above-mentioned spectral change between the two types of experiments, DD- and CP-MAS methods, is clearly ascribed to the conformational change of agarose from random coil to an ordered conformation (Table 1).

Several investigators[33-35] have questioned the validity of the double-helical junction-zones and proposed an alternative model of gel network containing extended single helices. In accordance with this line, we examined ^{13}C NMR spectra of dried agarose film and its hydrate prepared from N,N-dimethylacetamide solution at 80°C under anhydrous conditions with the expectation of an increased proportion of single chains.[36] We found that the ^{13}C NMR spectrum thus obtained is identical to that obtained from gel sample. Therefore, it is more likely that the network structure of agarose gel consists mainly of the single helical chains, as summarised in Table 1.

4. DYNAMIC FEATURE OF GEL NETWORK

It is well recognized that ^{13}C spin-lattice relaxation times in the laboratory frame is very sensitive to the presence of motions with correlation time of 10^{-8} s. On the contrary, ^{1}H spin-lattice relaxation time in the rotating frame is sensitive to motion whose correlation time is of the order of 10^{-4} s. Therefore, it seems worthwhile to perform a comparative study of various types of polysaccharide gels as studied in this work by means of such relaxation parameters.

It is very important to use completly sealed sample rotor for gel samples in order to obtain reliable relative peak-intensities essential for measurements of such relaxation parameters. Otherwise, the peak-intensities may be distorted by detuning of probe circuitry caused by loss of water molecules during rapid sample rotation. In the present work, we tightly glued a teflon cap to the sample rotor with rapid Araldite (Ciba-Geigy). In Figure 9, we illustrate a typical example of a set of stacked plots of ^{13}C NMR spectra of starch gel as a function of delay time, t, in order to obtain the spin-lattice relaxation times in the laboratory frame, T_1's. The T_1 values were calculated by means of the following formula.[37]

$$M_{net}(t) = M_{cp}(0)\exp(-t/T_1), \qquad (1)$$

where $M_{net}(t)$ and $M_{cp}(0)$ are the peak intensity at the delay time t and zero, respectively. In Table 2, we summarize the ^{13}C T_1' s of starch (potato) gel (33%) thus obtained (solid-like domain), together with those determined by DD-MAS technique (liquid-like domain), recorded by a Chemagnetics CMX-400 spectrometer (100 MHz). It is interesting to note that the ^{13}C T_1values of the liquid-like domain are in the vicinity of the T_1 minimum, $\omega_0 \approx 10^{-8}$ s ($\omega_0\tau_c = 1$), whereas those of the solid-like domain are in the lower temperature side of the T_1 minimum, $\tau_c > 10^{-8}$ s. Here, the values

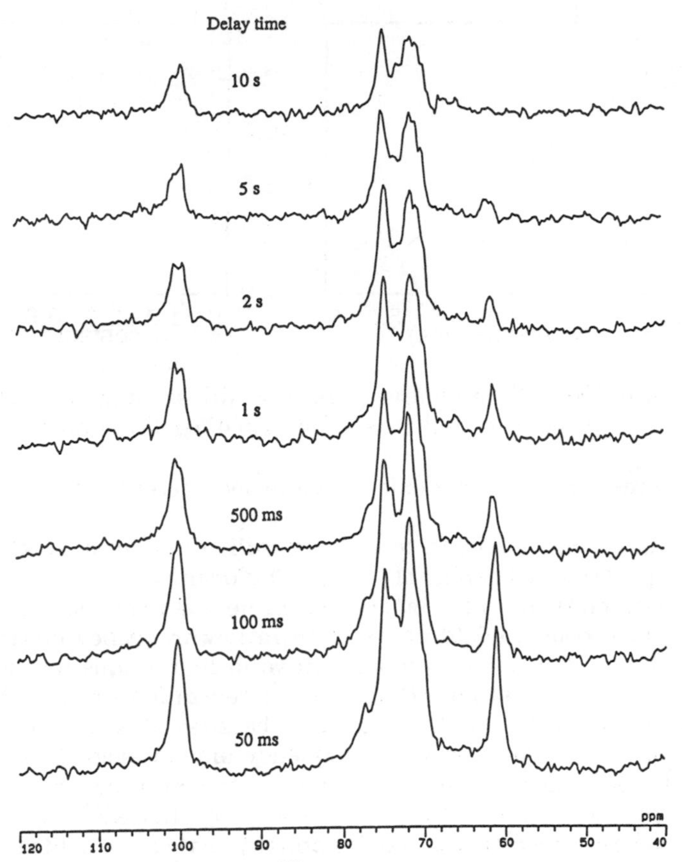

Figure 9 A stacked plot of ^{13}C NMR spectra of starch gel as a function of delay time.

Table 2 ^{13}C spin-lattice relaxation times of starch gel (33%) by DD- and CP-MAS NMR method (s)

	C-1	C-4	C-3	C-2	C-5	C-6
Liquid-like domain	0.36	0.29	0.30	0.32	0.29	0.16
Solid-like domain	9.2	11.8	11.9	11.9	11.9	2.1

ω_0 and τ_c denote the larmor frequency and correlation time, respectively. In fact, it is pointed out that the latter values are very close to those observed for a variety of (1->3)-β-D-glucans in the solid state.[8] Therefore, this finding is consistent with a view that the solid-like domain arising from the cross-linked region of starch gel is crystalline portion as obtained in the solid-state. On the contrary, the T_1 values from the liquid-like domain arose from flexible molecular chains taking random coil form, although their

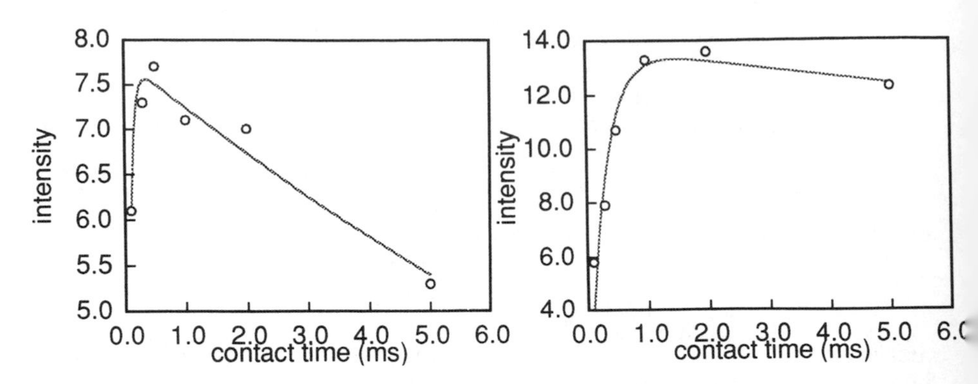

Figure 10 Plots of the relative peak-intensities of agarose (left) and strach gels (right) as a function of contact times.

mobility is restricted to some extent due to the presence of the cross-linked structure.

It turned out, however, to be very difficult to distinguish the single from multiple-stranded helices of (1->3)-β-D-glucans on the basis of the spin-lattice relaxation times in the laboratory frame (data not shown). It is likely that such distinction sholuld be detected in low frequency motions, if any, although the T_1 values are rather sensitive to high frequency motions. In such case, it is expected that ^1H spin-lattice relaxation times in the rotating frame might be useful for this purpose, because this parameter is very sensitive to the presence of lower frequency motion such as 10^{-4} s. ^{13}C resolved ^1H spin-lattice relaxation times in the rotating frame ($T_{1\rho}$) and cross polarization time (T_{CH}) were easily evaluated[38] by fitting the variation of an experimental peak-intensity $I(\tau)$ as recorded by varing contact times τ (Figure 10).

$$I(\tau) = (I_0/T_{CH})[\exp(-\tau/T_{1\rho}) - \exp(-\tau/T_{CH})]/(T_{CH} - T_{1\rho}) \qquad (2)$$

In particular, the observed peak decays exponentially after reaching the maximum intensity. This is clearly seen by examination of the decay

Table 3 The observed T_{CH} and $T_{1\rho}$ of linear and branched (1-3)-β-D-glucans

	C-1		C-3		C-5		C-2		C-4		C-6	
	T_{CH} μs	$T_{1\rho}$ ms	T_{CH} μs	$T_{1\rho}$ ms	T_{CH} μs	$T_{1\rho}$ ms	T_{CH} μs	$T_{1\rho}$ ms	T_{CH} μs	$T_{1\rho}$ ms	T_{CH} μs	$T_{1\rho}$ ms
curdlan	128	17.9	138	16.6	137	17.0	145	19.2	110	14.0	82.2	22.9
schizophyllan	67.3	3.32	62.8	4.18	53.1	4.30	56.4	4.09	32.0	5.69	22.1	10.0
HA-β glucan	56.6	5.04	35.4	5.38	64.4	6.27	47.8	5.34	50.7	5.80	53.0	7.28

Table 4 Typical T_{CH} (μs) and $T_{1\rho}$ (ms) values of starch (33%) and agarose (20%) gels

	Starch		Agarose	
	C-5	C-6	G-1	G-6
T_{CH} (μs)	290	320	64.8	128
$T_{1\rho}$ (ms)	46	53	13.7	39.4

pattern as a function of $\exp(-\tau/T_{1\rho})$. This means that the $T_{1\rho}$ of agarose gel (left) is much shorter than that of starch gel (right), as will be discussed later. In Table 3, we summarize the ^{13}C T_1 and ^1H $T_{1\rho}$ of (1->3)-β-D-glucans thus obtained. Clearly, it is seen from Table 3 that the T_{CH} and $T_{1\rho}$ values of curdlan taking the single helix conformation are significantly longer than those of schizophyllan and HA-β-glucan taking the triple helix conformation. This means that the single helical curdlan is able to afford low frequency motions in the solid-like domains, whereas the triple helical glucans are not. In fact, the $T_{1\rho}$ valules of the latter two in the gel state are very close to those observed in curdlan (single chain) and paramylon (triple helix) in the solid state (data not shown) in spite of their different conformations.

Table 4 summarizes the typical examples of the T_{CH} and $T_{1\rho}$ values of starch and agarose gels to gain insight into the dynamic feature of these gel network structures as compared with those of (1->3)-β-D-glucans. It is demonstrated from the data of Table 4 that the molecular chains of the solid-like domains of both the starch and agarose are more flexible than those of the single helical curdlan gel, as judged from the siginificantly longer T_{CH} and $T_{1\rho}$ values of these gels as compared with those of Table 3. In this connection, it should be taken into account that the major part of molecular chains in these starch and agarose gels are taking random coil conformation, in contrast to the cases of (1->3)-β-D-glucans (Table 1). Therefore, the flexibility of the molecular chains in these polysaccharide gels is decreased in the following order: starch > agarose ≈curdlan >schizophyllan ≈ HA-β-glucan. In addition, it is also pointed out that the amount of the multiple-stranded chain in starch and agarose gels is nominal, if any, on the basis of the relaxation parameters.

5 CONCLUDING REMARKS

We demonstrate here that conformation and dynamics of polysaccharide gels are very conveniently studied by high-resolution solid-state ^{13}C NMR spectroscopy. The major advantage of the NMR study is that the liquid- and solid-like domains can be examined separately by DD- and CP-MAS NMR technique, respectively, and their conformations are elucidated by an empirical manner on the basis of the conformation-dependent displacements of peaks, once ^{13}C chemical shifts of polymorphic structures of polysaccharides have been accumulated in advance. It turned out that a systematic study of conversion diagram among polymorphs by a series of physical treatments is especially useful to clarify whether the polysaccharide chain under consideration is taking either single or multiple-chain/helical conformations both in the solid and gel state.

Further, we found that both the ^{13}C and ^{1}H spin-lattice relaxation times in the laboratory and rotating frame, respectively, are very useful to examine the dynamic features of the network structures. In particular, the ^{1}H spin-lattice relaxation times in the rotating frame turned out to be very sensitive to the presence or absence of low frequency motions and are very useful to distinguish the single helical chains from the multiple-helical chains in the gel network.

References

1. D. A. Rees, Pure & Appl. Chem., 1981, **53**, 1.
2. A. H. Clark and S. B. Ross-Murphy, Adv. Polym. Sci., 1987, **83**, 57.
3. H. Saitô, in "Viscoelasticity of Biomaterials", ed. by W. Glasser and H. Hatakeyama, ACS Symp. Ser. 489, pp. 296-310, 1992.
4. H. Saitô, Magn. Res. Chem., 1986, **24**, 835.
5. H. Saitô and I. Ando, Ann. Rep. NMR Spectrosc. 1989, **21**, 209.
6. B. A. Stone and A. E. Clarke, "Chemistry and Biology of (1-3)-β-D-Glucans", 1992, La Trobe Univ. Press.
7. H. Saitô, M. Yokoi and Y. Yoshioka, Macromolecules, 1989, **22**, 3892.
8. H. Saitô, and M. Yokoi, Bull. Chem. Soc. Jpn., 1989, **62**, 392.
9. H. Saitô, R. Tabeta, M. Yokoi, and T. Erata, Bull. Chem. Soc., 1987, **60**, 4259.
10. Y. Yoshioka, N. Uehara, and H. Saitô, Chem. Pharm. Bull., 1992, **40**, 1221.
11. H. Saitô, J. Yamada, Y. Yoshioka, Y. Shibata and T. Erata, Biopolymers, 1991, **31**, 933.
12. C. T. Chuah, A. Sarko, Y. Deslandes, and R. H. Marchessault, Macromolecules, 1983, **16**, 1375.
13. K. Okuyama, A. Otsubo, Y. Fukuzawa, M. Ozawa, T. Harada, and N. Kasai, J. Carbohydr. Chem., 1991, **10**, 645.
14. H. Saitô, T. Ohki, and T. Sasaki, Biochemistry, 1977, **16**, 908.
15. A. Sarko and P. Zugenmaier, ACS Symp. Ser., 1980, **141**, 459.
16. D. French, "Starch: Chemistry and Technology", 2nd ed. ed. by R. L. Wistler and E. D. Pascall, Academic Press, New York, 1984, p. 183.
17. R. P. Veregin, C. A. Fyfe, R. H. Marchessault, and M. G. Taylor, Macromolecules, 1986, **19**, 1030.
18. M. J. Gidley and S. M. Bociek, J. Am. Chem. Soc., 1985, **107**, 7040.
19. F. Horii, H. Yamamoto, A. Hirai, and R. Kitamaru, Carbohydr. Res., 1987, **160**, 29.
20. H. Saitô, J. Yamada, T. Yukumoto, H. Yajima, and R. Endo, Bull. Chem. Soc. Jpn., 1991, **64**, 3528.
21. R. P. Veregin, C. A. Fyfe, and R. h. Marchessault, Macromolecules, 1987, **20**, 3007.

22. M. J. Gidley, and S. M. Bociek, J. Am. Chem. Soc., 1988, **110**, 3820.
23. F. Horii, A. Hirai, and R. Kitamaru, Macromolecules, 1986, **19**, 930.
24. F. R. Senti and L. P. Witnauer, J. Am. Chem. Soc., 1948, **70**, 1438.
25. J. Blackwell, A. Sarko, and R. H. Marchessault, J. Mol. Biol., 1969, **42**, 379.
26. H. C. H. Wu and A. Sarko, Carbohydr. Res., 61, **7**, 1978.,
27. A. Imberty and S. Perez, Biopolymers., 1988, **27**, 1205. 365.
28. H. Saitô, Y. Yoshioka, M. Yokoi, and J. Yamada, Biopolymers, 1990, **29,** 1689.
29. M. J. Gidley, Macromolecules, 1989, **22**, 351.
30. M. J. Miles, V. J. Morris, and S. G. Ring, Carbohydr. Res., 1985, **135**, 257.
31. S. Arnott, W. E. Scott, D. A. Rees, and C. G. A. McNab, J. Mol. Biol., 1974, **90**, 253.
32. A. I. Usov, Botanica Marina, 1984, **27**, 189.
33. M. R.Letherby and D. A. Young, J. C. S. Faraday I, 1981, 1953.
34. I. T. Norton, D.M.Goodall, K. R. I. Austin, E. R. Morris, and D. A. Rees, Biopolymers, 1986, **25**, 1009.
35. S. A. Foord and E. D. T. Atkins, Biopolymers, 1989, **28**, 1345.
36. H. Saitô, M. Yokoi, and J. Yamada, Carbohydr. Res., 1990, **199**, 1.
37. D. A. Torchia, J. Magn. Reson., 1978, **30**, 613.
38. M. Mehring, "High resolution NMR spectroscopy in solids", Springer, New York, 1983.

NMR Studies of Cereal Proteins

A. M. Gil

DEPARTMENT OF CHEMISTRY, UNIVERSITY OF AVEIRO, 3800 AVEIRO, PORTUGAL

1 INTRODUCTION

The prolamin storage cereal proteins account for nearly half of the total protein present in most mature seeds[1]. The original definition of prolamins was based on their amino acid composition and their solubility properties. These proteins are very rich in glutamine and proline amino acids and are characterized by unusual primary structures consisting of the repetition of well defined peptide sequences. Most of the prolamin proteins are insoluble in water and in salt solutions but some are soluble in alcohol/water mixtures [1-3].

Three main groups of prolamins are defined according to their average molecular weight as well as to their content of amino acid residues containing sulfur and it has been shown that similar fractions can be found in different cereals (Table 1)[1-3]. The properties of these proteins and particularly their response to hydration are believed to account for the characteristic viscosity and elasticity properties observed in these systems. The alcohol-insoluble fraction of wheat gluten (glutenins) is believed to be responsible for its elasticity properties whereas the alcohol-soluble fraction of gluten (gliadins) seems to account for its viscosity properties[1,3]. Such properties should also determine the behavior of dough as well as some of the final properties of baked products[1,3]. It is therefore important to investigate and characterize the relationship between the structural properties and the macroscopic properties of this type of systems. Such studies should help to inform the food processor about how to modify a particular property of his product to improve its quality or to adapt it for further uses, including applications in non-food areas.

Due to the general insolubility of these systems and difficulties associated with their extraction and purification, a relatively limited number of studies have been carried out so far on these proteins having been mainly devoted to solution state systems[1,3-5]. Since it is important to model and understand their behavior in a state that approximates real systems

Table 1 *Prolamin protein fractions found in Wheat, Barley and Rye*

Type of Fraction	WHEAT	BARLEY	RYE
HMW (high molecular units)	HMW subunits (Glutenins)	D-Hordein	HMW Secalins
S-Poor	ω-Gliadins	C-Hordein	ω-Secalins
S-Rich	γ-Gliadins	γ-Hordeins	γ-Secalins 40K
	α-Gliadins	---	---
	LMW subunits (Glutenins)	B-Hordein	---
	---	---	γ-Secalins 75K

such as dough, it is of interest to study the low water content systems. In order to carry out these studies, solid state analytical methods must be applied and the developments of techniques such as vibrational spectroscopy, Nuclear Magnetic Resonance (NMR) spectroscopy and X-ray diffraction have been of extreme importance.

Reference should be made to some of the published spectroscopic studies of gluten as a whole[6,7], of gluten HMW subunits[8] and of barley protein C-hordein[9,10]. This barley protein has been used as a suitable structural model system for wheat gluten because of its composition, similar to the one of ω-gliadins (Table 1), and its availability in a relatively pure and homogeneous form.

This work describes an NMR study aiming to clarify the structural implications of the hydration process of wheat gliadins. Some studies were carried out on the model protein C-Hordein (composed of the repeat motif: -[Pro-Gln-Gln-Pro-Phe-Pro-Gln-Gln]-)[1-3] and compared with some preliminary results obtained for the ω-gliadins system. Techniques such as proton relaxation measurements, ^{13}C Cross Polarization and Magic Angle Spinning (CP/MAS), Single Pulse Excitation (SPE) experiments and High Speed ^{1}H MAS experiments were used. In some cases, comparison of the effects of protonated and deuterated water enabled distinction between the effects of water-related nuclear magnetic mechanisms and protein plasticization processes upon hydration.

2 EXPERIMENTAL PROCEDURE

The cereal proteins C-hordein and ω-gliadins were prepared as described previously[9-11]. Hydrated samples were prepared by drying to constant weight under vacuum and standing over solutions of different relative humidities. The deuteriated samples were prepared by stirring the sample in D_2O for *ca.* 4 h, freeze-drying and drying under vacuum to constant weight. All hydration percentages indicated are expressed in terms of g water/100 g dry protein.

For all C-hordein studies at 2.3T a Bruker MSL100 spectrometer operating at a proton frequency of 100MHz was used. All experimental conditions have been described elsewhere[9,10]. The lineshape studies were carried out on the above spectrometer and on a Bruker MSL300 spectrometer, operating at a proton frequency of 300MHz. All experimental parameters relative to the studies at 300MHz have been described previously[10]. All NMR studies on ω-gliadins were carried out on a Bruker MSL400P operating at a proton frequency of 400MHz and using a 4mm MAS probe. CP/MAS spectra were obtained under the conditions stated in the figure caption. The SPE spectrum was obtained with a recycle time of 15 s. In these studies a carbon 90° pulse length of 4μs was used and for proton studies the value used for the 90° pulse length was 2.5μs .

3 RESULTS AND DISCUSSION

3.1 Proton T_1 measurements on C-hordein

The variation of T_1 relaxation times with temperature and water content for protonated dry samples shows a maximum at around 225 K and tends to a minimum at around 350 K (Figure 1a). The shape of this curve and the effect of increasing water content have been discussed elsewhere[10] and show that water addition causes more efficient relaxation and that the temperature at which the spectral density function is maximum moves to lower temperatures. This can be accounted for by plasticization of protein enhancing rapid motions (of the order of ω_0, 100MHz) and the effects of spin chemical and quantum

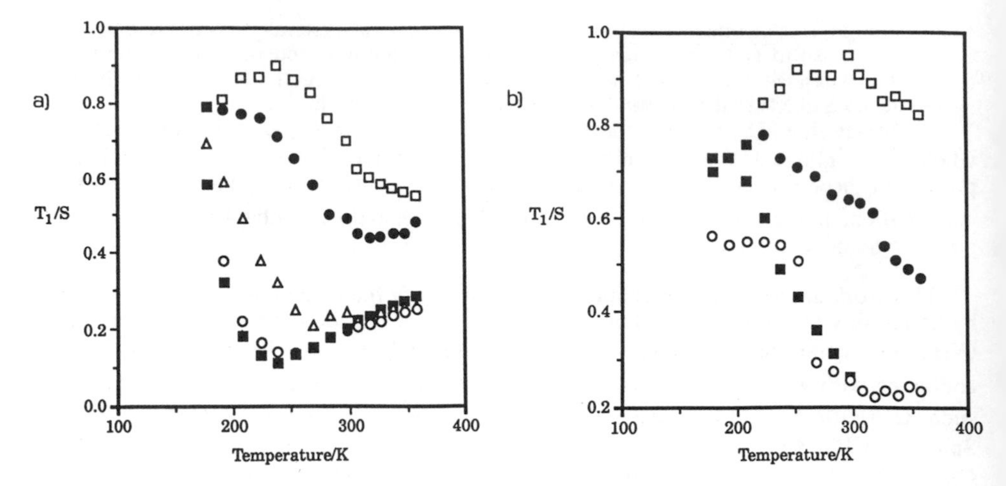

Figure 1 *Variation of proton spin-lattice relaxation time, T_1, for C-hordein, as a function of temperature and a) percentage H_2O content, (□) 0, (●) 7, (Δ) 16, (O) 49, (■) 67; and b) percentage of D_2O content, (□) 0, (●) 16, (■) 72, (O) 221.*

mechanical exchange[12,13].

Comparison of these results with the results obtained for the deuteriated samples (Figure 1b) enables the study of the system without the exchange mechanisms being involved. As the D_2O content is increased, relaxation is enhanced demonstrating the result of protein plasticization. Information about the role of the high number of glutamine sidechain amide groups may be obtained by comparison of the results shown in Figure 1. Since these groups should have exchanged fully with D_2O in the deuteriated samples, the relaxation enhancement can be attributed to backbone motions or other sidechain motions such as proline ring puckering or aromatic ring motions. However, comparison of both protonated and deuteriated dry samples shows significant differences, particularly the absence of a high temperature minimum in the deuteriated dry sample. This difference can be assigned to the deuteriation of the sidechain amide groups of glutamine. Rotation of NH_2 and NH_3^+ groups has been identified as a cause of T_1 relaxation[14]. In the protonated samples these groups should also contribute to relaxation through their participation in exchange processes with the water.

3.2 Proton $T_{1\rho}$ relaxation on C-hordein

The measurement of proton $T_{1\rho}$ relaxation times helps to identify significant slower motions functioning as $T_{1\rho}$ relaxation mechanisms (motions of the order of 50-60kHz, in this study). In the dry materials, relaxation in the protonated samples is more efficient than in the deuteriated samples (Figure 2b) implying that the exchangeable amide groups on the glutamine chains or the backbone have a role in the relaxation process. At high water contents the protonated samples show a sharp excursion towards longer relaxation times (Figure 2a). A possible cause is the increased efficiency of chemical exchange as the water content and temperature rise[15]. The likely candidates for exchange effects are water and sidechain amide groups the motion of which may also be affected and cause slower relaxation in the protonated material. An explanation for the sudden change in $T_{1\rho}$ in terms of the occurrence of a glassy to rubbery transition has been considered and discarded[10,16]. An alternative explanation consists of a conformational transition. At

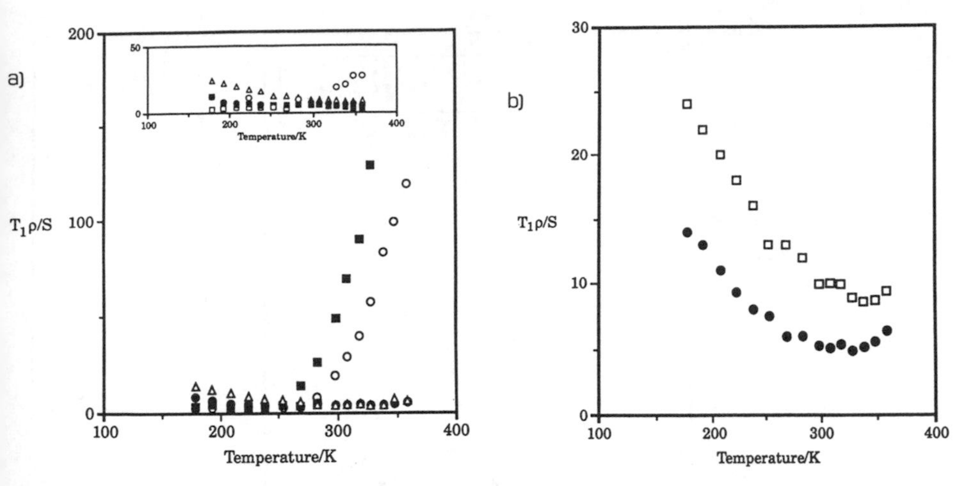

Figure 2 *Variation of proton spin-lattice relaxation time in the rotating frame, $T_{1\rho}$, for C-hordein as a function of temperature and a) percentage H_2O content, (Δ) 0, (●) 22, (O) 49, (■) 67; and (insert) percentage D_2O content, (Δ) 0, (■) 16, (□) 72, (O) 221. b) Plots for the dry protonated (●) and deuteriated (□) samples.*

higher temperatures and water contents there may be a transition from β-sheet-rich structure to a more mobile β-turn rich structure[5] with enhanced motion of the glutamine side chains and increased access to water to enable more rapid exchange.

3.3 Static proton lineshape variations

Figure 3 shows a typical proton spectrum of a hydrated protein sample showing two superimposed components. Evidence for the occurrence of a conformational transition should be visible in the corresponding proton spectra. These were obtained for the same protonated and deuteriated samples, at 2.3T field strength, and the corresponding linewidths of the broad and narrow components measured[10]. The narrow components show distinctly different behavior which depends upon deuteration (Figure 4). The protonated material shows considerable variation in the behavior of the narrow component which is naturally related with the water contribution for that signal. However, comparison of the components of the two driest protonated and deuteriated samples shows that there must be also a contribution from the proton exchangable groups, namely, the glutamine

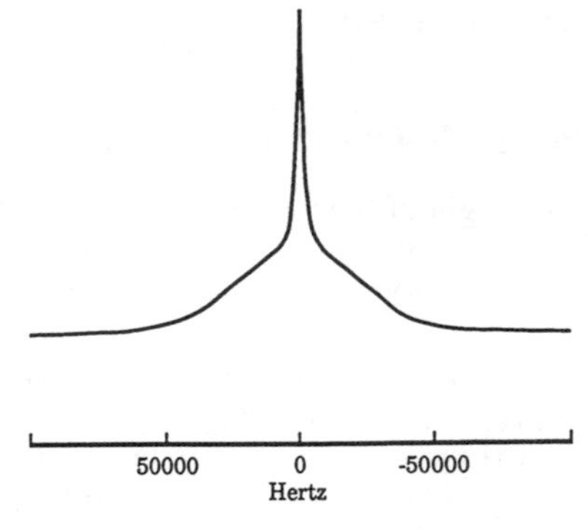

Figure 3 *Proton wide-line NMR spectrum of C-hordein. This sample contains 7% water .*

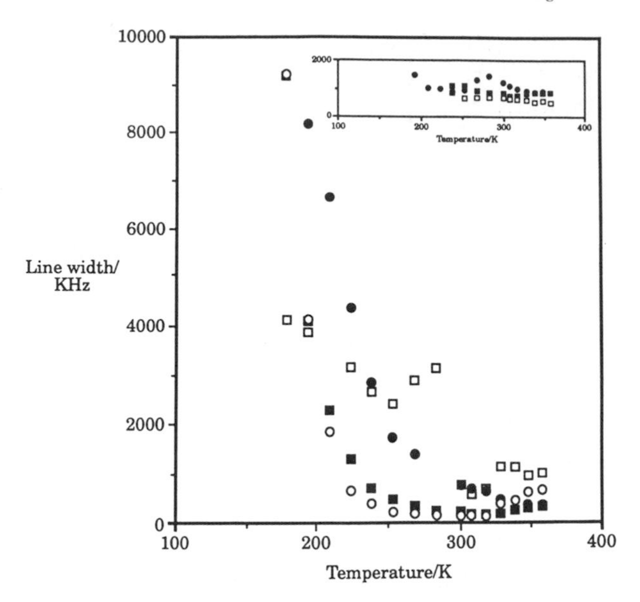

Figure 4 *Linewidth at half height of the narrow component of the proton spectra as a function of temperature and percentage H_2O content, (□) 7, (●) 16, (■) 49, (O) 67; and (insert) percentage D_2O content (□) 0, (■) 16, (●) 221.*

sidechain groups.

At higher water contents and temperatures, there is a point when the broad and narrow components become indistinguishable and, in the protonated samples, this coincides with the conditions for the sharp change in $T_{1\rho}$. This observation is again consistent with the occurrence of a conformational transition[10]. The linebroadening and spectral finestructure seen in such conditions (Figure 5) must also be the result of some change in conformation since the low temperature narrow component is narrower than the high temperature structured line.

Taken together with the relaxation measurements presented, the study of lineshape variations suggests an important involvement of glutamine side chains in the dynamic behavior of the system. In order to investigate further the role of those groups, as well as other sidechain groups, it is important to understand fully the origins of the proton spectra lineshape and of the finestructure observed.

3.4 On the origins of proton spectral lineshapes

In order to understand the origin of the narrow component in the proton spectra of C-hordein, the effects of variable field strength and MAS were investigated as well as those of the spin echo sequences $90^o{}_x$-τ-$180^o{}_y$ (Hahn echo) and $90^o{}_x$-τ-$90^o{}_y$ (solid echo).

The effect of varying the field strength from 2.3 T to 7 T was that linewidths increased at the higher field[9,17]. This implies that some sort of field-dependent interactions such as chemical shift anisotropy or magnetic susceptibility must be involved. Similar to the high water content lineshape at 2.3T (Figure 5), the 7T proton spectra show asymmetry at lower water contents and clear structure at higher water contents (Figures 6 and 7).

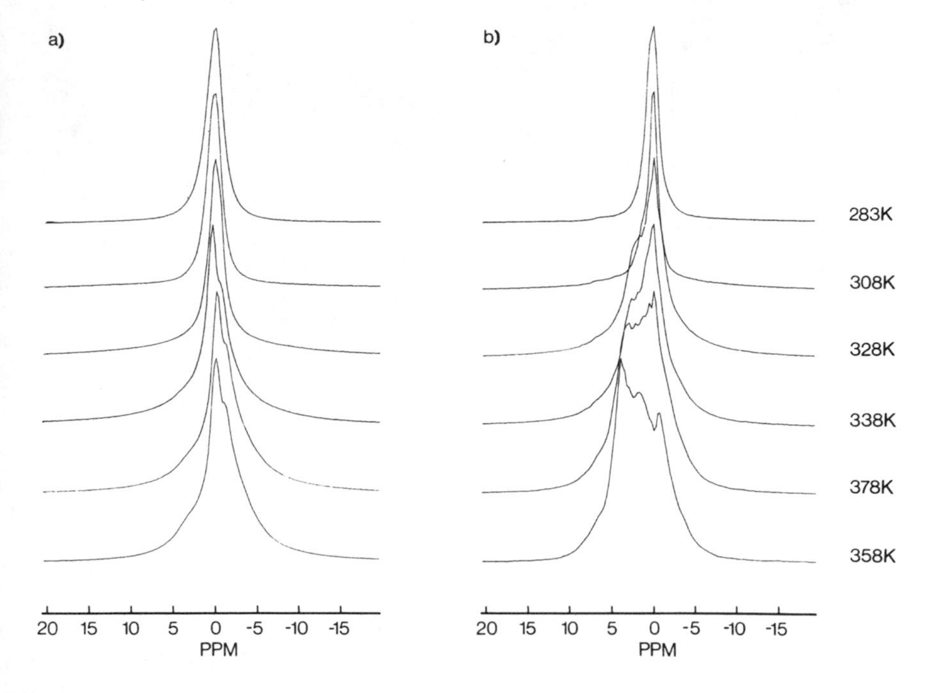

Figure 5 *Proton spectra of C-hordein samples with a) 49% and b) 67% of H_2O at different temperatures.*

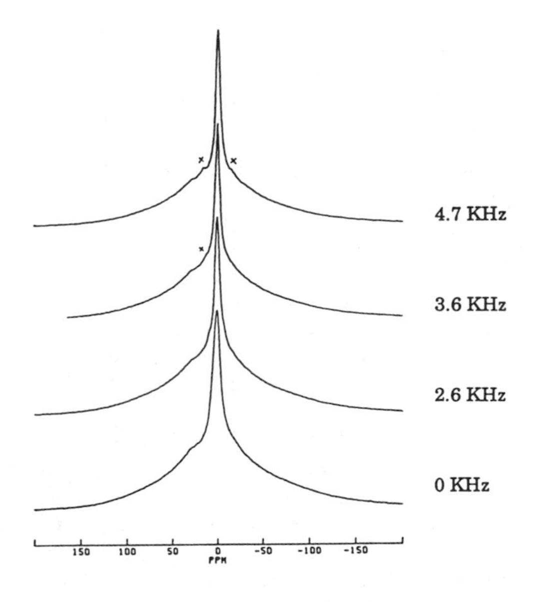

Figure 6 *Proton spectra, at 7 T, of a sample of C-hordein with 10% water, a) in static conditions and with MAS at spinning rates 2.6 , 3.6 and 4.7 kHz. x indicates the position of spinning sidebands (s.s.b.).*

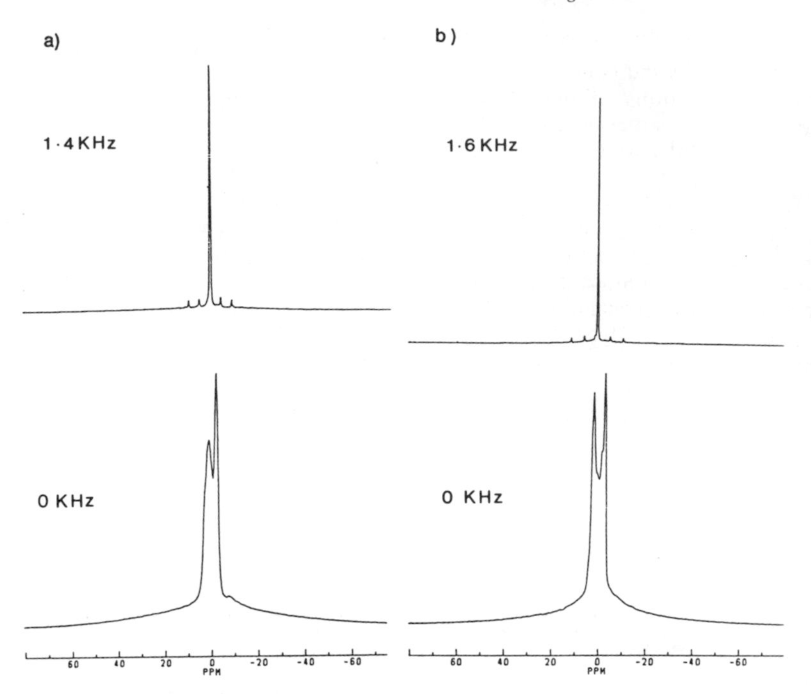

Figure 7 *Proton spectra, at 7 T, of C-hordein with a) 27% water and b) 30% water, under static conditions and sample spinning conditions.*

In order to understand the finestructure observed, the possibility of it being due to a distribution of isotropic chemical shifts must be investigated. This was done by studying the effect of MAS on the samples containing 10, 27 and 30% water (Figures 6 and 7). It is apparent that for all the three samples MAS affects the lineshape. The possibility that isotropic chemical shifts contribute to the lineshape can thus be dismissed since those would remain unchanged under MAS conditions. For the 10% sample, the effect of MAS is to narrow the line significantly and to generate spinning sidebands. The line narrowing is observed even at the lowest spinning speed of 2.6 kHz and the change, thereafter, with increasing spinning speed is very small. Since the static lineshape is 3.1 kHz, this implies that the line is a sum of a number of separate interactions[17]. The contribution of homonuclear proton dipolar interactions must be another important contribution for the narrow component of the sample hydrated to 10% water. In the more hydrated samples, however, considerable linenarrowing is achieved at spinning rates as low as 1.4 kHz (Figure 7) which may suggest that the contribution of static dipolar interactions is greatly decreased.

In order to investigate further the relative importance of dipolar effects at higher water contents it is possible to make use of the 90^o_x-τ–180^o_y and 90^o_x-τ–90^o_y spin-echo sequences. Both refocus field-dependent effects such as chemical shift and magnetic susceptibility but the solid echo sequence can also refocus static dipolar interactions arising from an isolated spin pair[18,19]. In the 90-180 sequence the spin pair interactions are not refocussed and the pulse sequence is twice as efficient at refocussing field dependent effects as the 90-90 sequence[18]. Therefore, in the absence of dipolar interactions the ration: $R = 2\,E_{90}/E_{180}$ is unity. E_{90} is the echo intensity at 2τ for the 90-90 sequence and E_{180} is that for the 90-180 sequence.

When R is plotted as a function of τ for liquid ethane-1,2-diol it is seen to be independent of τ and close to unity (*ca.* 1.2) which is consistent with the absence of static dipolar interactions (Figure 8a). When the same ratio is plotted for C-Hordein hydrated to 36% water, R is much larger than one at short times and falls to the same value of *ca.* 1.2 after about 2.5ms (Figure 8b). This demonstrates the presence of dipolar interactions.

If the part of the 90-180 decay after 2.5ms is subtracted from the whole decay[9] the remaining component may be fitted to an exponential function with a decay constant of 6.58×10^3 s^{-1}, equivalent to a linewidth of 2kHz. This represents the total static dipolar interaction contribution to the linewidth. Comparison of the intercepts of the two exponential components of the 90-180 echo decay also indicate that approximately 60% of the peak area comes from static dipolar interactions with the remaining 40% originating from other interactions including field-dependent effects[9].

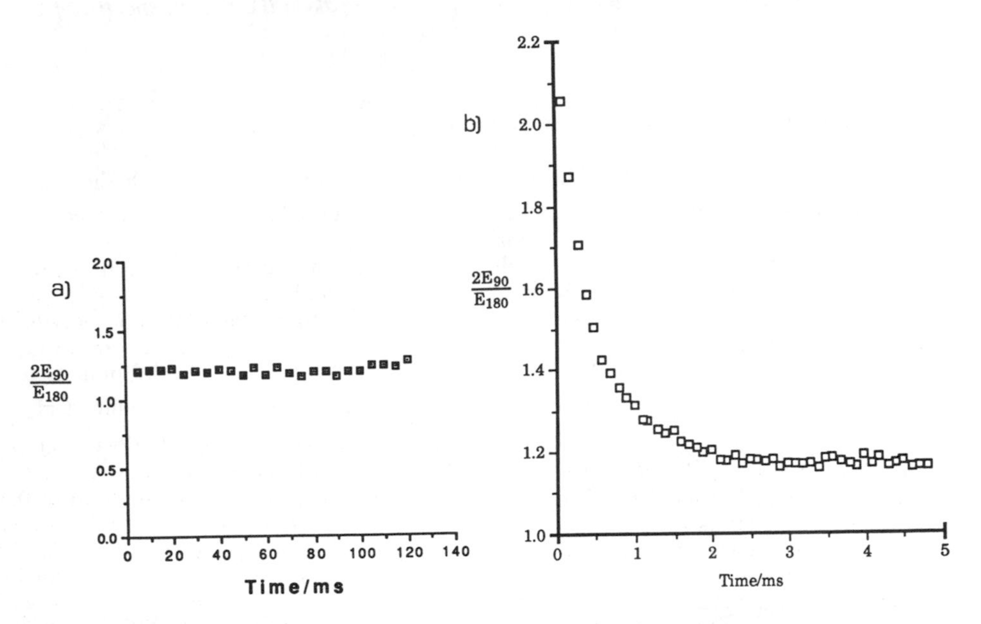

Figure 8 *Plot of the intensity ration, $R = 2E_{90}/E_{180}$, for a) liquid ethane-1,2-diol and for b) C-hordein with 35% water.*

These studies show that, contrary to what is generally assumed in the interpretation of these type of proton spectra, the motional narrowing regime can not be automatically associated with the narrow component of the spectrum. This work shows that both field-dependent and dipolar interactions contribute significantly for the reproducible doublet feature observed in the static spectra of C-Hordein.

The question of how to interpret the observed doublet shape remains and, at this stage, it is important to verify if the same shape is reproduced for the structurally similar system of wheat ω-gliadins. The observation of finestructure in the proton range of 10 ppm at the basis of the main peak under spinning conditions (Figure 9) also suggests that the effect of high speed MAS studies should be investigated since it may be of value to produce proton spectra with high resolution features.

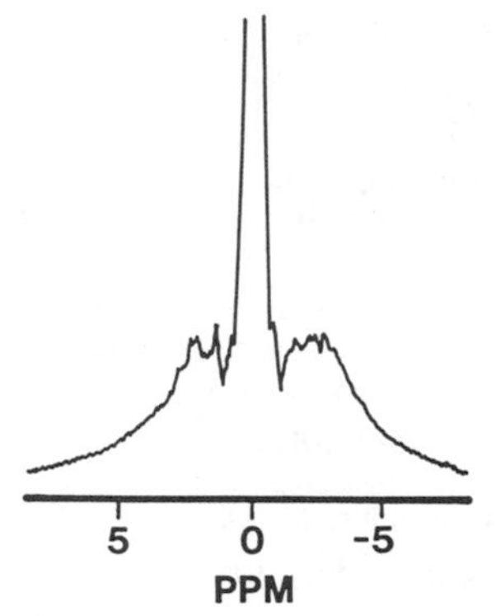

Figure 9 *Finestructure observed at the basis of intense peak in the proton spectra of C-hordein with 30% water, at a spinning speed of 1.6 kHz.*

3.5 NMR studies on the wheat fraction ω-gliadins

Before comparing the proton NMR results obtained for C-hordein with the ones obtained for ω-gliadins the ^{13}C spectra of gliadins were recorded to see if any evidence of increases in motion and conformational transitions is shown.

The CP/MAS spectra of a dry sample and a hydrated sample of gliadins show significant differences (Figure 10). An increase in resolution is clear particularly in the carbonyl region, where the well defined peak at *ca.* 177 ppm shows that a specific environment is more uniformly adopted by the glutamine sidechain groups as a result of the increased freedom of motion in the presence of water. The peak at 30.4 ppm also shows clear narrowing reflecting an increase in motion which affecs the glutamine γCH_2 and the proline βCH_2 carbons[20]. The SPE spectrum of ω-gliadins hydrated to 48% water confirms these effects (not shown). It should be noted that some aromatic groups (belonging mainly to phenylalanine residues) as well as some backbone carbonyl groups seem also to adopt enough motion to be visible under single pulse excitation conditions. Not surprisingly, the glutamine βCH_2 groups (at 25.4ppm) are relatively more immobilized.

MAS with speeds of up to 15 kHz was applied to a dried sample of ω-gliadins and to a sample hydrated to 48% water in order to investigate the reproducibility of the doublet pattern in static conditions and of the finestructure observed at higher spinning speeds. In the static spectrum of dry gliadins sample an asymmetric narrow component is observed (Figure 11) similar to that seen for the dry sample of C-hordein. A broad component is present and breaks up into spinning sidebands at about 10 kHz. The narrow component shows very clear structure in the spinning samples and it narrows slightly with increasing spinning speed. These results show that, although the more immobile part of the system is still characterized by relatively strong dipolar interactions (of the order of some tens of kHz), the more mobile component corresponds to a very weakly coupled environment corresponding of static interactions in the range of a few kHz since very faint spinning bands are observed at that spinning rate. The effect of hydration is shown in Figure 12. The doublet observed for the static spectra of hydrated C-hordein is also observed for this sample suggesting that its nature may be related with the similar primary structures of the two proteins and with a similar hydration mechanism. The interpretation of the pattern is not straightforward and is currently under investigation. As seen for the dry sample, spectral structure is seen for the spinning samples and the number of peaks resolved

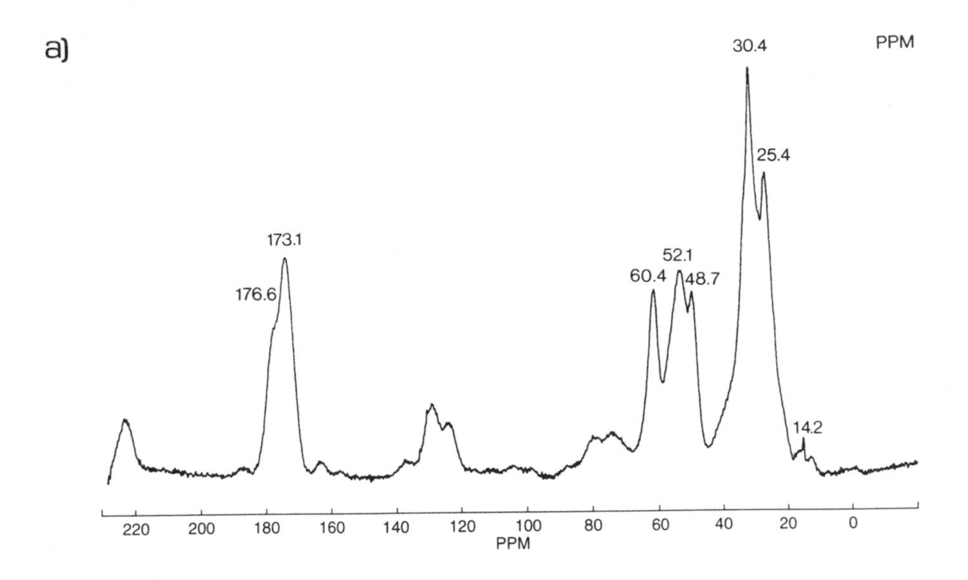

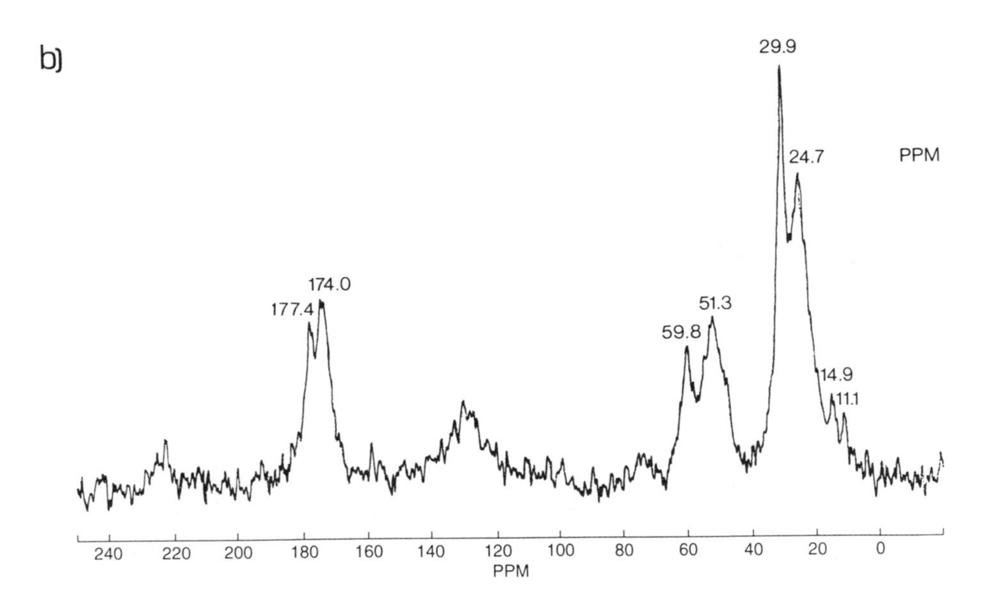

Figure 10 *^{13}C CP/MAS spectra of a) dry ω-gliadins and b) ω-gliadins with 48% water. Contact time 5ms, recycle 5s, 11 560 transients acquired in a) and 32 300 in b).*

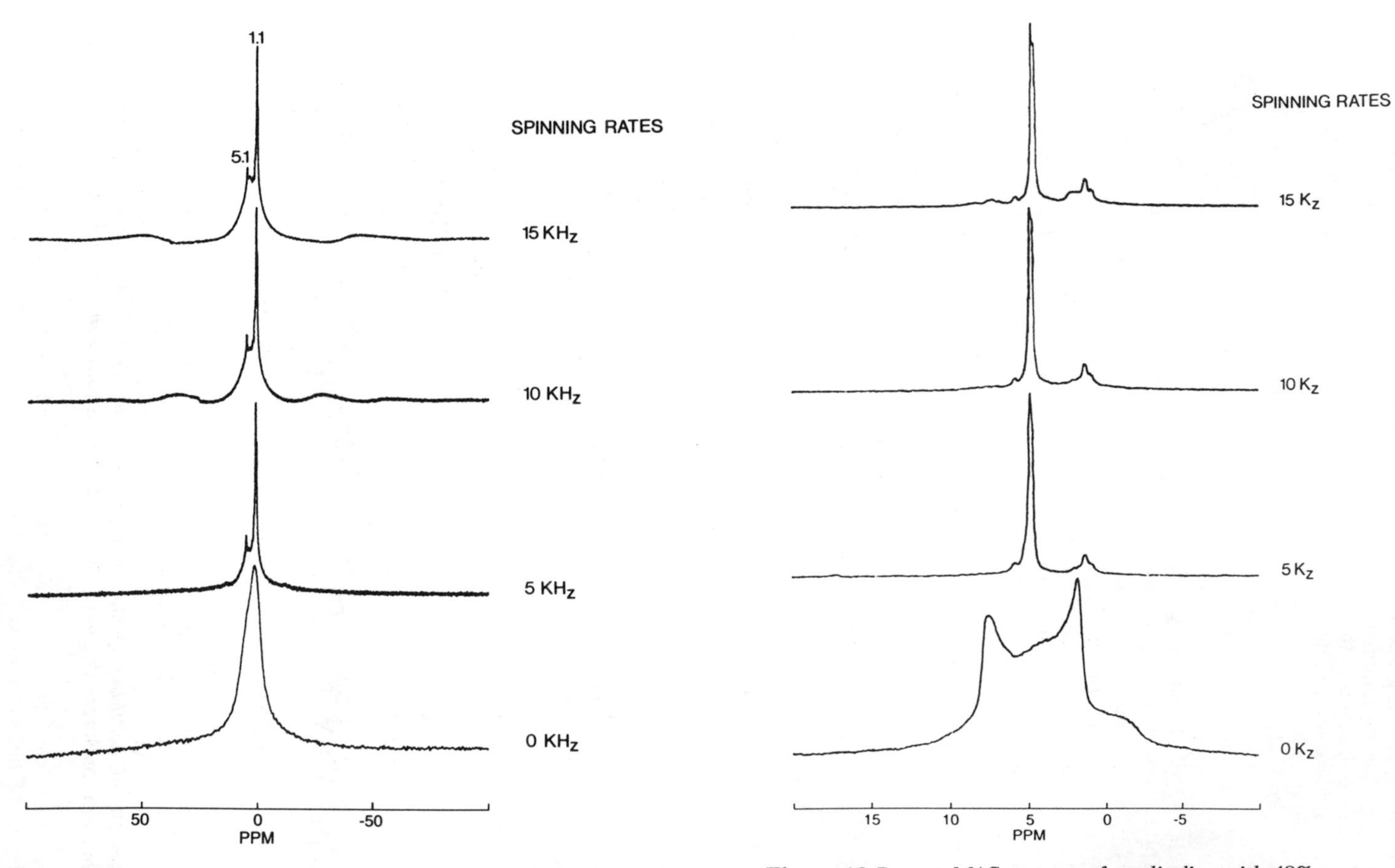

Figure 11 *Proton MAS spectra of dry ω-gliadins at different spinning speeds: 0, 5, 10 and 15 kHz.*

Figure 12 *Proton MAS spectra of ω-gliadins with 48% water at different spinning speeds: 0, 5, 10 and 15 kHz.*

increases with spinning speed until, at 15kHz spinning speed, a proton spectrum approaching high resolution standards is produced (Figure 13). Table 2 lists the proton chemical shifts observed in the MAS 15kHz spectra as well as the spin-lattice relaxation times measured for some of the peaks using the inversion-recovery pulse sequence.

For a complete assignment to be made it should also be born in mind that the observed patterns are naturally the result of the overlap of the different scalar coupling patterns. In this work, we present only a preliminary assignment of these spectra (Table 2). The peaks at 6.8-7.5 ppm, more intense than the ones at *ca.* 8.3 ppm, show that the glutamine sidechain amide groups adopt, as expected, a relatively more flexible conformation than the backbone NH's. However, the aliphatic peaks at 1.0-1.3 ppm are undoubtedly the more mobile protein moieties, even in the dry sample. These should correspond to the glutamine γCH_2 with some contribution from proline and phenylalanine residues[15]. The fact that these groups are more readily observed than the sidechain amide peaks, contrary to what was initially expected, is currently under investigation. One possible explanation may be that rapid exchange with water is occurring with the glutamine groups preventing their clear observation. Variable temperature studies will help to investigate this possibility.

Table 2 *^1H-MAS NMR of ω-Gliadins at spinning rate 15kHz (bold indicates most intense peaks).*

SAMPLE	Chemical Shift/ppm	T_1/ms	Assignment (from literature)
DRY	0.7	630.7	CH_3
	1.1	620.9	CH_2 (Gln, Pro)
	1.8	636.4	CH_2 (Gln, Pro)
	2.5	680.0	CH_2 ?
	5.1	707.6	-
HYDRATION 48%	0.9	469.4	CH_3
	1.3	448.9	CH_2 (Gln, Pro)
	1.9	528.2	CH_2 (Gln, Pro)
	2.5	473.2	CH_2 ?
	4.6	510.1	Water
	4.7	(not measured)	Water
	5.7	735.6	-
	6.8	(not measured)	Gln δNH_2
	7.3	(not measured)	Gln δNH_2, Phe ring H's
	7.5	(not measured)	Gln δNH_2
	8.3	(not measured)	backbone NH protons

Preliminary T_1 relaxation measurements of the different peaks showed that the general effect of hydration is to decrease T_1 values except for the peak at *ca.* 5 ppm. This peak has not been assigned but this relaxation measurement suggests that it belongs to a part of the system not easily affected by hydration, e.g. the protein backbone.

A self-consistent model of the behavior of the protein upon hydration may be advanced at this stage and must take account of the insolubility of the material in water and the high content of glutamine residues (40-50 mol%).

At lower water content, hydrogen bonding, both intra- and inter-molecular, can occur between the amide groups of the glutamine side chains and between glutamine residues and the amide backbone (Figure 14). As water is added, hydrogen bonds form between the amide groups and water. This increases amide groups rotation and will allow increased backbone motion. Therefore, both effects are observed on spin-lattice relaxation and lineshapes. As the water content increases the motional freedom is increased and at high water content a conformational change may occur. However, there must always remain

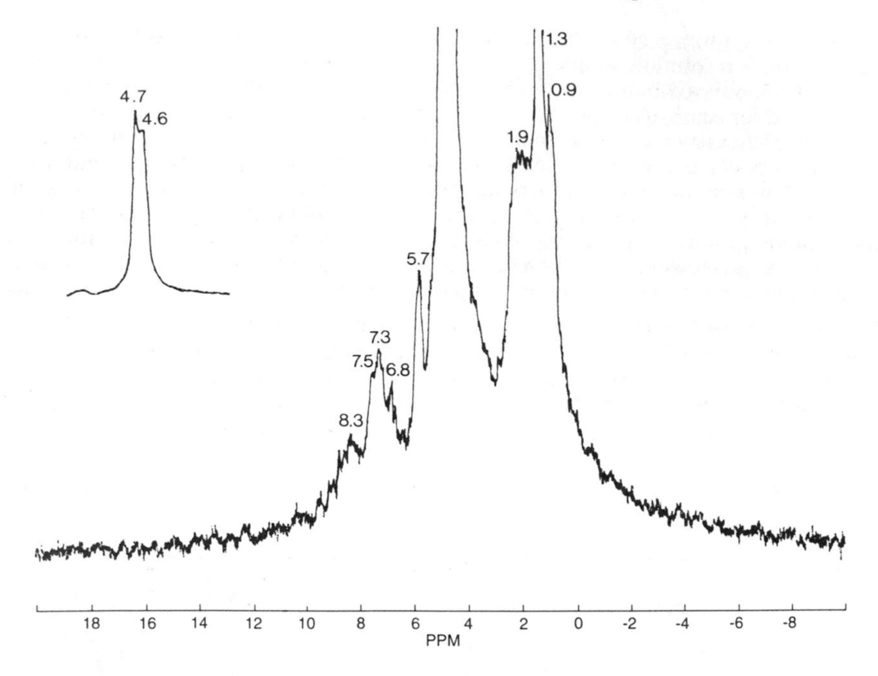

Figure 13 *Proton MAS spectrum of ω-gliadins with 48% water at 15 kHz spinning rate.*

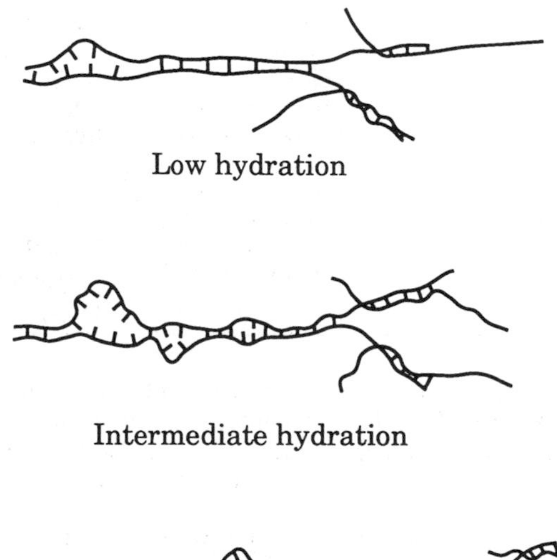

Figure 14 *Scheme of hydration model advanced for C-hordein and ω-gliadins.*

some interchain hydrogen bonds otherwise in the limit of excess water the protein would dissolve (Figure 14). The situation is analogous to polymers adsorbed at interfaces where, as the solvent quality increases, the polymer exhibits a greater proportion of loops and fewer trains[21]. There are always, however, sufficient numbers of surface contacts to ensure that dissolution does not occur.

Such a model is also consistent with scaling theories of transverse relaxation such as those proposed by Brereton[22]. As the water content increases the number of residues between crosslinks increases and the transverse relaxation process becomes generally slower, corresponding to reduced linewidth. In the case of very high water content samples, a conformational change causes a sharp change in the free chain length which results in rapid loss of the broad component and highly resolved spectra upon averaging of static interactions by high speed MAS.

Explanations have not yet been advanced for other interesting aspects observed such as the doublet structure of the water signal under MAS and the "powder pattern" observed in static conditions of the hydrated materials and these are currently being investigated.

4 CONCLUSIONS

Proton NMR techniques have been used to extract important dynamic and structural information on the hydration process of the cereal ,water-insoluble, proteins barley C-Hordein and wheat ω-Gliadins. Proton relaxation measurements and spectral lineshape studies have helped to identify specific groups where the motion is most affected by hydration. The high number of glutamine amide sidechains seem to have a dominant role in the changes occurring upon hydration, including the occurrence of a conformational transition at high water contents and temperatures.

Under these conditions finestructure is observed in the static spectra but studies of the origins of the lineshape showed that static structured lineshapes still have significant contributions from field-dependent effects as well as homonuclear dipolar interactions. Therefore, careful examination of the origins of the linewidth is necessary before any interpretation or assignment is attempted. MAS studies produced significant linenarrowing and fine structure becomes clear at high water contents and higher spinning rates.

High Speed MAS applied to gliadins produced exceptionally well resolved proton spectra in which a number of isotropic signals are readily observed. This result enables the application of a variety of high resolution techniques such as correlation spectroscopy which will help to clarify the effects of hydration in these systems. Chemical modification of the natural proteins such as deamidation to different extents is also being investigated in order to help the complete interpretation of the observed spectra.

A model for the hydration process can be proposed which involves in excess water a decreased number of interchain links, enabling a sharp change in the free chain length at high temperature.

A.M.G. thanks the Calouste Gulbenkian Foundation for funding. This work was funded in part by the Commission of the European Communities ECLAIR Programme Agre 0052. A.M.G. is grateful to Dr. A.S.Tatham for the gifts of C-hordein and ω-gliadins.

References

1. A.S.Tatham, P.R.Shewry, and P.S.Belton, in *Advances in Cereal Science and Technology*, ed. Y. Pomeranz, American Association of Cereal Chemistry, St. Paul, USA, 1991, vol.10, p.1.
2. P.R.Shewry and A.S.Tatham, *Biochem.J.*, 1990, **267**, 1.
3. P.R.Shewry and A.S.Tatham, *Comm. Agric. Food Chem.*, 1987, **1**, 71.
4. K.J.I'Anson, V.J.Morris, P.R.Shewry and A.S.Tatham, *Biochem.J.*, 1992, **287**, 183.

5. A.S.Tatham, A.F.Drake and P.R.Shewry, *Biochem.J.*, 1989, **259**, 471.
6. P.S.Belton, S.L.Duce, I.J.Colquhoun and A.S.Tatham, *Mag.Res.Chem.*, 1988, **26**, 245.
7. P.S.Belton, S.L.Duce and A.S.Tatham, *J. Cereal Sci.*, 1988, **7**, 113.
8. P.S.Belton, I.J.Colquhoun, J.M.Field, A.Grant, P.R.Shewry and A.S.Tatham, *J. Cereal Sci.*, 1994, **19**, 115.
9. P.S.Belton and A.M.Gil, *J.Chem.Soc. Faraday Trans.*, 1993, **89**, 4203.
10. P.S.Belton and A.M.Gil, *J.Chem.Soc. Faraday Trans.*, 1994, **90**, 1099.
11. A.S.Tatham and P.R.Shewry, *J.Cereal Sci.*, 1985, **3**, 103.
12. S.F.Tanner, B.P.Hills and R.Parker, *J.Chem.Soc., Faraday Trans.*, 1991, **87**, 2613.
13.P.S.Belton, *Commun.Agric.Food Chem.*, 1990, **2**, 179.
14. E.R.Andrew, D.N.Bone, D.J.Bryant, E.M.Cashell, R.Gaspar Jr. and Q.A.Meng, *Pure Appl. Chem*, 1982, **54**, 585,
15. K.Wutrich, 'NMR of Proteins and Nucleic Acids', Wiley, New York, 1986, p.24.
16. S.J.Ring and T.Noel, personal communication.
17. J.P.Yesinowski, H.Eckert and G.R. Rossman, *J. Am. Chem. Soc.*, 1988, **110**, 1367.
18. D.E.Woessner, *Mol. Phys.*, 1977, **34**, 899.
19. N.Boden and M.Mortimer, *Chem.Phys.*, 1973, **21**, 538.
20. I.C.Baianu, L.F.Johnson and D.K.Waddell, *J. Sci. Food Agric.* , 1982, **33**, 373.
21. E. Dickinson, 'An Introduction to Food Colloids', Oxford University Press, New York, 1992.
22. M.G.Brereton, *Macromolecules*, 1990, **23**, 1119.

Subject Index

C

D

E

F

G

H

I

K

L

M

N

O

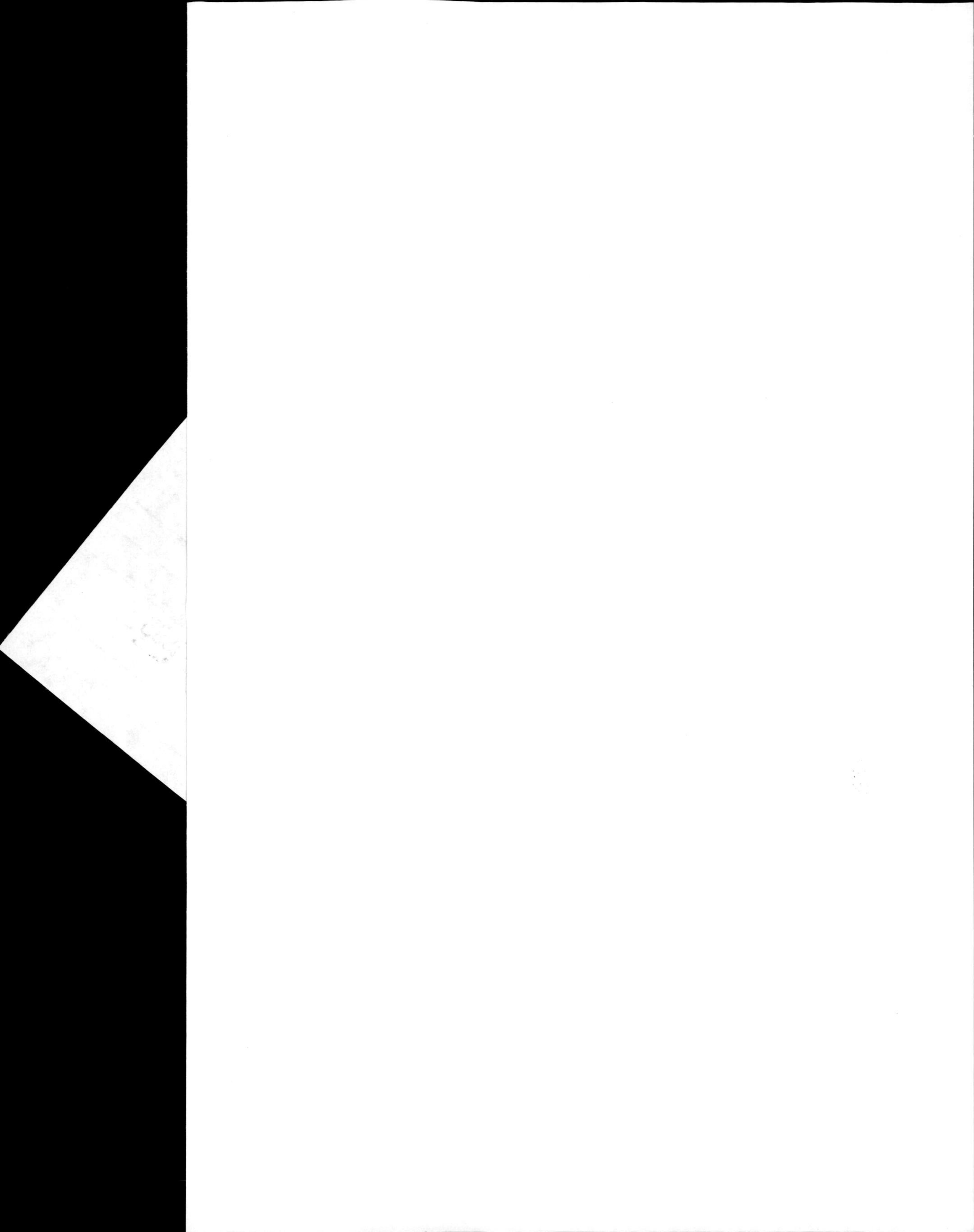

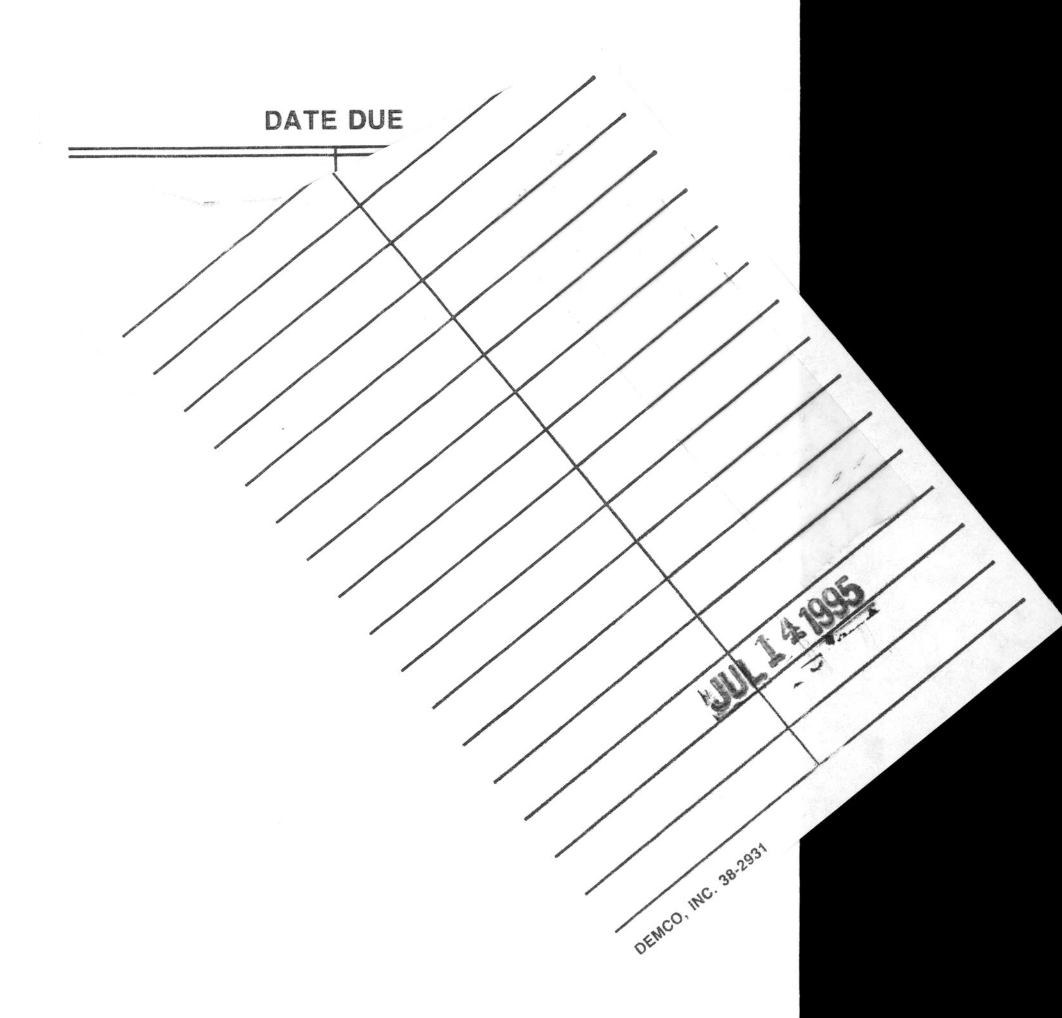
DATE DUE
JUL 14 1995
DEMCO, INC. 38-2931